Energy Technologies
and Conversion Systems

PRENTICE-HALL SERIES IN ENERGY

Wayne C. Turner and W. J. Kennedy, Jr., *editors*

Gibson, *Energy Graphics*
Hodge, *Analysis and Design of Energy Systems*
Kennedy and Turner, *Energy Management*
Kleinbach and Salvagin, *Energy Technologies and Conversion Systems*
Mills and Toké, *Energy, Economics, and the Environment*
Money, *Transportation Energy and the Future*
Murphy and Soyster, *Economic Behavior of Electric Utilities*
Myers, *Solar Applications in Industry and Commerce*
Plummer, Oatman, and Gupta, *Strategic Management and Planning for Electric Utilities*

Energy Technologies and Conversion Systems

Merlin H. Kleinbach
and
Carlton E. Salvagin

State University of New York
College at Oswego

PRENTICE-HALL, INC.
Englewood Cliffs, N.J. 07632

Library of Congress Cataloging in Publication Data

KLEINBACH, MERLIN H.
  Energy technologies and conversion systems.

  (Prentice-Hall series in energy)
  Bibliography: p.
  Includes index.
  1. Power resources. 2. Power (Mechanics)
  3. Energy conservation. I. Salvagin, Carlton E.
  II. Title. III. Series.
  TJ163.2.K57 1986   333.79   85-3369
  ISBN 0-13-277344-9

Editorial/production supervision and
  interior design: Martha Masterson and Lisa Halttunen
Cover design: Joe Curcio
Manufacturing buyer: Carol Bystrom

To Jessica and Kristin, for a more secure energy future

M.H.K.

To the future generations, who will bear
the burdens or the benefits of our energy decisions

C.E.S.

© 1986 by Prentice-Hall, Inc., Englewood Cliffs, New Jersey 07632

*All rights reserved. No part of this book may be
reproduced, in any form or by any means,
without permission in writing from the publisher.*

PRINTED IN THE UNITED STATES OF AMERICA

10 9 8 7 6 5 4 3 2 1

ISBN 0-13-277344-9 01

PRENTICE-HALL INTERNATIONAL (UK) LIMITED, *London*
PRENTICE-HALL OF AUSTRALIA PTY. LIMITED, *Sydney*
PRENTICE-HALL CANADA INC., *Toronto*
PRENTICE-HALL HISPANOAMERICANA, S.A., *Mexico*
PRENTICE-HALL OF INDIA PRIVATE LIMITED, *New Delhi*
PRENTICE-HALL OF JAPAN, INC., *Tokyo*
PRENTICE-HALL OF SOUTHEAST ASIA PTE. LTD., *Singapore*
EDITORA PRENTICE-HALL DO BRASIL, LTDA., *Rio de Janeiro*
WHITEHALL BOOKS LIMITED, *Wellington, New Zealand*

CONTENTS

Preface　　xi

INTRODUCTION: THE ENERGY PICTURE

1 **Perspectives: Where We Stand—A Point of View**　　1
Concepts 1　　Introduction 1　　National Security 3
Energy Options 4　　National Energy Policy 5
Environmental Degradation 6　　Conservation 7
The Future 8　　Summary 9　　Activities 9
Bibliography 10

2 **Society's Experiences with Energy: How We Use It**　　11
Concepts 11　　Glossary 11　　Introduction 12
Renewable Energy Use 15　　Nonrenewable Energy
Use 19　　Electricity Generation 27　　Waste as a
Recoverable Resource 29　　Conservation 29
Activities 30　　Bibliography 31

3 **Understanding Energy: What It Is, How It Works, Forms, and Properties**　　32
Concepts 32　　Glossary 32　　Introduction 33
Forms of Energy 33　　Sources of Energy 34
Energy Terms 35　　Energy Use 37　　The Laws of
Thermodynamics 39　　Summary 41　　Activities 41
Bibliography 42

WHERE WE STAND: FINITE ENERGY SOURCES

4 Coal: The Plentiful Fossil Resource   43
Concepts 43   Glossary 43   Introduction 44
Background of Coal Technology 44   Physical
Characteristics of Coal 45   Coal Reserves 48
Coal Extraction 49   Coal Preparation 56
Distribution Systems 57   Personnel Health and
Safety 57   The Uses for Coal 58   Synthetic
Fuels from Coal 62   Environmental Impacts 66
A Look at the Future 71   Summary 73
Activities 73   Bibliography 74

5 Petroleum and Natural Gas: Our Most Popular Fuels   75
Concepts 75   Glossary 76   Introduction 76
Petroleum: Origins and Components 77   Petroleum
as a Resource 78   Petroleum Reserves 79
Production 80   Exploration 81   Drilling
Structures 83   Extraction 85   Refining 88
Distribution and Storage 90   Environmental
Factors 92   Natural Gas 94   Background 94
Supplies of Natural Gas and Reserves 95
Summary 99   Activities 100   Bibliography 100

6 Nuclear Fission:
The Tarnished Star of Our Energy Future   102
Concepts 102   Glossary 103   Introduction 103
Nuclear Overview and Current Status 104   How
Nuclear Plants Work 106   The Fuel Cycle 113
Financial Costs 114   Radioactive Wastes 121
Public Opinion 124   Environmental Impact and
Safety 125   The Future of Nuclear Fission 129
Summary 131   Activities 131   Bibliography 132

THE DEVELOPING SCENE: RENEWABLE AND INEXHAUSTIBLE ENERGY SOURCES

7 Solar Energy: The Primary Energy Source   134
Concepts 134   Glossary 134   Introduction
and Historical Background 135   The Nature of
Solar Energy 137   Current Solar Applications 140
Domestic and Commercial Water Heating 141   Types
of Solar Collectors 142   Flat-Plate Collectors 143
Types of Solar Water Heaters 146   Sizing the
System 150   Focusing Collectors 154   Space
Heating Applications 157   Greenhouses/Solar Rooms/
Solar Spaces 163   Space Cooling Applications 164
Solar Thermal Storage Systems 166   Photovoltaics 170
The Future of Solar Energy 177   Summary 178
Activities 179   Bibliography 180

Contents vii

8 Wind Energy: Energy from Thin Air 181
Concepts 181 Glossary 181 Introduction 182
Principles 183 Site Characteristics 184 Blade
Design 185 Control Mechanisms 187 Electricity-
Producing Units 188 Generator Types 188
Inverters 189 "Storage" Systems 189 Small/
Intermediate Versus Megawatt Systems 191 Other
Rotor Designs 195 Institutional Barriers to
WECS 204 Environmental Impacts 204
Summary 205 Activities 206 Bibliography 207

9 Geothermal Energy: Heat from Down Under 208
Concepts 208 Glossary 208 Introduction 209
Background: Geothermal Heat as an Energy Source 209
Geothermal Site Locations 211 Geothermal
Features 215 Geothermal Energy Conversion
Technology 216 Technical Problems 220
Environmental Issues 221 The Geothermal
Future 223 Summary 225 Activities 225
Bibliography 226

10 Energy from Waste and Biomass: Trash to Energy 227
Concepts 227 Glossary 227 Introduction 228
Thermal Conversion of Wood 229 Forestry
Management and Environmental Impact 235 Urban
and Industrial Waste 237 Biomass Conversion:
Methane 245 Composting 249 Alcohol Fuels 251
Biomass Utilization 256 Summary 258
Activities 259 Bibliography 259

11 Hydropower: Energy from Flowing Water 261
Concepts 261 Glossary 262 Introduction 262
The Development of a Technology 262 Turbine
Design and Use 267 Other Considerations 270
Expanding Electricity Demand 273 Potential for
Hydroelectric Facilities 273 Resurgence of Small-
Scale Hydroelectric Power 274 Environmental
Issues 280 Future of Hydroelectric Power 281
Summary 282 Activities 282 Bibliography 283

12 Ocean Energy Systems: A Sea of Energy 284
Concepts 284 Glossary 284 Introduction 285
Tidal Energy 285 Ocean Current Energy 291
Ocean Thermal Energy Conversion 295 Wave
Energy 302 Salinity Gradient Power 309 Ocean
Bioconversion 311 Summary 312 Activities 312
Bibliography 312

CONSERVATION: EASIEST, FASTEST, LOWEST COST

13 **Conservation and Legislation: The Wise Use of Energy** — 314
Concepts 314 Glossary 314 Introduction 315
Energy Consumption 315 Personal Attitudes
toward Conservation 320 Recycling Used
Materials 321 Energy-Saving Technologies for
Industry 322 Factors Affecting Heating and
Cooling 324 Payback (or Return on
Investment) 325 Residential Conservation 326
Caulking and Weather Stripping 327 Insulation:
Types and Uses 329 Energy Audits and Appliances 331
Air Pollution 333 Legislation and Regulations 336
Legislation and Tax Incentives for Conservation 337
Summary 340 Activities 341 Bibliography 341

14 **Cogeneration and Heat Reclamation: Two Energy Forms for the Price of One?** — 343
Concepts 343 Glossary 343 Introduction 344
Background 344 Cogeneration Systems 345
High-Grade Heat Recovery 350 Low-Grade and
"Waste" Heat Recovery 351 Legislation 366
Environmental Impact 366 Cogeneration and Heat
Reclamation Potential 367 Summary 369
Activities 370 Bibliography 371

THE FUTURE: WHERE DO WE GO FROM HERE?

15 **Energy for the Future: From Fossil Resources toward a Mix of Renewables** — 372
Concepts 372 Glossary 372 Introduction 373
Overview 373 Energy Sources of the Future 375
Future Projections for Each of the Energy Resources 379
Implementing the Energy Mix 393 Future
Energy Technologies: A Crystal Ball View 395
Summary 409 Activities 410 Bibliography 410

EMPLOYMENT: A LOOK AT EMPLOYMENT OPPORTUNITIES

16 **Energy-Related Employment and Careers: Where the Jobs Are** — 412
Concepts 412 Glossary 412 Introduction 413
Occupations in Renewable Energy Fields 414
Renewable Energy, Employment, and the Economy 420
Education and Training Programs 425 Summary 426
Activities 428 Bibliography 428

APPENDIXES

A	Energy Units and Conversion Factors	430
B	Average Heat Value of Fuels	434
C	Heat Value of Various Wood Species	435
D	Power Consumption by Appliances	436
E	Home Energy Audit	438
F	Estimate Your Annual Exposure to Radiation	445
G	World Use of Renewable Energy: 1980, 2000, Change Factor, and Potential	446
H	U.S. Energy Production and Consumption by Source	447

Index   448

PREFACE

This book presents a comprehensive view of energy technology, including the fossil, renewable, and inexhaustible sources. We argue that it is not wise to extensively consume resources that took millions of years to form, saving little for our children and their children. Our wastefulness leaves future generations with a rapidly deteriorating environment and the responsibility of caring for our waste products.

The text encompasses a historical perspective of society's energy use and surveys the energy resources we have and those that may be available in the near or distant future. Each energy resource is comprehensively reviewed, and its advantages, disadvantages, and potential are addressed. Energy conservation and appropriate use are strongly emphasized. Employment opportunities in each of the energy fields are also included.

Energy, its availability, cost, and impact upon our daily lives, is an exciting area of study. We hope that this text will provide the reader with an understanding and appreciation of energy and all that it does to make our lives comfortable and rewarding.

Chapter 1
Perspectives
Where We Stand—
A Point of View

CONCEPTS

1. Domestic fossil fuel energy resources which were once acquired easily and economically have been depleted.
2. Unlimited utilization of fossil fuels poses serious economic, political, and environmental problems for the next generation.
3. Consuming our way to prosperity is not sustainable.
4. Energy will increasingly affect our lives and livelihoods.
5. The solution to the energy dilemma will be multifaceted.

INTRODUCTION

The industrialized nations of the world are facing a serious dilemma. Energy sources that have been utilized to heat our homes, power our industries, and fuel our transportation systems will be nearly depleted within the next generation. Fossil fuels, which have taken hundreds of thousands of years to form from organic materials, are in finite supply, and the attitude of many people is that high-technology magic will bail us out of the problem at little cost.

Our international dilemma cannot appropriately be described as a crisis. The sudden appearance of a specific time, occasion, stage, or point during which a short-term critical event is happening is often called a crisis. We are facing an ongoing, continually more serious problem which demonstrates only further deterioration. We only have to observe for a short time before realizing that government, the business community, and the media have been painting an ominous picture of America's energy predicament which projects throughout the balance of the century and perhaps beyond.

Energy is essential to the economy. It provides the propulsion to the processes that generate economic output, but we must remember that its economic cost must also be charged against that output. Our industrialized economic system depends heavily on fossil (nonrenewable) energy sources which are continually escalating in price. When price increases occur, it is generally the consumer who must bear the additional burden.

Proof of both depletion of resources and increase in costs may readily be found (see Table 1-1). We are having to go farther and farther away from our places of end use to find and acquire the necessary fuels for our economy. It is necessary to endure the climate of Alaska's North Slope to drill for, pump, pipe, and ship petroleum to the continental United States. We are fighting the winds and the waves of the seas while drilling thousands of feet below the surface for the life's blood of industry. We are even assisting unfriendly nations to acquire the most recent technologies so that we may purchase their natural gas. If other resources were readily available, would we not be using them rather than going to these extreme and expensive lengths? The domestic energy sources that were acquired easily and inexpensively have dwindled greatly.

TABLE 1-1 Average National Prices of Selected Energy Sources

Year	Electricity (cents/kWh)	No. 2 Fuel Oil (cents/gal)	Natural Gas (cents/1000 ft^3)	Gasoline (Regular) (cents/gal)
1973	2.54	22.8	108.2	38.8
1975	3.51	37.7	154.2	56.7
1976	3.73	40.6	184.6	59.0
1977	4.05	46.0	226.4	62.2
1978	4.31	49.4	262.6	62.6
1979	4.64	65.6	323.1	85.7
1980	5.36	97.8	394.6	119.1
1981	6.20	120.5	455.7	131.1
1982	6.58	115.3	489.2	120.6

U.S. Department of Commerce, *Statistical Abstract of the United States*, 1982-1983, p. 470.

It would appear that as industrial societies have consumed their way to prosperity, success, and affluence, little attention has been directed toward thoughts of the future. The industrial nations are now at war (economically) because the energy system on which they are dependent is cannibalizing the very economic system it is supposed to support. This fact is substantiated by observing the cost of energy forms, which increase at a rate above the general rate of inflation (see Table 1-2). As energy costs rise, more and more of the household income is diverted from the purchase of goods and services—reducing demand and consequently fueling economic recession. Domestic industrial products become less competitive with more economical imported goods.

NATIONAL SECURITY

Let's explore another problem of concern to all of us. America is very dependent on electricity. It heats many of our homes, supplies hot water, provides illumination, and is the foundation for many industrial processes, such as the making of iron, steel, and aluminum. The number of machine tools and pieces of equipment requiring electric motors is astronomical. Each of these examples converts electrical energy to another usable form, such as heat, light, or mechanical energy.

The source of electrical energy is most often a centrally located utility company which converts other forms of energy (chemical, nuclear, or gravitational) to electricity, which is then distributed via a complex system throughout the geographic area. This system is supported by an even more complex system called a grid, which interconnects one utility company with others for the purposes of safety, security, and overall support of the delivery system.

TABLE 1-2 Consumer Price Index: Annual Percent Change in Cost

Year	All Items	Fuel Oil and Coal	Electricity and Natural Gas
1973	6.2	14.8	4.9
1975	9.1	9.7	16.3
1976	5.8	6.6	11.4
1977	6.5	13.0	12.9
1978	7.7	5.3	9.0
1979	11.3	35.1	10.8
1980	13.5	37.9	17.1
1981	10.4	21.6	14.6
1982	6.7	−6.0	14.4

U.S. Department of Commerce, *Statistical Abstract of the United States*, 1982–1983, p. 461.

Our political leaders are very concerned about our national security, military superiority, and the strategic locations of defense installations—at the same time totally disregarding a very fragile, vulnerable energy supply system. If this system is sabotaged or otherwise interrupted, the disruption could easily incapacitate the entire nation in a matter of minutes. Consider the impact of the loss of a single key electrical relay that tripped, sending a wave of outages all across the northeastern United States on November 9, 1965. Another outage occurred in New York on July 13, 1977, and yet another in the New York City garment district during the summer of 1983. People were in the dark, with no heating or cooling, no refrigeration, and there was chaos in city streets. A similar electrical blackout on Wall Street in New York City lasting only a few days could create a serious threat to the national economy by closing down the major financial institutions. For some very thought-provoking reading, acquire a copy of Lovins and Lovins' book, *Brittle Power: Energy Strategy for National Security*. The Lovinses elaborate on several energy situations, including the vulnerable national distribution system for natural gas, 70% of which originates in the state of Louisiana.

Looking a bit deeper into America's energy supply picture, we find that oil- or petroleum-derived products fuel most of the country's energy conversion systems. Of our total national energy utilization, 45.5% is petroleum based, and most disturbing of all, 40% of this supply is imported (1982). One has only to review the United States balance of payments and the political and economic instability of the members of the Organization of Petroleum Exporting Countries (OPEC) to see that this affects not only our oil supplies but our entire energy system—oil, natural gas, nuclear, and electric power stations and grids—and national security. It places the United States in a position where it must defend these nations militarily in order to support our energy utilization.

ENERGY OPTIONS

We are at a crossroads where the energy choices we make now will determine what kind of natural heritage we leave to future generations. As the population of the world and nation expands, so does our annual need for energy. What can we do?

All of our energy options have several common denominators. Each requires financial investment. Each demands an increase in the use and further development of our current technical knowledge. Each will contribute to environmental pollution. All options have both positive and negative aspects (see Table 1-3). We are in a quandary. Maintaining the status quo is not possible. Which way do we jump?

TABLE 1-3 Energy Options

Option	Results
1. Increase production of fossil fuels: coal, oil, and natural gas	Increased rate of depletion of finite sources
	Increased environmental pollution
	Increased costs due to limited accessibility, distance, and availability
2. Increase the use of nuclear power	Additional problems of radioactive waste disposal and storage
	Exponential increases in costs of facilities construction
	Increased potential dangers inherent with nuclear power use
3. Development of manufactured fuels from fossil sources: shale oil, coal gasification, synthetics	Depletion of finite sources
	Increased environmental pollution, especially water
	Increased costs due to the manufacturing process
4. Develop renewable energy sources: solar, wind, biomass, geothermal, others	Requires funding
	Some sources are site-specific
	Many are not easily compatible with our current philosophy of centrally located utilities
	Environmental pollution
5. Reduce energy demand: conservation	Reduced profits to producers of fossil fuels
	Reduced dependency on foreign suppliers
	New businesses required
	Employee career changes
	Reduced costs
	Environmental pollution
	Reduced rate of energy utilization

NATIONAL ENERGY POLICY

Our national energy policy is strongly focused on dominance over Middle East oil supplies and support of a faltering nuclear industry. In addition, our policy shifts and sways with the ideology of each succeeding administration.

Efforts to conserve our resources and encourage the development and use of renewable energy resources, thereby reducing our foreign dependency, rise and fall according to the interests of the current political incumbent. One administration provides financial support for research and development, energy tax credits for individuals and industry, and an emphasis on conservation, while the next places emphasis on additional exploration, development, and exploitation of the remaining fossil fuel supplies while wishing to abolish the tax credit incentives for conservation and renewable energy established by the predecessor.

The United States must establish a sound, ongoing, future-oriented energy policy. This policy must consider the vulnerability of our nation's dependency on other countries' resources, our centralized systems for national energy distribution, improved efficiency of use, conservation of the remaining fossil-based assets, and the use of energy sources which are renewable rather than finite. The policy would also have to be resistant to changeable administrative ideologies, consistent with our national environmental goals, and still meet the nation's present and future energy needs.

As idealistic as all this sounds, the security and actual survival of our technological society depend on a national energy policy and its rapid establishment and implementation. An initial effort to attain this goal internationally was made in April 1983 at the first meeting of the United Nations permanent International Committee for Development and Utilization of New and Renewable Sources of Energy. It will be interesting to see if the representatives of several nations can establish a goal which thus far has been unattainable by most individual nations.

ENVIRONMENTAL DEGRADATION

Each of our energy options contributes to the degradation of the natural environment. This is one of the inherent characteristics of a technological society. As we shape and form natural materials to meet our needs, energy is used and chemical changes occur. Unfortunately, not all of these changes are positive.

If we take a close look at the environment, we will find that there are many changes taking place which are not attributable to human intervention. Scientific studies have proven that nature alone contributes significantly to air pollution: specifically, particulates, 55%; sulfur dioxide (SO_2), 65%; hydrocarbons (HC), 70%; carbon monoxide (CO), 93%; and carbon dioxide (CO_2) and oxides of nitrogen (NO_x), 99% (lightning fixes oxides and nitrogens). The problems arise when human activity adds to these natural levels.

Earth Day in April 1970 was an event that made both political officials and the general public aware of just how much disregard we had for the environment and that strong measures had to be taken immediately before reversal of the rampant degradation would no longer be possible. This day initiated a decade of environmental awareness which made many people aware that a clean environment is a fundamental component of quality life.

The condition of the environment is no longer viewed by many people as a crisis issue. A national survey, commissioned by federal agencies in 1980 and conducted by Resources for the Future (RFF), demonstrated that environmental protection enjoys continued strong support. Many people believe that energy development is compatible with environmental quality. The study, which asked participants to project toward the year 2000, revealed that 61% think that our efforts should be made in solar energy technologies, while one-third indicated that the least effort should be expended on nuclear energy. The need for, and value of, a strong environmental watchdog agency at the federal level was evident during the summer of 1983, when there was an upheaval in the Environmental Protection Agency (EPA). The turmoil was caused because of alleged failure to enforce environmental regulations and alleged shelving of studies that pointed to environmental degradation practices.

Energy development is not necessarily incompatible with environmental quality. Economic prosperity and environmental protection are not necessarily at opposite ends of a continuum. If a battle is ever staged between economics and ecology, you can be sure that ecology will win. The environment will absorb only a certain level of waste before the impact of one member of the economic system begins causing hardship on another member.

CONSERVATION

The conservation of energy has been termed by many people as a "key energy source." This can be compared with the notion that "a penny saved is better than a penny earned" (if you consider not having to work for it or pay taxes on the savings). Of course, you cannot actually consider conservation as a "source" of energy, but the concept of saving contributes to a longer available life for our finite fossil fuel energy sources.

Conservation as an energy "option" is not truly an option but a necessity. It is by far the fastest and most economical method for reducing our dependency on fossil fuels. When considering the four sectors of industry, transportation, commercial, and residential, the United States has been wasting 49% of the total energy it uses. It is the richest and the most wasteful nation in the world. The lack of conservation is due to the affluence of the American economy and the indulgent attitude of the society. In the past there appeared to be no necessity to conserve or use energy or raw materials efficiently as long as they were economical and in abundant supply. This scenario is changing as prices escalate and supplies become diminished. The changes are a result of economics, not the acquisition of a benevolent attitude.

The attractive characteristic of conservation is that efforts toward the reduction in use, or more efficient use, of energy have little impact on our life styles. For example, U.S. residential energy use could easily be reduced by 30% and with some modification or effort, 50 to 75%, without sacrificing comfort. The consumer just needs to know how and where to apply the efforts to attain these savings.

Conservation, no matter how strongly applied, is not sustainable by itself. It will provide "better mileage" from our resources. However, because of increasing population, our annual need for energy continues to increase. Conservation efforts will assist us by providing additional time to develop sustainable energy sources.

THE FUTURE

The 1980s and 1990s will be an exciting time of innovation, redesign for conservation, recycling, environmental improvement, decentralization of economics, and revival of smaller businesses. Macroeconomics is running into difficulty because cheap, dependable supplies of energy and raw materials are diminishing. Consuming our way to prosperity is not sustainable.

The United States, with the richest and most wasteful economy, is in the most advantageous position to solve the energy problem plaguing our political, economic, and social sectors. However, there is no "quick fix." The solution will not be easy, and no one option will be successful by itself in solving the dilemma.

Economic forces are attacking American complacency and are heading us in the direction of an energy solution.

> Imported oil is no longer cheap.
>
> Many new centrally located electricity-generating plants are too expensive to compete in the energy services market.
>
> Costs will force us toward the marketing of products that improve energy efficiency.
>
> Nonrenewable sources, which make up 95% of our total energy picture, are dwindling.

By depleting our petroleum and natural gas, we lose valuable resources that we need for purposes other than energy (i.e., lubricants, plastics, oil-based synthetic fibers, medicines, fertilizers, and other chemicals).

"Externalized" costs are being passed on to taxpayers, society at large, the environment, or future generations.

Americans want it all. We want a clean environment for our families, powered by energy sources that are economical and sustainable. These restrictions reduce the number of available options. The key to our dilemma will be multifaceted and will not be focused on large, centralized, high-tech use of fossil fuels, although there will be many facilities continuing into the next century. Instead, the solution will involve intensive conservation coupled with highly efficient, diverse, dispersed, smaller applications of renewable energy sources.

SUMMARY

The industrialized nations of the world established their systems on the availability of plentiful, economical fossil fuels. As the availability, ease of acquisition, and subsequent costs of these fuels become more and more a burden to society, changes are going to be necessary. There are a number of energy options available, and each will affect national security, the consumer costs of goods and services, and the condition of the environment.

An economy based on renewable resources and carefully managed for sustained yield and long-time productivity of all its resources can provide useful satisfying work and richly rewarding life styles for all its participants. All energy must be cost-effective—from acquisition through end-use applications and environmental impact. As we strive toward this goal, we need to invest in efficiency and conservation until the price of saving the next unit of energy is as high as the cost of producing it. It will take considerable time, effort, and financing, but as a result of our efforts we will establish "the age of inexhaustible fuels."

ACTIVITIES

1. Discuss national security and how it relates to international policy, alliances, and foreign trade.
2. Follow the steps that an energy form takes from its origin to its final disposition. Determine sources, acquisition, utilization, advantages, disadvantages,

potential pollution problems, people/agencies/nations involved, and related direct and indirect costs.
3. Discuss the international balance of payments and how it relates to inflation.
4. Review and discuss newspaper articles about environmental degradation, protection efforts, and cleanup costs. Who bears the costs?
5. Review periodicals containing articles about the most recent energy technology developments. Select a topic and
    a. Lead a discussion.
    b. Give a presentation.
    c. Prepare a written report.
6. Invite guest speakers to share their philosophies on energy with your group:
    a. Politician
    b. Utility company representative
    c. Environmentalist
    d. Homeowner
    e. Futurist
   Compare these philosophies. How are they the same? How are they different?
7. List society's possible energy options for the future. Discuss their feasibility, advantages, disadvantages, and environmental impact. Compare short-term costs with long-term costs of each option.
8. From the list of energy options in Activity 7, identify options that are feasible for your geographic location. Cite specific advantages and disadvantages of developing each option. Compare projected short- and long-term costs for each option.
9. *Scenario:* It is January and the distribution of electrical energy has just been totally interrupted. You receive an announcement over your battery-powered transistor radio (the radio station has its own emergency diesel-powered generator) that there will be no electricity in your area of the country for at least 4 full weeks. Plan what you should do to sustain life until the electricity is restored.
10. Referring to the scenario in Activity 9, plan what you would do if the radio announcement provided you with a 1-week warning before the blackout.

BIBLIOGRAPHY

Lovins, Amory B., and L. Hunter Lovins, *Brittle Power: Energy Strategy for National Security*. Andover, Mass.: Brick House, 1982.

Priest, Joseph, *Energy for a Technological Society: Principles, Problems, Alternatives*. Reading, Mass.: Addison-Wesley, 1979.

Steinhart, Carol, and John Steinhart, *Energy: Sources, Use, and Role in Human Affairs*. North Scituate, Mass.: Duxbury Press, 1974.

U.S. Department of Commerce, Bureau of the Census, *Statistical Abstract of the United States, 1982-83*, 103rd ed. Washington, D.C.: U.S. Government Printing Office, 1982.

Chapter 2
Society's Experiences with Energy
How We Use It

CONCEPTS

1. There is no energy shortage, only a shortage of applications of known resources.
2. Current energy resources are restricted as to type and with regard to flexibility for consumers.
3. Economic development is predicated on the use of increased amounts of energy.
4. Energy sources have shifted from labor-intensive to more convenient, less labor-intensive fuels.
5. Energy has become more expensive as fossil resources have become increasingly more difficult to extract.

GLOSSARY

Btu (BRITISH THERMAL UNIT)—the amount of energy necessary to raise the temperature of 1 pound of water 1 degree Fahrenheit.
DOE (DEPARTMENT OF ENERGY)—an agency of the federal government set up to administer federal energy programs.

KILOCALORIE (KCal)—the physics equivalent to the calorie used in discussions of food value, equivalent to 3.968 Btu.
KILOWATT (kW)—a unit of rate of energy use or production equal to 0.947 Btu per second.
MEGAWATT (MW)—1000 kW.
PAYBACK TIME—period of time it takes for the investment cost of energy conservation measures to equal the cost of energy not used.
QUAD—a unit of measure used for very large amounts of energy, equal to 1 quadrillion (10^{15}) Btu.
WECS—wind energy conversion system.

INTRODUCTION

Our ancestors' needs for energy developed as early as the first use of fire for heat and light, protection from wild beasts, and the preparation of food. The first fuel was undoubtedly wood, an abundant, easily gathered, easily utilized fuel. The relative comfort derived from the satisfaction of basic needs and simple creature comforts led to the recognition of other available fuels. It was discovered that a black rock, coal, would burn and provide more heat than wood. Liquid fuels—first animal fats and oils, and in more recent times petroleum products—became the source of heat and light, as each was perceived to be more convenient, less expensive, readily available, or cleaner than other fuels. Natural gas, first considered a dangerous by-product of coal mines and oil wells, was recognized for its fuel properties. In recent decades, other materials, such as uranium, have been exploited as energy sources.

It is believed that early human beings used as little as 2200 kcal of energy per day (less than 9000 Btu) averaged over a lifetime. The food consumed was gathered from wild plants and animals. The recognition of the need for a more consistent source of food energy was undoubtedly a major impetus to the development of agriculture and the domestication of animals.

Toolmaking helped to tame both land and animals and began the ever-increasing use of larger and larger amounts of energy and materials. As tool use increased, more use was made of animal power to perform tasks beyond the limits of human strength and endurance. Each of these steps increased the per capita energy consumption level as it increased the energy control exercised over the environment.

The search for easier and "better" means to provide what we feel are the needs and wants of life has led to greater and greater use of energy. The Bronze Age, beginning about 3000 B.C., and the Iron Age, beginning about 1000 B.C., both increased the level of energy consumption several times over that of earlier eras. Not only was there a need

TABLE 2-1 Chronology of Energy Development

Stone Age 1,000,000 years ago	Human muscle	Tools improvised from bones and sticks	0.1 hp
Up to 200,000 years ago	Human muscle	Fire-hardened tools, crafted tools, carrying devices	0.1 hp
Up to 40,000 years ago	Human muscle	Fire-making kits, dressed skins, compound tools	0.12 hp
Up to 12,000 years ago	Human muscle	Spear points, decorated artifacts, clay lamps	0.15 hp
Circa 6000 B.C.	Human muscle, small dogs (?) as pack animals	Paddles, dugout canoes, nets, bow drill	0.18 hp
Bronze Age— 3000 B.C.	Draft animals, sails	Plant and animal husbandry, trade, first metallurgy	0.2 hp+
Iron Age— 1000 B.C.	Water, wind, and animal power		Several hp+
Industrial Revolution— seventeenth and eighteenth centuries	Heat engines		100s of hp
Electric Age— late nineteenth century	Electric motors, storage and control of multiple forms of energy		1000s of hp
Information Age—late twentieth century	Use of electronics, computers, transmission of information and power globally and beyond		Unlimited hp

for more energy, but the techniques used were very inefficient; the "return on investment" was very low. It was not until the Industrial Revolution, in the eighteenth century in Western Europe and nearly a century later in the United States, that energy demands began to make serious inroads into supplies.

Improvements in products enabled human beings to travel farther, faster, and at greater energy expense than ever before. Energy use at the beginning of the Industrial Revolution was still heavily reliant on renewable resources: wood, wind, and water. As greater demands were

TABLE 2-2 Energy Sources Through Human History

Date	Source	Use
100,000+ years ago	Human (muscle)	Basic survival
100,000 years ago	Fire from wood	Heat, cooking, protection
10,000 years ago	Domestication of animals	Oxen, horses
2000 B.C.	Coal for metalworking	Greater heat
1700 B.C.	Wind	Windmills for pumping water
100 B.C. (China)	Coal	Domestic heating
	Waterwheels	Mechanical operations
A.D. 1200	Windmills	Mechanical operations
	Commercial coal mining	Domestic heating
1710	Steam engine	Pumping water from mines
1750	Steam engine	Transportation
1792 (Scotland)	Coal gas	Illumination
1808 (England)	Coal gas	Illumination and heating
1850	Coal replaces wood as major energy source	Industrial process heat
1859	First petroleum from "oil wells"	Industrial process heat
1880	Internal combustion engine	Increased heat to mechanical energy
1882	Hydroelectric plants	Lighting and motors
1892	Wind driven electric generators	Local lighting needs
	First American internal combustion vehicle	Increased mobility
1905	Geothermal energy	Electricity generating and water heating
1910 (Germany)	Fischer-Tropsch process (synthetic fuels)	Oil and gas from coal
1930s	Pier, Lurgi processes	Oil and gas from coal
1950s (USSR)	Underground coal gas electricity generation	Reduced cost of shipping coal
1954	First nuclear plant on line	"Cheap" electricity
1968	Tidal generators	Electricity from nonpolluting source
1970s	Photovoltaic generation (commercial)	Direct electric generation
1980s(?)	Fusion reactors	Net generation of electricity (?)

made on production, other sources of fuel had to be found. People were often considered to be a resource to be exploited. Colonization had as one of its goals the location of new sources of raw materials. Another goal was the relocation of numbers of people, some drawn by the promise of riches, some by the pressure of personal beliefs, some by exile, some by choice. The wider dissemination of peoples who were tied together by trade and commerce, even though separated by greater and greater distances, meant greater demands on energy resources.

RENEWABLE ENERGY USE

Solar energy has been recognized as an important part of life and as the primary source of energy by many cultures, including the Anacazi Indians of the Southwestern United States. Their cliff dwellings were designed to take advantage of the natural rock formation overhangs to block the high summer sun yet to admit the low winter sun for heat and light. The cultures of desert dwellers in the Middle East have used the principle of the "solar chimney" to increase air movement in their homes and other buildings. The solar chimney uses the upward movement of heated air coupled with air movement from underground passages through earth tubes, where the subsurface temperatures are substantially lower than air temperatures surrounding the structure, resulting in a cooler, more comfortable building. The air movement is provided by the same source that people seek to avoid during the hottest hours of the day—the sun.

The effects of the sun (and the moon) are not limited to collectors and other direct and indirect solar-gain devices and systems. The combined effect of both these space travelers is found in the tides along every coastline. Although variable at any location, many sites have proved suitable for dam building, blocking tidal flows at their high point, then allowing the trapped water to flow back through waterwheels (turbines) when the tide has receded. Tidal power has been used along the Atlantic coast since colonial times for grinding grain and sawing wood. Although the technology to use the tides to produce electricity is not yet fully developed, feasibility studies have shown that this is destined to become an important alternative energy source for some countries.

The use of energy from other renewable sources is not new. Falling water and wind power have long found great favor in performing many small-scale tasks requiring mechanical energy. The use of water mills of various designs dates back some 2000 years to the eastern Mediterranean basin. Most mills were used to grind grain into flour, but there were many other uses, such as fulling hand-loomed cloth, raising water for irrigation, crushing olives, carding wool, and sawing lumber. Western expansion in the United States was aided by water for transportation but also as an energy source using the same types of mills found in Europe. The lack of a well-developed road system tended to restrict agricultural and community development to distances of less than 25 miles (40 km), a 1-day trip with horse and wagon, from these waterways.

The federal census of 1840 reported nearly 60,000 rural water mills: gristmills, sawmills, and fulling mills. The size ranged from under 1 hp to as much as 2 or 3 hp, with efficiencies in the range of 15 to 20%. As recently as 1920, more than 10,000 mills were still operating.

Falling water was the principal power source for the generation of electricity before the extensive use of coal, oil, natural gas, and nuclear-fueled facilities. Many of these smaller hydroelectric plants are still in place but unused and in disrepair, because fossil fuel products have been easily available and relatively inexpensive and because hydro plants often are not sited or sized to provide the large-scale electricity-generating capacities demanded by extensive geographic distribution systems. They remain, however, a feasible source of electricity production and an alternative to the dwindling finite sources of energy based on fossil remains.

Recent improvements in the design and efficiency of hydro-powered equipment, as well as escalating fossil fuel prices, have again made these plants the most economical method of generating electricity. Studies of the hydroelectric generating potential in the northeastern United States have shown that if it were harnessed by known (existing) technology, hydroelectric could compensate for the energy equivalent now supplied by uncertain and expensive overseas petroleum supplies.

Wind systems are as varied as the countries that make use of this "solar" resource. "Wind catchers" have been used from the earliest known times as a propulsion system for boats and as a means to raise water from wells, lakes, and rivers. The best known illustration is that of the Dutch windmill used to "create" farmland from swampland adjacent to the sea. Other uses include the use of windmills to grind grain and saw wood.

Wind energy was used during the 1920s and 1930s all across the mid-United States for pumping water from shallow and deep wells for farmers and their livestock. The wind was also used to generate electrical energy using thousands of small wind electric plants. Banks of storage batteries were used to meet the relatively low electric needs on farms when the winds did not blow. Wind turbines have more recently been designed to be used in large "wind farms" such as the systems in the Altamont and Tehachapi Passes in California, where there are more than 20 wind-farm sites. (The turbines in use there range in size from

25 kW to over 100 kW each and have a combined capacity of over 220 MW.)

The electricity produced is sold to public utilities within the state. Plans are being developed to have more than 2000 generators erected in San Gorgnio Pass near Palm Springs, California, by 1990. The output is expected to be 1000 MW, an amount in excess of that produced by Hoover Dam. The National Aeronautics and Space Administration (NASA), in cooperation with the DOE, has undertaken an extensive study of large-scale wind generators. Considerable debate exists over the relative merits of medium-scale wind-energy conversion systems versus megawatt-size wind units.

Wood as a fuel currently contributes approximately 8% of the world's energy supply. It was the fuel of choice long before recorded history, being replaced as a major fuel in the United States only in recent history. Peat and coal, with their higher Btu content per weight,

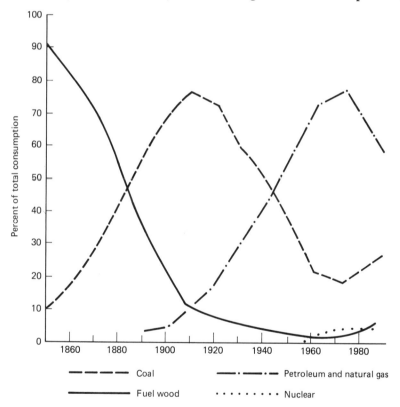

Figure 2-1 Transitions in major U.S. energy sources, 1850–1980. [Adapted from Richard C. Dorf, *The Energy Fact Book* (New York: McGraw-Hill, 1981), p. 25.]

replaced wood in the late eighteenth century. Oil in turn replaced coal in the latter part of the nineteenth century, being easier to transport to its ultimate use point and requiring no waste or residue handling. By 1900 wood accounted for less than 25% of the energy consumption in this country and by 1972 reached a low of 2%. Fuel wood use is now showing an increase as the result of escalating fossil fuel prices.

Wood is easily found, has a relatively high Btu content for its weight, and is easily gathered—so easily gathered, in fact, that entire forested areas have systematically been stripped to the degree that regrowth can take place only by extensive reforestation practices. Accompanying the loss of forests are the loss of ground cover and nutrients in soils which are in many cases marginal for agricultural purposes. The loss of soil due to erosion from both wind and water, and loss of habitat for entire ecological systems are also serious problems.

In spite of our recognition of the ecological damage resulting from increased use of wood as a fuel, the recent trend is to use it in larger and larger amounts. Industry now accounts for more than 60% of the 150 million tons (136,000 kg) burned every year in the United States.

Wood can become a more feasible energy resource than before. Together with cogeneration processes, wood can again be competitive with other fuels, especially for on-site or near-site use. These conditions can often make marginally profitable industries into money makers, saving and often increasing the job opportunities in industries hard hit by conventional fuel prices.

Dried dung, the manure of any species of the bovine family, has been used as a source of fuel by the people of many nations. American Indians used the dung of the buffalo for fuel, since there was little wood on the Great Plains until the white man planted trees to help keep the light soil from blowing away. The early settlers in the westward expansion took their cue from those who had lived off the land for centuries and used dried manure from their oxen to fuel the cooking fires on the long trek toward the west coast. Dung is a major source of fuel for those living on the subcontinent of India, and its use gives cause for concern with the declining fertility of the soil in that region.

Geothermal energy is energy directly from the heat of the earth. Geysers and volcanic flows are witness to the fact that the deeper one goes into the earth, the hotter the environment becomes. The therapeutic values of mineral water have been known and used for centuries for personal rejuvenation as well as for medicinal purposes. Hot springs provided thermal energy, especially for heating water for domestic purposes.

Early in the twentieth century, geothermal wells were used in the United States to produce electricity as well as to heat water. This process was not competitive with other ways to produce electricity, and techniques to deal with the high temperature and often corrosive fluids had yet to be overcome.

To make use of the geothermal principle requires only that one drill deep enough to reach the heat, tap it directly or by some type of heat exchanger, and bring that heat to the surface. Each step in the process is costly. Add to this the problem of the corrosive effects of heated, chemically laden fluids, and other alternatives may be more economically feasible. Nevertheless, many sites in the United States and in other parts of the world are currently being developed.

NONRENEWABLE ENERGY USE

The rapidly rising cost of fossil fuels has made people all over the world recognize that fossil fuel reserves are dwindling faster than new deposits can be found. There has been an upsurge in exploring new ways to supply various forms of energy from sources that were once thought to be too expensive or too exotic to consider.

The world energy supply during the recent past has been predominantly from fossil (nonrenewable) sources and is expected to continue to be from those sources until at least the year 2000. Figure 2-2 indicates the relative shares of the principal energy sources for the past two decades and the decade to come. Petroleum is indicated as decreasing rapidly as a major energy source, with coal increasing, but at a slower rate.

Nuclear and hydro, used primarily for generating electricity, have been less than 20% until recently but will increase toward the end of the century. Geothermal, wood, waste, and wind energy (and photovoltaics) usually termed "other sources," have shown rapid growth to this point and will continue to increase in importance.

In less than 125 years, we have consumed the majority of fossil fuel resources, which took millions of years to produce by natural means. Many of the deposits of oil and natural gas were relatively easy to access, develop, and distribute. The fuels derived were initially inexpensive compared to other fuels, often more convenient, and thought to be less harmful to the environment. This last misconception, as well as the "need" to supply a rapidly expanding population with an ever-growing appetite for the "better things in life," led to a rapid increase in the rate of consumption of all types of goods.

We were not content to consume only those natural resources found in our own country, but went to great lengths to import petroleum and many other mineral products from other countries. Our dependence on imports has created a strain on our economy as well as a drain on the resources of other countries.

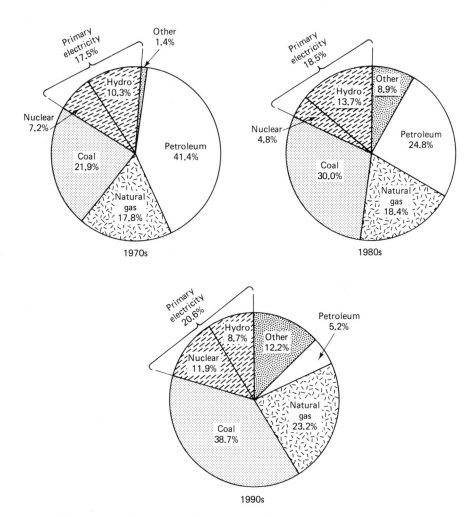

Figure 2-2 Percentage shares of world energy supply, 1970-2000. (From Duncan L. Gibson, *Energy Graphics*. Copyright © 1983. Reprinted by permission of Prentice-Hall, Inc., Englewood Cliffs, N.J.)

Nuclear energy, once thought to be the "ultimate solution," has been found to have too many associated problems and is too expensive to rely on as the major alternative to fossil fuels. Nuclear reactors for generating electricity have increased in cost and in construction time to the point where serious questions must be asked regarding their ultimate usefulness. Although the technology is little more than 30 years old, a multitude of changes in federal regulations have made the industry a high-risk proposition for investors and developers. There have been many problems, such as too many unproven design factors, too little

TABLE 2-3 Per Capita Energy Consumption, United States

Year	Total Energy Consumed (quads)	Total Energy per Capita (millions Btu)
1850	2.5	105
1900	8.3	110
1925	20.8	180
1950	34.1	226
1970	67.4	330
1972	72.1	345
1973	75.6	351
1974	73.9	346
1975	70.7	330[a]
1976	74.4	345
1977	76.3	352[a]
1978	78.0	357[c]
1979	78.9[b]	351
1980	75.9	337[d]
1981	74.1	324[a]
1982	70.9	308[a]
1983	70.6	301

[a] Estimated value.
[b] Total peak energy consumption.
[c] Per capita peak energy consumption.
[d] Conservation efforts expanded.
Source: U.S. Department of Energy.

concern for the disposal of radioactive waste products, and the lack of acceptable techniques for the decommissioning of "old" plants. Electricity that was once "too cheap to meter" may now be too costly to consider as an energy source for the future.

The consumption of energy in the free world is expected to increase from 202 quads used in 1979 to 266 quads by 1995, according to the DOE. That portion of the diagrams shown in Figure 2-3 noted as "other" includes developing countries in several parts of the world. Their demand for energy is expected to increase at a more rapid rate than that of the "developed" or industrialized nations. Their share of the world's 1979 energy "budget" was 40 quads, but this is expected to increase to 72.8 quads in 1995, an amount about equal to energy-use levels in the United States in 1983.

The energy growth rate in the United States is projected to grow by about 1.5% per year according to one estimate of the DOE. As shown in Figure 2-4, the proportions of both oil and natural gas will decrease substantially, and coal use is expected to increase by a factor of about 3.

An increase in nuclear and "other" sources of energy is expected

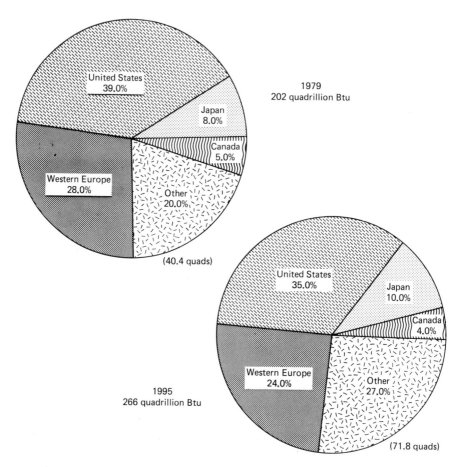

Figure 2-3 Consumption of energy by region, 1979 and 1995. (From Duncan L. Gibson, *Energy Graphics.* Copyright © 1983. Reprinted by permission of Prentice-Hall, Inc., Englewood Cliffs, N.J.)

as well. The rate of increase for nuclear will depend largely on the licensing of nuclear facilities, an area of much concern and controversy over the past decade. The role of "other" will include increased amounts of energy from renewable sources.

Several scenarios have been developed by the DOE to determine the growth/decline in various energy sources to the first decade of the twenty-first century. Figure 2-5 shows those energy sources with the largest expected growth to be coal and renewables, with nuclear remaining level after 1990, and natural gas and oil declining as indicated by other projections. Shale oil, although much too costly in today's economy, is shown to make some growth after the turn of the century. Coal for synthetics is expected to become a factor after 1995.

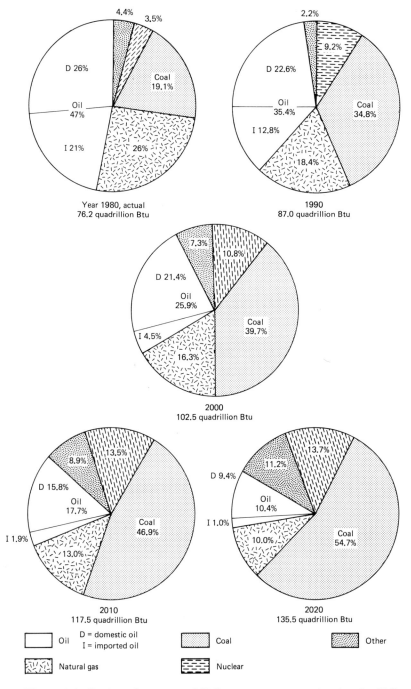

Figure 2-4 Projected sources of U.S. energy consumption by the U.S. Department of Energy (1980 and 1981). (From Duncan L. Gibson, *Energy Graphics*. Copyright © 1983. Reprinted by permission of Prentice-Hall, Inc., Englewood Cliffs, N.J.)

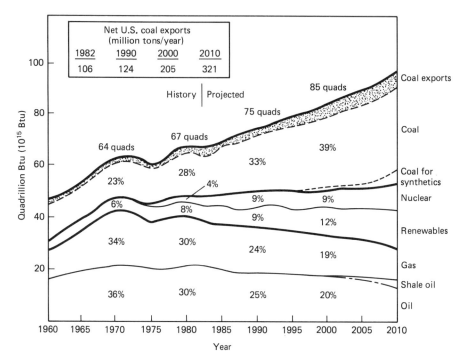

Figure 2-5 U.S. energy production: scenario B. (From the U.S. Department of Energy.)

The trend of rapidly increased energy use began in the early 1970s but then fell off to a more sustainable rate in recent years, as shown in Figure 2-6. Some of this reduced demand is believed to be due to conservation efforts as well as to the increased cost of energy. The results of the economic downturn undoubtedly helped keep energy consumption down as well. A more important factor is the rate of energy consumption per dollar of gross national product (as shown by the shaded bars). Energy consumption need not increase in direct proportion to economic growth (i.e., greater productivity is possible with lowered energy expenditure).

All this is encouraging news compared with the knowledge that the known fields of oil and natural gas are being depleted more rapidly than they are being replaced with newly discovered reserves. Comparisons of known reserves, estimated reserves, and potential reserves are difficult to determine. The best estimates of crude oil potential were made by M. King Hubbert during the 1960s. His projections indicated a discovery/production pattern which assumed the shape of the well-known "normal curve" beloved by statisticians. Hubbert's projections have been close to actual production rates in spite of dramatic changes

Nonrenewable Energy Use

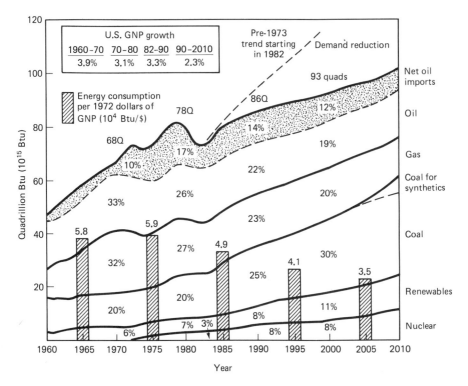

Figure 2-6 U.S. energy consumption: scenario B. (From the U.S. Department of Energy.)

in oil prices, changing production technologies, and reductions in demand. Figure 2-7 indicates that the actual crude oil production rate during the decades between 1960 and 1980, although sagging below the projection early and rising above it later, is still "on target." Most important the point of peak production occurred almost exactly (in the early 1970s) as anticipated.

The United States is importing much less oil than in the past two decades. This factor may help to extend the year when "peak production" of oil occurs for the world. Production is still increasing for much of the known fields and was recently projected to peak about 2000, as shown in Figure 2-8.

Coal is expected to have a much extended period of world production, extending for several hundred years based on known reserves (Figure 2-9). The reserves of the United States are expected to provide energy for at least 285 years. The use of coal is expected to increase several environmental problems, including additional amounts of sulfur in the air and therefore greater amounts of acid rain. Increased

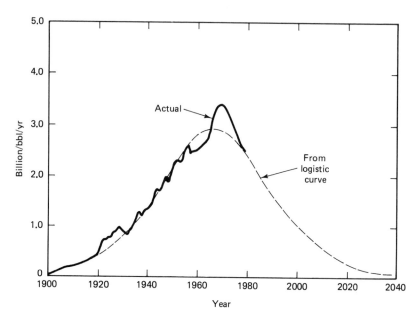

Figure 2-7 U.S. crude oil production rate (excluding Alaska). (From Jerrold H. Krenz, *Energy: Conversion and Utilization*, Second Edition. Copyright © 1984 by Allyn and Bacon, Inc. Used with permission.)

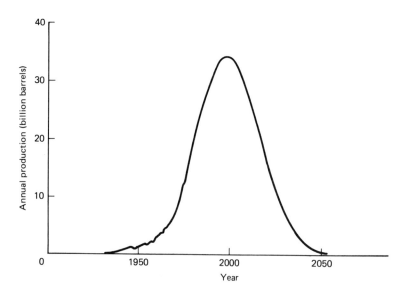

Figure 2-8 Projected world oil production. (Adapted from S. F. Singer, *Energy: Readings from Scientific American*, p. 209. Copyright © 1979 by Scientific American, Inc. All rights reserved.)

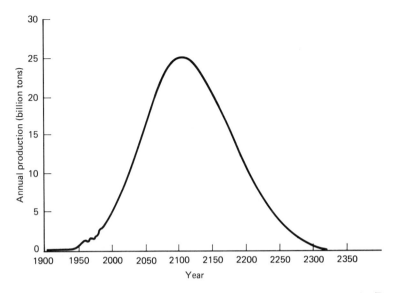

Figure 2-9 Projected world coal production. (Adapted from S. F. Singer, *Energy: Readings from Scientific American*, p. 209. Copyright © 1979 by Scientific American, Inc. All rights reserved.)

coal consumption could also raise the level of CO_2 in the atmosphere, increasing temperatures through the "greenhouse effect."

See Appendix H for tables of recent U.S. energy production and consumption.

ELECTRICITY GENERATION

As shown in Chapter 3, the generation of electricity from other energy forms is in most processes very inefficient. Still, we continue to rely more and more on this very versatile energy form. It remains the easiest to "transport," and it is difficult to match for its multiple end-use applications—heat, light, and ease of conversion to mechanical energy. The initially high cost of electricity produced using photovoltaic cells has been reduced to the point where it is competitive with electricity generated by means of nuclear processes. Although there is ample electricity-generating capacity in the United States at the present time, many of the facilities are reaching the point where major repair or replacement will be needed. This is a particularly pressing issue when it is noted that the trend is to rely more and more on electricity as a major energy source.

Electricity use in the United States grew at an annual rate of 7% per year until the mid-1970s. Since then the increase has been less than

0.5% per year, due in part to economic conditions and conservation efforts. As the economy recovers, a greater rate of increase is anticipated—in the range of 3% per year through 1991 according to the North American Reliability Council.

In view of this relatively modest increase, there are concerns about the ability of the electrical industry to meet the demands of the commercial, industrial, and domestic sectors by 1990. If orders for new nuclear power plants continue to be canceled and costs of those under construction continue to escalate, costs for electricity could increase substantially, while reliability could be reduced.

Electrical generation in 1983 was supplied primarily from coal-fired plants (54.5%), with much less costly hydro providing the next largest amount (14.3%) as indicated in Figure 2-10. Hydro sites have been located in many states and are being developed in substantial numbers.

Nuclear sources are expected to provide about the same proportion in the near future, but gas and oil are projected to decrease. Renewables are represented as part of "other," which also includes geothermal and waste. Although the contribution of this group is quite small at the present time, it is expected to increase substantially in the next 20 years. Wind generators installed in the years 1980-1983 in California alone have produced as much electricity as one nuclear plant, at substantially less cost and in one-third the construction time. It should

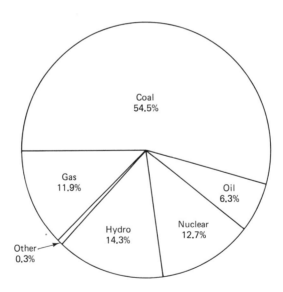

Figure 2-10 U.S. electricity generation by primary source. (From Energy Information Administration, *1983 EIA-Monthly Energy Review*, February 1984, p. 70.)

be noted that not all sites are suitable for wind generators. Other renewable sources, such as geothermal and hydro, are site-specific as well.

WASTE AS A RECOVERABLE RESOURCE

Americans produce about 1 ton (900 kg) of waste per capita each year. The demand for products for living and working in a technological age produces waste and by-products that are highly complex and often dangerous. Life in a less technologically-oriented society produces waste material that is easily (and often naturally) recycled.

"Technical progress" has made products with more physical durability but not necessarily with greater "workability." Mining and manufacturing, chemical processing, and development of artificial materials such as polymers have multiplied the problem of disposal of products and by-products of the Industrial Age.

Conventional practices make use of some by-products, as distinguished from "waste," because productive uses have been found for them. If we begin to view waste as a resource, for recycling or as an energy source, it begins to take on value and becomes a "natural resource."

The amount of waste that is recycled in the United States is increasing because of a shortage of landfill areas and because of an awareness of the value of recycling to the energy and environmental picture. Europeans have done much more than Americans to take advantage of recycling technologies, primarily because land is too scarce to use it for a waste receptacle and their costs of fossil fuels are substantially higher than costs in this country.

CONSERVATION

Until the several "energy crises" of the 1970s, conservation of energy was not popular. Energy was "too cheap" to conserve! "There is always more where that came from!" Costs of insulation were considered to be higher than the fuel such material would save. The importance of upward mobility was "fueled" by the perceived need to acquire more goods and services, without consideration for our dwindling resources.

During the past three or four decades the booming birthrate worldwide and a corresponding increase in the demand for goods and services, coupled with other disruptive economic and political factors, all seemed to occur at the same time. Higher costs for all energy products, based primarily on artificially higher prices for petroleum products, suddenly made an impact on individuals and on the business and industrial sectors in 1973. Smaller cars, more efficient appliances, insulation, and any other measure that can be expected to reduce consumption are now in vogue.

Conservation techniques, ranging from "no-cost" procedures such as turning off lights to keeping doors and windows of unused rooms closed and thermostats turned lower in cold weather and higher in warm weather, have saved precious energy resources. Low-cost measures include water-flow reducers, automatic-setback thermostats, and insulation both for buildings and water heaters. Higher-cost practices, generally measured in "payback time," have found wide acceptance in commercial and industrial firms. Computer-controlled systems manage heat, ventilation, and light usage. Reuse of "waste" process heat, previously vented to the atmosphere, is now sent through heat exchangers to preheat incoming air and water, saving billions of Btu. Energy resources, their impacts, and their potential are addressed in detail in the following chapters.

The depletion of our nonrenewable resources has resulted largely from our unwillingness to examine energy sources based on our knowledge of the principles and laws of physics. We must organize and manage all sources of energy wisely. By doing so we will be able to increase standards of living for all people throughout the world.

There is no energy shortage, only a shortage of applications of known resources, knowledge, and technologies.

ACTIVITIES

1. Make two lists of energy-consuming devices found in your home or that of a friend, using categories of "essential" (used to prepare food, for heat, and for light), and "nonessential" (for entertainment, convenience, vanity, or appearance). Determine the Btu or wattage consumed by each appliance over a period of time—a day, week, or month. Which list uses larger amounts of electricity?
2. Which of the appliances listed in Activity 1 use nonrenewable forms of energy? Which use renewable forms?
3. Review the historical use of energy as a general resource to provide for human needs. Consider especially the many sources, methods of use, and quantities used, and how emphasis of use has changed from source to source.
4. Human beings began using natural, renewable energy sources, turned to fossil fuels, and are now returning to renewables. Discuss this cycle of events, reflecting on the social conditions, quality of life, economic factors, and environmental impact that have caused and are the result of these events.
5. What renewable energy resources are major sources of energy for your community?

6. Which of the renewable energy resources can be used in your community? Which ones would not be usable, or usable for only part of the year?

BIBLIOGRAPHY

Brown, Lester R., and Pamela Shaw, *Six Steps to a Sustainable Society*, Worldwatch paper 48. Washington, D.C.: Worldwatch Institute, March 1982.

Deudney, Daniel, *Rivers of Energy: The Hydropower Potential*, Worldwatch paper 44. Washington, D.C.: Worldwatch Institute, June 1981.

Deudney, Daniel, and Christopher Flavin, *Renewable Energy: The Power to Choose*. New York: W. W. Norton, 1983.

Diamond, Henry L., Chairman, Citizens' Advisory Committee on Environmental Quality. Washington, D.C.: The Committee, 1974.

Federal Republic of Germany, *New and Renewable Sources of Energy*. Bonn: The Federal Minister for Research and Technology, August 1981.

Flavin, Christopher, *The Future of Synthetic Materials: The Petroleum Connection*, Worldwatch paper 36. Washington, D.C.: Worldwatch Institute, April 1980.

Gibbons, John H., and William U. Chandler, *Energy: The Conservation Revolution*. New York: Plenum Press, 1981.

Hayes, Denis, *Energy for Development: Third World Options*, Worldwatch paper 15. Washington, D.C.: Worldwatch Institute, December 1977.

———, *Repairs, Reuse, Recycling—First Steps toward a Sustainable Society*, Worldwatch paper 23. Washington, D.C.: Worldwatch Institute, September 1978.

Krenz, Jerrold H., *Energy: Conversion and Utilization*. Boston: Allyn and Bacon, 1976.

Norman, Colin, *Soft Technologies, Hard Choices*, Worldwatch paper 21. Washington, D.C.: Worldwatch Institute, June 1978.

Priest, Joseph, *Energy for a Technological Society*. Reading, Mass.: Addison-Wesley, 1979.

Pryde, Philip R., *Nonconventional Energy Resources*. New York: Wiley, 1983.

Smith, Nigel, *Wood: An Ancient Fuel with a New Future*, Worldwatch paper 42. Washington, D.C.: Worldwatch Institute, January 1981.

Steinhart, Carol E., and John Steinhart, *Energy Sources, Use, and Role in Human Affairs*. North Scituate, Mass.: Duxbury Press, 1974.

Chapter 3
Understanding Energy
What It Is, How It Works, Forms, and Properties

CONCEPTS

1. Energy is required to do work.
2. In any energy conversion, the total amount of energy is the same; it just changes form.
3. We do not actually "consume" energy; it is simply changed or converted to other forms. Once changed, energy cannot be used again in the original form.
4. Some energy is too diffused or disorganized for use.
5. Some sources of energy are finite, others are renewable.
6. Efficient use and appropriate application of energy is necessary to emerge from our energy dilemma.
7. Steady growth in a finite environment is impossible.

GLOSSARY

DOUBLING TIME—the time required for a quantity to double its initial size.
EXPONENTIAL GROWTH—when a quantity has a fixed growth rate, it is undergoing exponential growth.

FINITE—having measurable or definite limits; not renewable.
FOSSIL FUEL—any naturally occurring fuel of an organic nature, such as coal, natural gas, and crude oil, which requires thousands of years to form.
INEXHAUSTIBLE—sources of energy which are regenerative, such as geothermal.
KINETIC ENERGY—energy in motion.
PETROLEUM—a naturally occurring material, which is a chemical compound of carbon and hydrogen.
POTENTIAL ENERGY—energy at rest.
RENEWABLE—sources of energy which are replaced regularly, such as solar, wood, and wind.
TEMPERATURE—a relative quantity of heat.

INTRODUCTION

What is energy? We can see the effects of energy after a severe windstorm as we look at damage to trees, homes, or personal property. We can see the damage that has been done to an automobile that has been in an accident. In many cases we cannot actually see the energy itself, but we are often able to observe as the energy is being changed from one form to another, and we can often see the results of the change, as in the wind-damaged trees or the smashed automobile.

We know that energy is necessary to warm our homes, heat water, move cars, and produce products. Energy is also required by our bodies. The food we eat is actually a form of chemical energy (fuel) which heats our bodies to 98.6 degrees Fahrenheit (°F) [37 degrees Celsius (°C)] and allows us to participate in daily activities. Our bodies change or convert the chemical energy in the food to heat energy and muscle operation (motion energy). Our bodies are conversion systems that take one form of energy and convert it to other, more usable forms.

All forms of energy have the ability to do something or to cause something to happen. The potential of or the result from energy performing a task is called work. Energy may therefore be defined as the measure of ability or capacity to do work.

FORMS OF ENERGY

There are several forms of energy on which our society depends. Some are used as they are found, while others must be converted to forms that are suitable to the task. These energy forms are:

CHEMICAL—the composition of fuels. Easily converted to heat.
ELECTRICAL—the movement of electrons. Often converted to heat, light, or magnetic energy.

GRAVITY—draws all things toward the center of the earth.
HEAT—cooks our food, warms our homes, available from the conversion of fuels.
LIGHT—provides illumination; causes photosynthesis.
MAGNETIC—the energy form surrounding a magnet.
MECHANICAL—the energy of motion. Objects in motion also have kinetic energy.
NUCLEAR—mass energy: fission and fusion.
SOUND—vibrations; may be audible or inaudible.

There are two additional states of energy: kinetic energy, which is motion energy, and potential energy, which is energy at rest. A moving body, such as a baseball bat, can exert a force on another body, a baseball, and the transfer of energy can result in a home run. Potential energy, on the other hand, is relative to position or composition. A heavy object placed on a table has potential energy in relation to the floor. Gasoline, a chemical form of energy, has potential energy due to its composition. This energy can be released as heat with the aid of a spark of electrical energy. If this release of energy occurs within the confines of an engine, the rapidly expanding gases can be directed to perform work.

SOURCES OF ENERGY

Energy is a natural resource and therefore its various sources are limited as to total quantity available or rate of delivery. For example, petroleum, a form of chemical energy, is a finite resource. There is only a given amount available, and when that is used there will be no more. We can use petroleum rapidly, slowly, or at any rate we desire.

Other sources of energy, such as solar (light energy) or geothermal (heat energy), are available at only a given rate of delivery. The sun's rate of energy output cannot be increased, is available only during specific hours, and can be interrupted by stormy or cloudy weather.

All sources of energy can be classified in one of two major categories, fossil or renewable (Table 3-1). Fossil energy sources have been developed by natural forces over a period of many thousands of years. These forces have not stopped, but the rate at which they function is so slow that we cannot expect any increase in energy availability from this category. Energy sources that are continually replaced on a cyclical basis with no foreseeable termination of supply are classified as renewable. A third category, inexhaustible, is often included to encompass those energy sources which are neither finite nor truly renewable on a cyclical basis. Geothermal energy is often placed in this category.

TABLE 3-1 Energy Sources and Their Classifications

Fossil	Renewable	Inexhaustible
Coal	Solar	Geothermal
Petroleum	Wind	
Natural gas	Hydropower	
Nuclear: fission	Nuclear: fusion	
Synthetic fuels	Biomass	
	Ocean energies	
	Tides	
	Waves	
	Currents	
	Thermal differences	
	Salinity gradients	
	Synthetic fuels	
	Hydrogen fuels	

ENERGY TERMS

To discuss energy intelligently, including its various forms and methods of conversion, it is necessary to become familiar with some of the terms. Many of these terms are expressed in units of measurement so that quantities can be calculated and the scientific interrelationships understood.

Force

We generally think of force as a push or pull. Force, for example, is necessary to remove a stuck automobile from a ditch. The force may be applied humanly or with the assistance of another vehicle, such as a tow truck.

There are four fundamental kinds of forces: gravity, electromagnetic, weak interaction (as in radioactive decay), and strong interaction (as that which holds atoms together). Sir Isaac Newton was responsible for finding the force exerted on a body and the equation

$$\text{force} = \text{mass} \times \text{acceleration}$$

$$F = m \times a$$

Force is expressed in units of newtons (N) or foot/pound/seconds (ft/lb/sec).

Work

From a physics standpoint, work is the application of force in relation to the distance that force has acted. Work is actually energy in transition. It may be expressed by the equation

$$\text{work} = \text{force} \times \text{distance}$$

$$\text{work} = F \times d$$

Work is expressed in units of joules (J) or pounds per second per square foot (lb/sec/ft^2).

Power

The rate at which work is done is relatively important when dealing with energy. If you were to carry a chair across the room in 6 seconds and your friend were to return it in 24 seconds, both of you would have done the same amount of work. That is, you both provided the necessary force to move the chair an equal distance. However, since you moved the chair in one-fourth of the time, you expended four times the power of your friend. Power is expressed as the rate of energy conversion, the rate of energy dissipation, or the rate of doing work:

$$\text{power} = \frac{\text{work}}{\text{time}}$$

$$P = \frac{\text{work}}{t}$$

Power is expressed in units of joules per second (J/sec), watts (W), foot-pounds per second (ft-lb/sec), or horsepower (hp).

An interrelationship may be seen when we consider that the work done by a body, which is equal to the energy given up by the body in doing the work, is equal to the power generated times the time necessary to do the work. This is expressed by the equation

$$\text{work} = E = P \times t$$

To understand this concept, think of electrical energy. We use and pay for this energy form in watts over a period of time. Since watts are relatively small units, we multiply them by 1000 to obtain kilowatts (kW). Our bills are based on the kilowatt-hour (kWh), the unit of electrical energy used for 1 hour.

Efficiency

The efficient use of energy is mandatory if we are to emerge from our energy dilemma. When we convert one form of energy to another, not all the energy available is converted to the desired form. In fact, each time we change or convert one form of energy to another, some energy escapes, often in the form of heat. Some energy conversions are more efficient than others. The calculation of efficiency is relatively easy and can be expressed by the equation

$$\text{efficiency} = \frac{\text{useful energy (work) out}}{\text{energy in}} \times 100$$

Energy efficiency is generally expressed as a percentage, so by multiplying the quotient by 100 the resulting efficiency will be properly symbolized.

Coefficient of Performance

When studying heat pumps in Chapter 14 you will be exposed to energy conversion units which use electricity to enhance low-level thermal energy to higher, usable thermal levels. In calculating the efficiency of devices such as heat pumps, we must compare electrical energy input to heat energy output. Since 1 watt (W) of electrical energy is equal to 3.414 Btu, the formula for determining the coefficient of performance (COP) of these devices is

$$\text{COP} = \frac{\text{Btu output}}{W \times 3.414}$$

A heat pump that has a COP of 2 delivers twice as much usable Btu as is required in an equivalent energy form to run the pump.

Energy units and conversion factors for calculating English and metric (SI) quantities are given in Appendix A.

ENERGY USE

Many energy sources, whether fossil or renewable, are not initially available in a form that is easily used. For example, we need to heat our homes, but since heat is not easily transported, we must select an energy source which can be transported and then convert that source, often a chemical form, to usable heat energy.

It is most important that people understand what really occurs when we involve ourselves in the transportation and conversion of energy sources. Each time we move an energy source from where it is found to where it is used, we lose because the transporation of energy requires energy, both to manufacture the transportation device (truck, train, pipeline) and the energy to make the device function (gasoline, fuel oil, electricity). In addition, the conversion of energy from one form to another is never 100% efficient; some is always lost in the transition. Listed below are some energy uses and respective efficiencies. A more comprehensive list is provided in Table 14-1.

USE	EFFICIENCY (%)
Mechanical to electrical	98
Gas furnace to heat	60–80
Fossil fuel to electricity plant	35–40

When determining the overall or end-use efficiency of an energy conversion system, the efficiency of each step must be multiplied by the end-use efficiency of the previous step. Table 3-2 shows the resulting efficiency of using electricity from a coal-fired plant to light an incandescent bulb.

TABLE 3-2 End-Use Energy Conversion Efficiency (%)[a,b]

Step	Efficiency of Step	End-Use Efficiency
Production of coal	96	96
Transportation of coal	97	93
Burning of coal to generate electricity	38	35
Transmission of electricity	85	30
Incandescent bulb	5	1.5

[a]These are typical efficiencies.
[b]Example: $0.96 \times 0.97 = 0.93$.

To provide light, energy had to be changed from chemical form to heat energy to mechanical energy to electrical energy to light energy. Does it surprise you that 98.5% of the energy available in the coal is actually "wasted" so that we can artificially illuminate our environment?

To obtain a clearer picture of our energy problem, it is necessary to understand the concept of doubling time and exponential growth. Energy use is actually a product of activity and efficiency. The level of activity is related directly to the size of the population and the level of industrial sophistication that the population has attained. Efficiency refers to the rate at which energy is converted as well as the energy that is simply wasted by that population.

When a quantity has a fixed growth rate it is undergoing exponential growth. Table 3-3 shows how rapidly energy use will increase if there is a constant growth rate.

TABLE 3-3 Doubling Times for Selected Yearly Growth Rates

Yearly Growth Rate (%)	Doubling Time (years)
1	69.3
2	34.7
3	23.1
4	17.65
5	14.25
7	10.25
10	7.25

For a yearly growth rate of 7% (such as the rate for electrical energy demand in the early 1970s), the doubling time is 9.9 years. Since electricity-generating plants take so long to build, you can appreciate how utility companies, in planning to double their generating capacity in less than 10 years, became involved in the construction of large nuclear-powered generating facilities.

Let's follow the exponential growth concept a bit further. You have just inherited $1000 from a rich uncle, and rather than spend it foolishly, you deposit it in a bank at a fixed 10% annual interest so that you can watch it grow. Table 3-4 illustrates what will happen during the first 15 years of your investment.

You did well! It took a little more than 7 years and 3 months to double your initial investment. Five years and 2 months later you doubled it again and 2 years and 6 months later you doubled it a third time. The problem is that this same exponential growth theory applies to energy use and population increases as well as money.

THE LAWS OF THERMODYNAMICS

"Thermo" implies heat; "dynamics" implies movement or change. There are laws that relate specifically to energy conversions. The First Law addresses the conservation of energy and the fact that in the universe no energy is lost in transferring from one energy form to another. In other words, the energy quantity in the universe cannot be altered

TABLE 3-4 Investment Doubling Time

Year	Balance
0	$1000 (deposit)
1	1100
2	1210
3	1331
4	1464
5	1610
6	1771
7	1948 (doubled)
8	2143
9	2357
10	2593
11	2852
12	3137 (tripled)
13	3451
14	3796
15	4176 (quadrupled)

even though the forms are changed. As we reviewed the use of chemical energy and the necessary sequence to convert it to light energy, heat energy was present. Heat is often a result of energy conversion, but it is frequently dissipated or at temperatures too low for practical use.

The First Law is stated: When a control mass is altered adiabatically, the amount of work is fixed by the end states of the system and is independent of the details of the process.

Stated so that it can be better understood, the First Law says: The total energy in a system is equal to the quantity of the heat added minus the work done by the system. An equation for this would be

$$E_i - E_f = \Delta Q - \Delta W$$

where E_i and E_f represent the initial and final energy, ΔQ is the quantity of heat added, and ΔW is the work done by the system. The Second Law of thermodynamics will provide additional explanation.

Heat energy is transferred due to a change or difference in temperature. There are three methods of heat transfer: conduction, convection, and radiation.

Conduction is heat transfer through a material by the transfer of energy from molecule to molecule due to molecular collisions. The ability of a material to conduct heat is dependent on the physical properties of the material. For example, most metals conduct heat very well compared to materials such as wood or polystyrene foam.

Convection is heat transfer due to the flow or movement of

molecules. In this method of transfer, the medium must be fluid, such as air or water, where the molecules are able to move according to differences in temperature. For example, warm air is less dense (lighter) than cold air and therefore rises. Cold air is relatively more dense and falls. This is why the temperature in a room is always cooler near the floor and warmer near the ceiling.

Radiation is heat transfer by electromagnetic waves. For example, heat energy from the sun can reach the earth only by radiation. Air or space is not heated by radiation, and the energy is not transferred until the waves are intercepted by surfaces or objects. The warmth that you feel when standing near a wood stove or fireplace reaches you primarily due to radiation.

The Second Law of thermodynamics provides us with a way of comparing the effects of two forces. One statement of the Second Law is: In any isolated system there is always an increase in the entropy. Entropy is the amount of disorder, mixing, or confusion. For example, when energy forms are converted, heat often results. The problem is that this heat is often of low temperature and of negligible value, dispersed, or is not easily or economically retained for use. Another postulate of the Second Law is: Any system isolated from its surroundings evolves until it reaches a stable equilibrium state. This is the reason that perpetual-motion devices are not succcessful.

SUMMARY

Energy is the measure of the ability or capacity to do work. You have become familiar with terms such as force, work, power, and efficiency and can calculate each as they interrelate in our daily lives. Through the laws of thermodynamics you have seen how conduction, convection, and radiation transfer heat when there is a temperature differential. You have also seen how exponential growth rates affect the demand for available energy.

As we continue our study of energy forms and sources, the understanding that some sources are renewable and others are not will be important. The next chapters address each of the energy sources, the current availability of these sources, and provide projections—many of which will strongly affect your future.

ACTIVITIES

1. List each of the energy forms and give examples of how each was used by you today.

2. Make a table listing the sources of energy your family uses. Indicate whether each is in the renewable or the fossil classification. Also indicate the form of each before and after use. Example:

Energy Source	Renewable or/Fossil	Form Before Use	Form After Use

3. Suspend a weight from a string. Allow it to swing freely (like a pendulum) and measure the distance of its swing, initially, then after 1 minute.
   a. Does the swing distance change?
   b. Why?
   c. Where along the swing does one state of energy change to another?
   d. Will the weight ever completely stop swinging?
4. Follow the steps that an energy form takes from its origin to its final use. Calculate the end-use efficiency.
5. Compare two or more energy sources and the steps required to convert them to usable energy forms. Describe ways that one type of source could be substituted for another and the advantages and disadvantages of such a change.
6. Study the three methods of heat transfer. Experiment with various methods or techniques that might be used to reduce or increase heat transfer from various materials to the environment.

BIBLIOGRAPHY

Crawley, Gerard M., *Energy.* New York: Macmillan, 1975.

Hunt, V. Daniel, *Energy Dictionary.* New York: Van Nostrand, Reinhold, 1979.

Priest, Joseph, *Energy for a Technological Society.* Reading, Mass.: Addison-Wesley, 1979.

Schwaller, Anthony E., *Energy Technology: Sources of Power.* Worcester, Mass.: Davis, 1980.

Chapter 4
Coal
The Plentiful Fossil Resource

CONCEPTS

1. Coal was formed millions of years ago from buried plants.
2. Coal has been rejected as a major fuel source for decades.
3. Diminishing supplies and high prices for oil and natural gas are causing renewed interest in coal.
4. Of all the nations, the United States has the most extensive recoverable coal deposits.
5. Coal will play an important role as a future energy source.

GLOSSARY

AQUIFER—a subsurface water-bearing zone.
CULM—an impure form of "worthless" carbon found in coal seams.
FEEDSTOCK—fossil fuels used for their chemical properties to produce other products.
GREENHOUSE EFFECT—the name given to the atmospheric buildup of carbon dioxide (CO_2) gases which act like the glass in a greenhouse by allowing the sun's rays through to warm the earth and then retaining the heat.
MACERALS—organic substances derived from plant tissues and cells which have

decayed and have been altered physically and chemically by natural, geological processes.

OVERBURDEN—soil, rock, and other material located between the top of a coal seam and ground level.

SYNTHETIC FUEL—liquid or gaseous fuels resulting from conversion processes.

INTRODUCTION

Coal, the fuel that helped initiate the Industrial Revolution, is rapidly regaining a prominent role in the U.S. energy picture. Due to its bulk, inconvenience, and lack of cleanliness, coal became less popular with the development of oil and natural gas. The primary reason for this renaissance is economic. With petroleum and natural gas in comparatively short supply and with their prices rising, coal, which is available in abundant quantity, will again be important as a fossil fuel. The many types, varied chemical characteristics, and geographic locations as a natural resource will have considerable influence on the extraction, preparation, use or conversion, and environmental impacts as it regains a position of significance in our society.

BACKGROUND OF COAL TECHNOLOGY

Coal is an organic sedimentary rock that was originally formed millions of years ago. Black or brown in color, coal contains more than 50% carbonaceous material by weight and was formed by the accumulation and burial of plant matter acted on by temperature and pressure changes.

When burned, the chemical energy contained in coal is converted to heat energy that can be used to warm buildings, drive machines, or produce various products. Coal was once the main source of energy in all the industrialized nations. Its relatively low cost, high Btu content, comparative "cleanliness," and convenience aided its replacement of wood as a fuel. Coal-burning steam engines provided most of the power in these countries from the early nineteenth to the early twentieth century, and coal-fired furnaces met most of the industrial process and heating needs until the 1940s, when oil and natural gas came into prominence.

Although coal has been rejected as a primary fuel by most industries and homeowners since the 1940s, it has played a key role in the generating of electricity and the production of iron and steel. In 1983, 54% of all the electricity generated in the United States was obtained from coal-burning facilities. The smelting of iron ore to make iron and steel requires coke, a processed form of coal, for both its thermal and chemical properties. The rejection of coal by many of its former users was due to the associated handling, storage, and waste products. Oil, natural gas, and electricity reduced these problems while providing greater convenience and fewer obligations for the consumers.

The resurgence in the use of coal is just beginning. Limited supplies of oil and natural gas and the associated increase in prices are the primary reasons, together with the availability of an abundant, reliable U.S. coal supply. This trend toward reduction in consumption of oil and natural gas is placing pressure on the coal industry not only to supply coal in appropriate quantities to meet the needs of society but also to develop coal technologies that reduce the burden of handling, storage, and waste problems as well as reduce associated environmental impacts. It is the opinion of many high-technology proponents that, with the uncertainties of nuclear energy and oil imports, the expanded use of coal is a necessity.

PHYSICAL CHARACTERISTICS OF COAL

Coal is a nonuniform mixture of carbon (C), hydrogen (H), oxygen (O), sulfur (S), and other elements. Since it developed from a diversity of plant species that died 1 million to 440 million years ago, we are actually dealing with a variety of coals of differing physical and chemical qualities.

The plants that ultimately became coal are assumed to have grown in swamps or bogs. As the plants died they formed a thick layer of organic materials that decayed and were altered physically and chemically by natural geological processes. These organic materials, called macerals, became mixed with various inorganic materials during the sedimentation process. Since various geographic areas have had differing climates, plant species, and soil chemical structures, and the time for formation is so extensive and variable, each geographic area has its own coal "fingerprint." Any given sample of coal can be analyzed and the exact geographic location from which it came determined because of these differences. Additionally, coal extracted from the same seam or layer often has more than one physical and chemical composition within the fingerprint due to these same variables.

There are three classes of macerals found in all coals, which are identified according to specific physical and chemical properties: vitrinite, liptinite, and inertinite. Vitrinite is the material derived from plant woody substances, liptinite from the exines, and inertinite from the resins. Since lignin and cellulose are the dominant components of plants, vitrinite is the most abundant coal component. Because of its prominence and ease of isolation, the vitrinite maceral is used for most coal analyses. Coal chemical analyses are necessary so that combustion or conversion systems can be designed for maximum efficiency with minimum environmental impact.

The large variation in chemical composition has required that ranks be established to grade or identify coals. These ranks are based on the fixed carbon content of the coal, which corresponds to the stage of metamorphic development. Listed in order from lowest to highest, these ranks are: peat, lignite, subbituminous, bituminous, and anthracite.

Peat is not truly a coal but actually a precoal material. Peat is the remains of dead plants buried for thousands of years. It is dark brown, spongy, and looks much like decayed wood. Many people throughout the world dry and burn peat to heat their homes.

As time passes, water, additional decaying matter, and sediments exert pressure on the peat, pressing it into thinner layers. The combination of time, temperature, and pressure turns peat into lignite. Although lignite is harder than peat and provides more thermal energy when burned, it still has a high moisture content and the lowest carbon level of the coal ranks.

With the additional passing of time and added pressure, lignite progresses through subbituminous, bituminous, and finally to the anthracite rank. Bituminous coal is called soft coal because it can be broken easily into various sizes for use. Anthracite, which began to form as long as 440 million years ago, is hard, shiny black coal that burns without smoke. Figure 4-1 shows the various stages or ranks of coal. Note the loosely compressed particles of the peat.

Many studies have been and are being conducted on selected properties of coals. Since both time and pressure vary the composition and characteristics, there are considerable differences among the ranks. For example, the moisture capacity and volatile matter decrease and the carbon and Btu content increase according to the pressure that has been placed on the coal formation. Table 4-1 provides a comparative listing of the coal ranks and properties associated with the vitrinite found within those ranks.

Additional scientific classifications are recommended by Richard C. Neavel of Exxon Research and Engineering Company due to the multivariate nature of coal. These classifications would be based on:

1. The nature of hydrogen bonding
2. The amount of hydrogen
3. Coal structure
4. Oxygen and sulfur content
5. Organic and inorganic interactions [Gorbaty and Ouchi, pp. 9-10].

Figure 4-1 Ranks of coal. (Courtesy of the National Coal Association.)

TABLE 4-1 Selected Properties of Coals of Different Ranks

Property	Lignite	Subbituminous	Bituminous	Anthracite
Moisture capacity[a]	40	25	10	5
Carbon[a]	69	74.6	83	94
Hydrogen[a]	5.0	5.1	5.5	3.0
Oxygen[a]	24	18.5	10	2.5
Volatile matter[a]	53	48	38	6
Density	1.43	1.39	1.30	1.5
Btu/lb	11,600	12,700	14,700	15,200

[a]Percent by weight.

Source: adapted from a table by Richard C. Neavel, Exxon Research & Engineering Company, which appeared in Martin L. Gorbaty and K. Ouchi, eds., *Coal Structure*, Advances in Chemistry Series 192 (Washington, D.C.: American Chemical Society, 1981).

All of this classification and scientific analysis may appear unnecessary to the reader, but it is required in order to attain chemically consistent liquids and gases from the many different coals.

COAL RESERVES

Coal is found throughout the world in varying quantities and ranks. According to the World Energy Conference Survey of Energy Resources, the United States has 30.8% of the world's recoverable coal reserves. The USSR has 23.1%; Europe, 21.4%; and China, 13.5%. The remaining countries have single-digit quantities or less [Edison Electric Institute].

The estimated quantity of coal, as with other fossil fuels, available in the United States is difficult to determine due to the variation in sources, inclusions, exclusions, accuracy, and honesty of those reporting or compiling the data. There are also extensive areas of the United States which have been neither mapped nor explored for their coal deposits. According to the Energy Information Administration (EIA), as of January 1979 our country had approximately 4 trillion short tons of coal resources ["America's Coal Reserve Base"]. Of this amount, only one-eighth, approximately 474 billion short tons, is proven mineable due to excessive depth, geographic and geologic considerations, and current extraction systems technology. The proven available coal is called the demonstrated reserve base (DRB), but since mining safety, practices, and technology conservatively result in only about 50% recovery, our estimated current coal supply is actually 237 billion short tons, enough for about 285 years at current consumption rates.

Coal deposits, existing in 38 states, are considered mineable or worth mining in 31, but extraction is being conducted currently in only 26 states (see Figure 4-2). Of the DRB, 45% is located in states east of the Mississippi River and 55% in western states and Alaska. The ranks of coal comprising the DRB are: bituminous, 51%; subbituminous, 38%; lignite, 9%; and anthracite, 2%. Lignite and subbituminous ranks are found primarily in the West, while bituminous and anthracite deposits occur in the East. Most anthracite is located in Pennsylvania. In descending order of mineable tonnage, the states of Montana, Wyoming, Illinois, West Virginia, Kentucky, and Pennsylvania possess 76% of all the country's coal. Montana and Wyoming have 40% between them, but most is lower-rank subbituminous coal ["America's Coal Reserve Base"]. The U.S. Geological Survey (USGS) estimates that 57.5% of U.S. coal reserves have a sulfur (S) content of 1% or less. Western states boast 84% of this low-sulfur coal, but with most of the low-sulfur demand in the Midwest and East, massive transportation systems are necessary if this source is to be used extensively. Although eastern coal has a higher Btu content, much of it possesses 2 to 4% sulfur.

U.S. miners extracted approximately 830 million short tons of coal from 6425 mines in 1980. ["U.S. Energy Review"]. Nearly 80% was mined east of the Mississippi River in the Appalachian region from underground and surface mines with coal seams 2.5 to 8 ft (0.63 to 2.44 m) thick. By comparison, western coal is nearer the surface, permitting more surface or strip mining, and the seams are 20 to 250 ft (6.0 to 76.2 m) thick. The largest deposit of coal in the world is the Fort Union Formation, which encompasses the Powder River basin and the western part of the Williston basin in Montana, Wyoming, and the Dakotas. The high-quality subbituminous and lignite coals from this area are found in seams 50 to 75 ft (14.2 to 22.9 m) thick under 30 to 40 ft (9.1 to 12.2 m) of overburden. These coals are in demand because they perform well below the limit set by the Environmental Protection Agency (EPA) combustion standards for air pollution.

COAL EXTRACTION

The type of coal mine is determined primarily by the location of the coal seam in relation to ground level and the topography of the area. Underground and surface mines are the two major categories. In 1980, 40% of the total annual coal production was acquired from underground

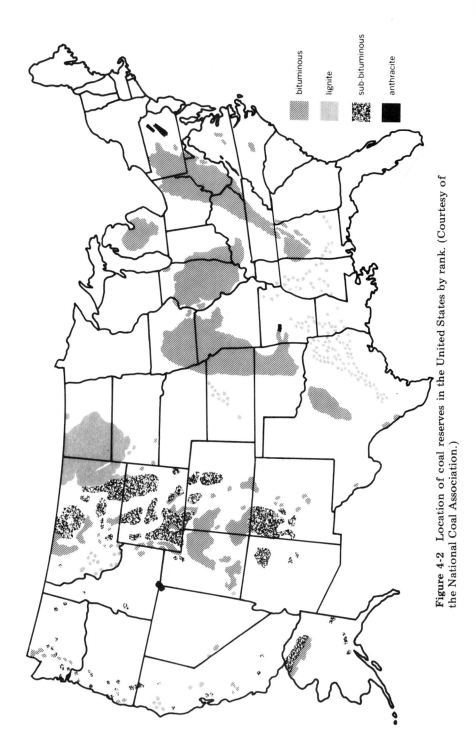

Figure 4-2 Location of coal reserves in the United States by rank. (Courtesy of the National Coal Association.)

mines and 60% from surface mines ["U.S. Energy Review"]. It is projected that of the total volume of coal that is feasibly recoverable from the DRB, 67% will be from underground mines, while 33% will be surface mined ["America's Coal Reserve Base"].

There are three types of underground mines: shaft, drift, and slope. Figure 4-3 illustrates how the topography and coal seam location dictate the mine design and processing plant location.

Extraction techniques have developed considerably since the days of pick-and-shovel mining. Cost-effective, highly specialized machines and transportation equipment as well as efficient extraction methods have greatly improved personnel safety and coal productivity. Approximately 95% of the coal from U.S. underground mines is extracted using the room-and-pillar method, although panel mining, popular in Europe, holds high potential for increasing production, worker safety, and decreased environmental impacts. Room-and-pillar mining is a method of extraction that requires leaving large roof-supporting pillars or columns of coal in place while excavating a pattern of intersecting and parallel tunnels 14 to 20 ft (4.3 to 6 m) wide by cutting and removing the coal. This method requires an interrupted cycle of cutting, loading, roof support, and equipment movement, permitting a maximum production level of only 15 short tons per minute. Room-and-pillar mining is not applicable for coal seams more than 20 ft (6 m) thick and the massive size of the support pillars permits a recovery rate of about only 30% of the available resource.

Removal of coal from the seam in room-and-pillar mining is performed by either drilling and blasting or continuous-mining machines. In the drill-and-blast technique, high-speed mobile drilling machines bore holes into the coal face. Explosives, placed in the holes, blast the coal from the seam for loading. Continuous-mining machines, shown at work in Figure 4-4, have many rotating cutters which tear the coal from the face of the seam. An integral conveyor system moves the coal from the cutter to the rear of the machine, where it is loaded into a shuttle car or mine conveyor system. Nearly 60% of all U.S. coal extracted from underground mines is by continuous-mining machines.

Panel mining using both short- and longwall systems, is being used increasingly in the United States where geological and underground mine conditions permit. This technique involves slicing large rectangular blocks of coal from the sides of the seam, facilitating high production levels and up to an 85% recovery rate. The shortwall system uses a continuous-mining machine which cuts and loads from the end of a rectangular panel of coal while the self-advancing roof supports provide protection for machine and operator.

The longwall system of panel mining was developed in Germany. This system is similar to the shortwall system with the exception that

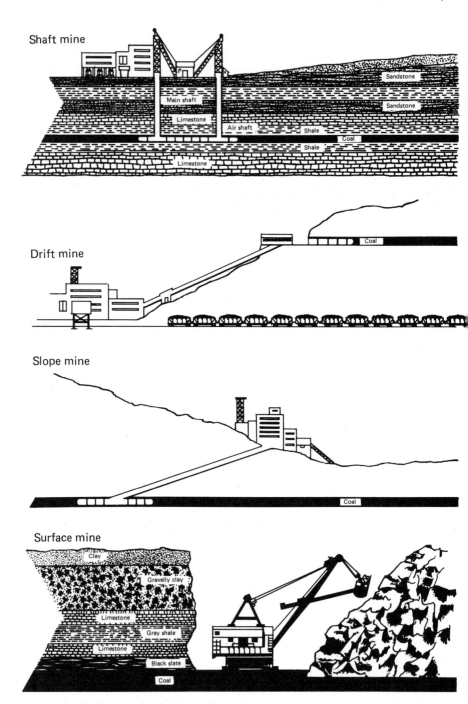

Figure 4-3 Principal types of coal mines. (Courtesy of the National Coal Association.)

Coal Extraction

Figure 4-4 Continuous-mining machine. (Courtesy of Joy Manufacturing Company.)

blocks of coal up to 500 ft (152 m) wide and 6000 ft (1829 m) long are removed. Retreat panel mining, both long- and shortwall, allows the roof material to cave in behind the advancing mining machine rather than leaving pillars of coal or other roof supports behind. Specialized supports protect the machine and operator, then advance to permit the roof to fall in a controlled manner.

Many innovative techniques and mechanisms are being developed to make underground mining safer and more productive. A combination mining and transportation system is being developed and tested which uses high-pressure jets of water to cut coal from the seam while pumps and pipes transport the resulting coal slurry to the surface. A hybrid continuous miner/bolter has been designed which drills holes in the roof of the mine and installs safety plates and bolts while continually advancing along the seam, removing coal. Remote-controlled mining machines and programmable roof-bolting systems are also improving conditions for human health and safety. New rotary drilling developments are permitting the boring of vertical shafts of up to 20 ft (6 m) in diameter and 2000 ft (610 m) deep using equipment located on the surface. Each of these developments and others are dedicated to the fast, safe, efficient, and cost-effective removal of coal from underground sources.

Surface or strip mines are used to remove coal from seams which are located relatively near ground level. Strip mining is characterized

Figure 4-5 Longwall mining machine. (Courtesy of Joy Manufacturing Company.)

by lower production costs, increased productivity, very large equipment, safer working conditions, but greater environmental impact than underground mining techniques. Strip-mining techniques are used in the Unites States to extract nearly one-half of the eastern and over 90% of the western annual coal production. The state of Wyoming is first in the production of coal using the strip-mining process.

Many of the coal-bearing areas of the western states are characterized by relatively flat topography, a thin layer of overburden, and thick seams of coal. By contrast, many eastern areas are rocky and mountainous, have considerable overburden and thin coal seams. Strip-mine production rates nearly eight times those of underground mines are derived from this ease of accessibility of coal resources and the use of extremely large equipment as well as the fact that nearly 90% of the coal can be removed. The coal seam must be located within approximately 225 ft (68.6 m) of the surface to make strip-mining methods more profitable than underground mining.

Whether an open cast or contour mine, strip mining generally involves a six-step mining process. First, large scrapers or bulldozers remove the topsoil and place it in a pile adjacent to the mining area.

Next, other loose or easily removed materials are moved by large draglines and/or power shovels and placed into a separate "spoils" pile. If the coal seam is not yet exposed, the strata overlying the coal are drilled and blasted and the dragline adds this material to the spoils pile. All the material located above the coal seam is termed *overburden*. The last three steps involve drilling and blasting the coal, loading trucks and transporting the coal from the mine, and finally returning first the spoils and then the topsoil, so that reseeding and transplanting efforts can restore the area for future use. Figure 4-6 shows a strip mine incorporating these features.

Figure 4-7 shows the world's largest dragline at work removing overburden. It is operated at the Central Ohio Coal Company strip mine near Zanesville, Ohio, and has a reach the length of a football field. The 220-cubic yard (168.2-m^3) bucket can move 325 short tons at each pass. To appreciate the size of this machine, compare the size of the people and the car with Big Muskie, which works 24 hours a day.

Open-cast strip mining is conducted in one of two ways, depending on the depth of overburden and the width of the coal seam. Where the seam is narrow or close to the surface, a single strip may be excavated to gain access to the coal. The overburden is piled to one side and replaced after the coal is extracted. More often, however, the coal seam

Figure 4-6 Strip mine. (Courtesy of the National Coal Association.)

Figure 4-7 World's largest dragline. (Courtesy of the National Coal Association.)

is wide or covered to a considerable depth with overburden, necessitating the use of modified open-pit mining techniques. This method involves excavating a series of adjacent strips, creating steps or benches, each deeper than the previous one, until the coal is reached and removed. By using relatively narrow steps or benches, overburden from the upper level can be placed in the lowest mined strip, thus eliminating rehandling.

Contour surface mining is frequently used in the Appalachian region due to the complex topography. This method involves excavating a strip mine around a hill or mountain following the contour. Overburden can be handled in one of two ways. It can be trucked to the rear of the pit and deposited where the coal has been removed, or by using a series of contour strips forming steps or benches, draglines can remove it from the upper level and place it in the lower, mined level.

COAL PREPARATION

Coal, as it comes from the mine, may contain up to one-third of its weight in dirt, pieces of rock, or other impurities that were in the seam or picked up during the mining process. The size of the individual pieces of coal may vary from extremely large chunks to chips and powder. Approximately 50% of all U.S. output is processed at facilities near the mine to provide a consistent product since many coal-fired combustion units require a uniform-size feedstock.

The preparation of coal involves crushing, sizing, and cleaning. After crushing, the coal is sorted through various screens of standardized sizes. In addition to the coal dust, rock, and dirt, modern cleaning operations use magnetic separators to remove pyrite. Chemical cleaning techniques, expected to be available commercially by 1987, will reduce the sulfur content by modifying the chemical properties of the coal. There are 10 processes or variations currently being developed which are projected to reduce air pollution caused by the coal-combustion process.

DISTRIBUTION SYSTEMS

Due to the weight and bulk of coal, large equipment is necessary for its transportation. Truck transportation is appropriate for small quantities such as those required by residential or small commercial users but is totally inadequate for large users such as electric utilities. The United States currently has a railroad network which accommodates special coal trains, but this system is inadequate for today's use, let alone the increasing need for coal in the future. The use of water transportation by barge or ship is functional on the Great Lakes, major river systems, and coastal areas, but this method also will need extensive revitalization as coal demands increase. There are currently two slurry pipelines which transport a coal—H_2O mixture from coal fields to the point of use. One of these is 175 miles (282 km) long and connects the Black Mesa coal field in Arizona with the Mohave Power Project in Nevada. At this time the coal transportation system in the United States is considered inefficient and inadequate, and will require considerable near-future investment. There are several leaders in the field who profess that a national pipeline distribution system similar to that currently used by the petroleum and natural gas industries is the only feasible means for the national distribution of coal.

PERSONNEL HEALTH AND SAFETY

The coal industry provides the least safe environment for workers of all the energy supply systems. Disease, primarily black lung, and accidents have been common, especially for personnel working in underground mines.

Extensive efforts are being made to improve health and safety conditions for miners. The nineteenth-century use of canaries to detect lethal, odorless fumes and gases has been replaced by sophisticated sensing/alarm systems which detect a multitude of gases, including explosive methane. Techniques for removing methane prior to coal-mining activities are undergoing feasibility study (see Chapter 5). The number of collapsing mine shafts and tunnels has been substantially reduced by new reinforcing and automated roof-support systems. Mining methods and equipment are constantly being designed and improved to safeguard the workers. Regulations established by the coal companies, the states, and the federal Coal Mine Health and Safety Act are primarily responsible for this improvement.

THE USES FOR COAL

Most of the coal mined in the United States is converted to thermal energy. Although the iron and steel industry uses a great quantity of coal and coke, industrial and process steam, and space and water heating by industrial and commercial users, require many tons per year. The expanding electric utilities consume 81% of the total annual production. They are by far the largest consumers ["Coal: The Primary Fuel Source for Electric Utilities"].

Reduced demand for coal between 1955 and 1970 is attributed to the availability of plentiful supplies of natural gas, oil, and economical electricity, coupled with their respective low prices and convenience of use. Nuclear energy was projected to supplant or at least greatly reduce coal use for producing electricity, but the many related concerns about its use and consequential slowdowns and terminations of nuclear-fueled electric generation facilities are expected to ensure an increase in the use of coal for maintaining and increasing generating capacity.

Coal consumption has increased 6.4% annually since 1971 (see Figure 4-8). Based on this increase, the consumption of coal for the production of electricity is projected to be 730 million tons in 1984 and 777 million tons in 1985 ["Coal: The Primary Fuel Source for Electric Utilities"]. Figure 4-9 shows the steps that are required to convert the chemical energy in fossil fuels to electrical energy.

Even though the United States has extensive coal reserves and exported 91.7 million tons in 1980, it also imported 1.2 million tons in the same year, mostly for conversion to electrical energy. South Africa, Poland, and Australia are the countries from which we import coal to provide a balance of trade. These imports also keep our ships from returning to the United States empty after delivering exported cargos of goods and grains.

There is considerable engineering effort being directed toward better coal-conversion processes in the attempt to improve efficiency

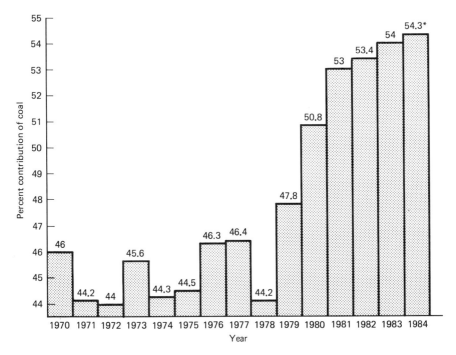

Figure 4-8 Electricity production from coal combustion by percent and year. (From the Energy Information Administration.)

*Estimated

and reduce environmental impacts. Often referred to as the "clean coal technologies," these efforts deal with the direct combustion, liquefaction, or gasification of coals. More than 90% of the coal used in the United States is burned in direct combustion systems. As with other fossil fuel systems, direct combustion units used to produce electricity are about 35% efficient. In addition, direct combustion of high-sulfur coals produces SO_2, NO_x, and particulates which when expelled to the atmosphere cause severe environmental impacts.

With the general failure of postcombustion cleanup systems, efforts have been directed toward new concepts in combustion systems and component parts which can safely utilize a wide range of coals and sulfur levels. Laboratory and engineering studies, experiments, research and development projects, and economic analyses have shown that many of the efforts to use coal-oil mixtures, various additives, chars, slurries, gases, liquids, and newly conceived equipment such as the improved combined cycle for generating electricity are superior to conventional combustion units.

Laboratory experiments have demonstrated that when limestone is burned with coal, SO_2 emissions are reduced to acceptable levels because

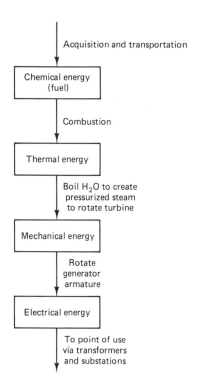

Figure 4-9 Steps in the conversion of chemical energy to electricity.

the limestone absorbs a majority of the SO_2. It also reduces the NO_x. The problem to be solved was that adding the limestone either in solid or powdered form reduced even further the poor combustion efficiency of the burning units. Limestone-injected multistage burners have recently been developed and are proven to the point where they are ready for retrofit installation in most electric power plants.

Experimentation with coal-combustion innovations has proven that greater thermal efficiency and cleaner emissions result from systems that provide forced air beneath the combustion area. As a result, fluidized-bed (FB) coal-combustion units have been developed. In these units, coal is mixed with limestone and burned above a bed of high-temperature inert particles which are "suspended" or fluidized by high-pressure air forced up from below. The forced air agitates the fuel particles and provides additional oxygen, thereby achieving more efficient combustion. During the combustion process, the limestone combines with the sulfur in the coal, eliminating or greatly reducing SO_2 emissions. Some advantages of fluidized-bed combustion units are reduced fuel consumption, smaller plants, less capital investment, no need for electrostatic precipitators, and greater combustion efficiency. The disposal of the ash bed and fly ash remains a problem.

Georgetown University in Washington, D.C. installed the first fluidized-bed combustion unit (Figure 4-10) in 1979, and another has been in operation since 1981 at the Great Lakes Naval Training Station in Illinois. Industrial-size units are now ready for the marketplace, but the larger units necessary for commercial use, such as electric utilities, require additional research and development to acquire higher combustion efficiencies, durable components, reliability of operation, and systems to burn high-sulfur coal directly with minimum environmental effects.

There are two types of fluidized-bed coal-combustion units, atmospheric (AFB) and pressurized (PFB). Atmospheric experimental units were initiated in 1967 and proved to use coal, char, and coal waste as acceptable fuels. In most AFB units, crushed coal and limestone are combined in a 4:1 ratio and blown into the combustion area while air is forced from beneath the fire. Both limestone and dolomite calcines have been found to react with SO_2 in oxygen to form a solid sulfate material that can be regenerated and marketed or disposed of with the ash as an inert granular landfill. The Monongahela Power Company of Riversville, West Virginia, took the first major step toward commercial-size units with their 30-MW demonstration project. This unit burns 300,000 lb (136,080 kg) of coal at one time in a multicell

Figure 4-10 Georgetown University's atmospheric fluidized-bed boiler plant extension. The right section of the building is the original oil/gas-fired plant. (Courtesy of Pope Engineers.)

Figure 4-11 Pressurized fluidized-bed coal-fired gas turbine pilot electricity-generating plant. (Courtesy of Curtiss-Wright Corporation.)

AFB boiler. Future electric utility installations are expected to be in the size range 200 to 800 MW.

The experimental use of coal waste to provide usable energy has been progressing at Shamokin, Pennsylvania, since August 1981. Sponsored by the DOE, an AFB boiler has been burning 100 tons of culm per day and providing steam for local businesses. Culm is an impure and seemingly worthless form of carbon dispersed throughout coal seams. Electricity is not presently being generated at this plant, although the capability exists.

Combustion units using PFB components are currently experimental, and commercialization is expected in the near future. Characterized by lower SO_2, NO_x, and CO emissions, lower projected cost, smaller size for equal output, and even greater efficiency than the AFBs, these units, combined with gas turbines (Figure 4-11) and steam turbines to form a combined-cycle power system, are expected to provide the cleanest, most cost-competitive method for combusting high-sulfur coal in producing electrical energy.

SYNTHETIC FUELS FROM COAL

Coal is not an ideal fuel. It is difficult to transport, burns less cleanly than gas and oil, leaves a bulky residue, and is not directly applicable to vehicular use. Because of its abundance, however, coal will probably be the main feedstock for an emerging synthetic fuels industry which will provide our society with fuels as oil and natural gas resources diminish. A major problem confronting development is the great variation in coals, requiring constant alteration of the conversion process components in order to obtain a consistent end product. Considerable effort has been directed toward developing a technology that will economically, cleanly, and easily provide liquid and gaseous fuels on a large scale.

All fossil fuels are combinations of carbon (C) and hydrogen (H). When one atom of each of these elements is combined, the result is coal, CH. Adding one atom of hydrogen produces oil, CH_2, and even more hydrogen results in natural gas, CH_4. In coal, the hydrogen and carbon atoms are in a ratio of 0.8:1, while oil has a ratio of 1.75:1 [*Scientific American*]. The additional hydrogen required to produce synthetic fuels is acquired from water (H_2O) when coal is reacted with an oxidizing agent and steam.

Significant advances have been made in coal conversion and utilization. Several small pilot and demonstration plants derive crude oil, fuel oil, distillates, chemical feedstocks, high- and low-Btu gas, and other products such as benzene, xylene, and toluene, from coal. The principles of the technology have been proven for decades. The problem is to be able to do the conversions economically.

Coal Liquefaction

Initial efforts to liquefy coal were conducted in Germany in the 1920s and 1930s, with some small-scale study of Germany's techniques by the United States in the 1930s. In 1944 Congress passed the Synthetic Liquid Fuels Act, which provided funding of research efforts through 1955.

Large-scale production of gasoline from coal was done by Germany during World War II when the Allied forces restricted petroleum supplies going to Germany in the attempt to terminate the war. Today only South Africa has commercial-size synthetic fuel facilities. They built their first plant in 1955, and currently 3500 tons of coal per day are converted to a gas and then to liquids that are similar to petroleum.

One ton of coal has the energy equivalent of approximately 4 barrels of crude oil. However, when the energy required for the conversion process is deducted using current technology, the resultant volume is 2.8 barrels. This figure does not include the energy required to extract or transport the coal to the conversion facility.

Basically there are four processes, most using solvents and catalysts, to derive a petroleum-equivalent liquid fuel from coal. The carbonization process involves heating the coal in the absence of air to acquire tar, gas, and coke. This is effectively done in a fluidized bed, after which the solid char is burned with air and steam, resulting in tar and pipeline gas. Substitute natural gas and high-quality (low-sulfur) fuel oil can be obtained by heating the coal in the presence of hydrogen at a pressure of 600 lb/in.^{2}

The Bergius process, originated in 1913 and used initially in Germany's war effort, is a hydrogenation process in which a pressurized reaction with coal, air, and steam produces a mixture of methane and synthetic gas consisting of carbon monoxide (CO) and hydrogen. Reacting this gas with a catalyst of cobalt molybdenum produces methanol—a transportation fuel or gasoline additive. Additional catalytic treatment is necessary for making a gasoline substitute.

The extraction process involves the partial or total dissolving of coal. One system uses an organic solvent in the presence of hydrogen gas under low pressure, and another requires high pressure to produce a low-sulfur oil. SRC-II (solvent-refined coal) is now undergoing combustion tests.

Fischer-Tropsch synthesis requires liquefying the coal. Coal is burned in the presence of O_2 and steam, producing a gas compound of CO and H. Blending with a catalyst produces methanol, waxes, and oils. This process, based upon hydrogenation, was used by Germany in the later portion of the war and is the process now being used in South Africa. This process is said to be economically competitive, as it yields 2 barrels of liquid and 10,000 ft^3 of gas per ton of coal.

Liquefied coal has two major markets: the production of electricity and industrial steam. Low-grade liquid fuels would be used in these industries, while high-grades would find their way into the transportation, heating, and chemical industries.

Coal Gasification

The process to convert coal to gas was used commercially in the United States nearly 100 years before the development of the first liquefaction process. During the 1820s, low-Btu-value gasified coal was used in many larger cities of the Northeast and Midwest for illumination and cooking. The process to acquire the gas involved the use of pyrolysis or heating the coal in the absence of air. Natural gas with a high Btu content replaced gasified coal after World War II, when the national pipeline distribution system was installed. Although low-Btu-value gas would provide a clean fuel for heating, it will not mix with commercially available high-value natural gas and would require that all burner units be changed.

Many of today's energy systems require the use of high-Btu-level fuels. Additional research and development is required to make

Synthetic Fuels From Coal

Figure 4-12 Exxon donor solvent (EDS) coal liquefaction pilot plant. (Courtesy of Exxon Corporation.)

commercial-level production of this gas from coal economically feasible. The heat and energy required for the pretreatment of the coal—grading, classifying, forming, drying, the process itself, and finally gas cleaning and separation—greatly reduce the net energy output of the coal gasification process. Ruedisili and Firebaugh report that there are 10 processes currently known for the production of high-Btu gas from coal, three for medium-Btu gas, and eight for low-Btu gas. Several of these processes involve sophisticated alterations but are based on those processes reviewed previously under coal liquefaction, since most of the liquefaction processes require initial gasification. Germany currently has pilot plants producing gas from a feedstock of Rhenisk brown coal. In the United States, pilot plants located near Alton and Chicago, Illinois, Rapid City, South Dakota, and Homer City and Bruceton, Pennsylvania, are involved in the preparations for commercialization of pipeline-quality gas from coal (see Figure 4-13).

The concept to gasify coal underground was initially proposed in 1868. After World War II, interest was high and experiments were conducted until 1960, when it was determined that the process was not economically competitive with plentifully available natural gas. The only commercial coal gasification facility known was reported in 1956 by the USSR.

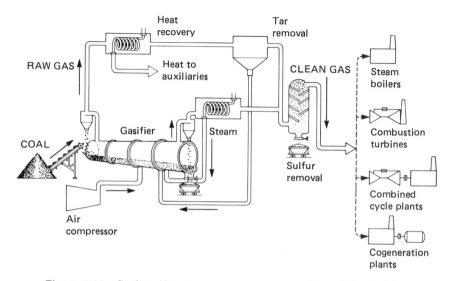

Figure 4-13 Coal gasification system schematic, Alton, Illinois. (Courtesy of Allis-Chalmers.)

Experimentation to gasify coal underground in Wyoming has been under way since 1973. A facility established by the Bureau of Mines burns 15 tons of coal per day to obtain gas from a fractured coal seam.

Synthetic fuels are not yet commercially available in the United States. Under the Carter administration, when a synthetic fuels program was strongly pursued, the construction of small-scale coal-conversion plants was projected for the 1980s with large-scale commercial plants, producing 1 to 1.5 million barrels of crude oil per day, being built during the 1990s. The Reagan administration has deemphasized commercialization of synthetic fuels, and a delay of 5 to 10 years or more is forecast by most leaders in the field. Another barrier to commercialization is that the synthesis processes require a considerable quantity of energy, resulting in expensive, inefficient conversion compared to using the naturally available fossil oil and gas resources. A dramatic economic shift will be required, which is expected to be met with much resistance from various sectors of the fossil fuel industry.

ENVIRONMENTAL IMPACTS

The utilization of coal creates environmental impacts from the time of extraction to long after it is burned. Each step along the way causes us to reflect on the impending importance of this abundant fossil fuel in our national economy and how long our environment will accept the related contamination of conventional coal-conversion systems. The physical characteristics and chemical composition of coal contribute to the impact, as do the methods and equipment for extraction, handling, and use.

There are several environmental issues related to underground mining, ranging from the working environment to the impact of mining activity on the site and adjacent area. As discussed earlier, miners are exposed to many health and safety risks, from noxious, explosive, and toxic gases to cave-ins and black lung disease. Water, whether used in the extraction process or found naturally in the mine, must be removed. The expulsion of this water, often laden with sulfur, acids, coal dust, and other contaminates, poses a serious threat to pure water supplies and erosion of the adjacent area. Removal of naturally occurring water also tends to lower the water table of the area. Excavation of the coal resource often leaves a problem of adequate support for the earth above. Sinkholes are common, and occasionally portions of mining towns are destroyed where mining activity has been conducted directly below the buildings.

Surface mining, although safer for the workers, holds high potential for environmental degradation. Since this extraction method involves disturbance of vast areas, the loss of agricultural, grazing, forest, and other uses is often characteristic. Even with extensive reclamation efforts, complete restoration is an unrealistic goal, especially in arid and semiarid climates. The lack of topsoil and rainfall, and the presence of alkaline soils, often lead to repeated failure and ultimate abandonment by the company. Some areas, such as the arid, mesa topography of the Four Corners region, where Utah, Arizona, New Mexico, and Colorado join, will be impossible to reclaim. Surface mining in this region would require it to be established as a national "sacrifice" area, which is not acceptable. In such cases mining should not be conducted.

When reclamation is planned from the beginning, the restoration of surface-mined areas is often successful. This involves segregating topsoil from the spoil, burying toxic materials, grading closely behind mining operations, and planting and seeding before erosion takes place. The end result can be an area equal or superior to that which was present prior to mining. Figure 4-14 shows an area that has been successfully reclaimed after surface mining operations. It is not uncommon for restoration efforts to exceed the value of the land. A report by the National Academy of Sciences for the Ford Foundation showed that reclamation costs vary between $500 and $5000 per acre.

The coal cleaning and processing operation that is conducted near the mine after extraction is another source of environmental concern. The grading, crushing, sizing, and cleaning operations require large

Figure 4-14 Reclaimed surface mine. (Courtesy of the National Coal Association.)

quantities of water, which can pollute the regional aquifer. The culm piles that result from this processing are also a continuing source of water degradation, erosion, dust, and potential fire hazard.

Burning coal to convert its chemical energy to heat energy causes still further environmental degradation. The massive piles of coal stored adjacent to a facility are an aesthetic concern as well as a source of water pollution, as precipitation leaches through the piles and merges with local groundwater. Additionally, approximately two-thirds of the heat energy contained in fossil fuels such as coal is released into the environment as a result of the inefficiency of current combustion units. Of far greater concern are the pollutants created by the combustion process.

The two major by-products of coal combustion, sulfur dioxide (SO_2) and particulate matter, have been of environmental concern for many years. SO_2 is formed when the sulfur in the coal is combined with the oxygen in the air during combustion. When released up the stack to the atmosphere, the SO_2 combines with water vapor (H_2O) to form sulfuric acid (H_2SO_4), commonly called acid rain. Nitric acid (NO_x), produced from the heating of air during combustion, is also a major concern, together with carcinogenic hydrocarbons and heavy metals.

Acid rain is a worldwide concern. It was reported in 1983 by Agriculture Minister Ignaz Kiechle that 35% (6.2 million acres) of West

German forests have been damaged by acid rain. Sulfurous emissions are also a major concern between countries such as the United States and Canada, where acid rain from the Ohio valley and midwestern U.S. coal-burning facilities is damaging forests and polluting watersheds in eastern Canadian provinces. Contention between states within countries is also strong where prevailing winds carry acid-laden moisture from these facilities to other locations, causing forest damage and virtually eliminating aquatic life in many lakes and ponds.

In the United States the call for action to curtail acid rain is being met with recommendations for long-range research and studies to determine exact relationships between SO_2 and NO_x atmospheric releases and the extent of increased acidity and biological damage to downwind areas. Demands for proof that most of the acid is from the coal-combustion industries have been met and substantiated by various groups, including the National Academy of Sciences and the White House Office of Science and Technology Policy. There is no agreement on the rate at which acidity has been increasing, how much damage is actually caused by how much acid rain, and how much reduction in emissions will cause how much improvement. In 1980 the U.S. Council on Environmental Quality embarked on a 10-year study of the problem.

Various techniques have been tried to reduce the acidity of lakes and ponds that are dead or dying due to industrial emissions. Although powdered limestone will work at varying levels of effectiveness, soda ash has been applied successfully to several Adirondack lakes in New York State and research is continuing. Of course, efforts toward treatment of the resulting problem are futile unless efforts are taken to alleviate the cause of acid rain at its source.

It has been suggested that excessive CO_2 production may cause a permanent global warming trend as a result of the "greenhouse effect." In an Associated Press release on October 10, 1983, the Environmental Protection Agency projected that within a decade major changes will be evident. This report also indicated that the CO_2 is produced by the burning of fossil fuels and that the effect is inevitable even if we were to take immediate steps to reduce emissions.

Particulates or fly-ash emissions are primarily unburned materials from the coal which is exhausted up the stack. Current federal regulations require that more than 99% of the ash particles be removed or scrubbed from the combustion gases before they are ejected from modern plants. This is done through the use of electrostatic precipitators or filtering in a baghouse.

In electrostatic precipitators, the stack gases containing dust and ash are given an electrical charge. This permits removal and collection of the charged particulates through electromagnetic separation. The baghouse operates much like a vacuum cleaner. Combustion gases are passed through a series of filter bags which remove the particles of dust and ash. The collected particulates are generally used as landfill. As of 1982, 84 coal-fired electricity generating plants had scrubbers. This amounts to about 10% of all plants, as the legislation is directed toward "modern" or newly constructed facilities. Stack-gas scrubbing technology is expensive, requires considerable maintenance, and is not constantly reliable.

The unburned waste or ash bed is yet another source of environmental pollution. Due to its physical characteristics and chemical content, this material is often used as landfill because it possesses nothing of industrial value. Transportation from the plant to the landfill site poses problems, as some of the material can be blown by the wind. Considerable effort is being expended to find a way to dispose of ash in an environmentally acceptable manner. One method being studied by the State University Marine Sciences Research Center at Stony Brook, New York, is to make blocks of coal waste which are bonded together with a chemical bonding agent. An artificial reef in the Atlantic Ocean off Fire Island, New York, is proving to be an environmentally safe repository while contributing to the fishing industry.

Legislation affecting environmental pollution has been passed by federal, state, and local governments. Its effectiveness, or lack thereof, depends on the variables of enforcement, and as might be expected, compliance costs someone money. The mining of coal today is closely monitored to avoid adverse effects on air and water quality, wildlife, and surrounding communities. In the past, reclamation efforts were attempted seriously only after environmental problems became evident. Today mining companies are much more aware and concerned about the future effects of their practices.

The Clean Air Act of 1967 established control regions in the United States for all sources of pollution according to the pollution level. A 1970 amendment shortened these procedures and shortened the time for compliance. In April 1971 the EPA was instructed to establish national primary ambient-air quality standards for SO_x, NO_x, CO, hydrocarbons, and photochemical oxidants. These standards were to be met by July 1975. Since the Clean Air Act of 1977, the emphasis has been on SO_x, NO_x, and particulates.

Violations of laws often result in the levying of fines, but most often the cost of the fine is much less than that of appropriate pollution control. In reality, fines become just another cost of doing business for some unscrupulous entrepreneurs.

Environmental protection legislation is expensive. It ultimately results in higher costs for the product, and if the related costs are too stringent, a decline in the industry will result. The Clean Air Act and succeeding amendments are said to have reduced our efforts toward energy self-sufficiency and increased foreign dependency due to restrictions on the use of large areas of our coal reserves, notably high-sulfur coals. Until combustion units are available commercially which will meet or exceed stringent legislation, either continued high-level use of oil and natural gas must persist, or relaxation of the current laws must take place. Future standards are expected to be even higher as we learn more about pollutants and their long-term effects on the environment and the health and welfare of living plants and animals.

A LOOK AT THE FUTURE

Looking at the total U.S. energy use for all purposes, coal provided 18% in 1978, 19% in 1980, and 20% in 1982. Various conservative projections indicate that by 1990 approximately 25% of our total energy will be derived from coal, requiring that nearly 1.5 billion tons be extracted annually from the earth. Other sources project that this figure could be substantially higher with the conversion of electric generating plants from oil and natural gas to coal and a stabilization or reduction in the number of nuclear-fueled plants. Since the high-technology commercialization of solar energy is not yet available, it appears that coal will be necessary to meet our needs at least for the immediate future.

Two groups of fuels are presently being addressed as safe, feasible replacements for oil and natural gas: the renewable biologically based fuels derived from agricultural and forest residue and industrial wastes and the fossil-coal-based fuels. Whether one or both become prominent in the future depends on a multitude of interacting factors, including appropriate handling and transportation facilities and the development of commercial-size systems to convert these chemical energy forms efficiently to usable energy in an environmentally acceptable manner.

The mining of coal will require continued development in order to meet the increasing demand. Underground mining, which is very labor intensive, will need considerably more automation, coupled with machine reliability and improved efficiency. Techniques that maximize extraction safety, recover and use methane gas, control effluents and other waste products, and reduce socioeconomic impacts are the major areas of concern. Surface mining, especially in the western states, is expected to increase steadily. Areas of arid climate will be held in reserve, while emphasis on the semiarid, more easily reclaimable lands will be strong. Mining of large quantities of coal from a small number of areas is projected rather than small quantities from many areas. The expected result is reduced reclamation expense and resulting consumer savings.

Transportation systems will be expanded and developed to move the coal from the mines to major market areas. Unit coal trains and a national network of slurry pipelines can be expected.

Continued research and development of combustion technology will result in methods to desulfurize coal, use coal wastes as fuel, and burn coal more cleanly and efficiently. Pelletizing, sintering, and coking techniques, as well as solvent refining methods, will reduce the potentially enormous environmental impact, as will conversion systems located adjacent to the mines. Liquefaction and gasification will be necessary so that coal-based fuels can be utilized in the transportation and other sectors where solid fuels are impractical. Conversion will also assist in the reduction of environmental impacts.

Liquefaction of coal is expected to provide low-sulfur fuels. The production of methanol from coal can result in boiler fuels, transportation fuels, synthetic gas, and methanol from a single process. Solvent-refined liquid fuels were successfully tested by Consolidated Edison in 1978 at their 74th Street electric generation station in New York City. Continued developmental efforts and foreign oil price increases are expected to make these processes economically competitive.

Synthetic gas from coal is becoming competitive as the price of natural gas increases. By 1985 the five pilot plants mentioned earlier should be in production. Six more are expected by 1990, and a total of 20 should be in operation by the year 2000. In the United States, 176 sites have been identified as having adequate coal and water supplies for commercial-size gasification plants. As research and development continues, high-Btu gas that is essentially free of sulfur and other pollutants should become available.

Since much of our nation's coal resource is not economically recoverable by either conventional underground or strip mining, in-situ coal gasification and liquefaction have been proposed. Neil Cochran has proposed a system that combines the hydrogenation, extraction, and Fischer-Tropsch synthesis processes for acquiring synthetic fuels from in-ground coal. If fully developed, this system may make the use of coal deposits located in arid areas both environmentally and economically attractive.

Both fuel cells and magnetohydrodynamics (MHD) systems using coal as a feedstock are expected to be available commercially by the late 1980s and early 1990s. These systems are expected to be highly efficient, modular, flexible, economically attractive, and environmentally acceptable. These systems are described elsewhere in the text.

Several economic and political problems are anticipated as coal development continues. Prices are expected to increase and then stabilize as the new technologies become standardized and environmental impacts become less severe. Ownership of coal deposits and water-use rights hold great potential for controversy in the near future, especially in the West. Coal deposits "belong" to the federal government, state, railroads, Indian tribes, and some private owners, while much of the surface land has been acquired through the Homestead Act or purchased by individuals. For example, approximately one-third of Montana's strippable coal is on Indian lands. The "rights" of all parties concerned are expected to result in considerable litigation regarding both coal deposits and slurry pipeline corridors.

SUMMARY

The availability of low-cost natural resources within the United States has been a major factor in our country's social and economic growth. Coal, which was a primary fuel in the past, is regaining attention as the availability of oil and natural gas is reduced and prices escalate.

There are four ranks of coal, each with its own physical and chemical characteristics. Although coal deposits are found in 38 states, mining is conducted in 26 states, where the quantity and quality are highest and/or access for extraction easiest. Coal is removed from the seams using either underground or surface-mining techniques.

Nearly all coal mined in the United States is used in direct combustion systems which convert its chemical energy to thermal energy. Since coal is a dirty fuel, considerable research and development is being directed toward new techniques which have less environmental impact. Innovative efforts are directed toward fluidized-bed combustion units, while others stress conversion of coal to a liquid or gas to make its use more flexible and less environmentally degrading.

ACTIVITIES

1. Discuss the environmental impact of surface mining in arid and semiarid regions.
2. Discuss the social and economic impacts of large-scale development of our nation's western coal deposits.
3. Investigate chemicals, products, and materials that can be made with coal as a feedstock.
4. Conduct additional study of some of the recent advances and research and development efforts described in this chapter: specifically, AFB, PFB, coal-oil mixtures, coal liquefaction, and coal gasification.
5. With your classmates, prepare a crossword or hidden-word puzzle using words associated with coal technology: bituminous, macerals, sulfur, and others. Exchange and solve the puzzles.
6. Read newspaper and magazine articles concerning the most recent events related to the environmental impact caused by the utilization of coal. Report or discuss your findings.

BIBLIOGRAPHY

"America's Coal Reserve Base," Energy Fact Sheet No. 17. U.S. Department of Energy. Washington, D.C.: U.S. Government Printing Office, 1981.

Associated Press, "World Will Get Warmer, New EPA Report Warns," *Syracuse Post Standard*, October 19, 1983.

"Coal," *The World Book Encyclopedia*. Chicago: World Book, 1979, pp. 566-582.

"Coal: Energy Source of the Past and Future," Energy Fact Sheet No. 3. U.S. Department of Energy. Washington, D.C.: U.S. Government Printing Office, 1980.

"Coal: The Primary Fuel Source for Electric Utilities," Energy Fact Sheet No. 21. U.S. Department of Energy. Washington, D.C.: U.S. Government Printing Office, 1981.

Denton, J. C., and N. H. Afgan, *Future Energy Production Systems*, Vol. 2. New York: Academic Press, 1976.

Edison Electric Institute, *Coal—Answers to Your Questions*. Washington, D.C.: Edison Electric Institute, 1980.

Gorbaty, Martin L., and K. Ouchi, eds., *Coal Structure*, Advances in Chemistry Series 192. Washington, D.C.: American Chemical Society, 1981.

Gould, Robert F., ed., *Coal Science*, Advances in Chemistry Series 55. Washington, D.C.: American Chemical Society, 1966.

Hunt, V. Daniel, *Handbook of Energy Technology*. New York: Van Nostrand Reinhold, 1982.

Jimeson, Robert M., and Roderick S. Spindt, eds., *Pollution Control and Energy Needs*, Advances in Chemistry Series 127. Washington, D.C.: American Chemical Society, 1973.

Massey, Lester G., ed., *Coal Gasification*, Advances in Chemistry Series 131. Washington, D.C.: American Chemical Society, 1974.

National Coal Association, *The Power of Coal*. Washington, D.C.: National Coal Association.

Peterson, Ivars, "The Dirty Face of Coal," *Science News*, September 17, 1983.

Ruedisili, Lon C., and Morris W. Firebaugh, eds., *Perspectives on Energy—Issues, Ideas, and Environmental Dilemmas*. New York: Oxford University Press, 1978.

Scientific American, *Energy*. San Francisco: W. H. Freeman, 1979.

Thomson, Sharon, "Coal—The Fiction and Facts," *The Energist*. Salt Lake City: Energy and Man's Environment, 1982.

"U.S. Energy Review," Energy Fact Sheet No. 24. U.S. Department of Energy, Washington, D.C.: U.S. Government Printing Office, 1981.

Chapter 5
Petroleum and Natural Gas
Our Most Popular Fuels

CONCEPTS

1. The economies and life styles of developed and developing nations are currently heavily dependent on petroleum and natural gas.
2. At present rates of use, petroleum reserves from both proved and potential sites are expected to be depleted soon after the beginning of the twenty-first century.
3. The majority of unproven petroleum reserves are thought to be in regions of harsh climatic and/or unstable political locations.
4. Natural gas is our cleanest fossil fuel energy source.
5. Natural gas has been relatively cheap due to "artificial" pricing systems.
6. As petroleum and natural gas reserves diminish, costs of extraction will increase and alternative energy sources will be needed to replace them.
7. The processes used to extract petroleum by primary and secondary processes is increasingly abusive to the environment and the economy.
8. Extracting natural gas from deep deposits under the sea may result in severe ecological and geological damage.

GLOSSARY

AROMATIC HYDROCARBONS—hydrocarbons characterized by unsaturated ring structures of carbon atoms, such as benzene, toluene, and xylene.

CATALYST—a material that assists in a chemical reaction but does not take part in it.

CATALYTIC CRACKING—the refining process of breaking down larger, heavier, and more complex hydrocarbon molecules into simpler and lighter molecules.

CRUDE OIL—a mixture of hydrocarbons that exists in liquid phase in underground reservoirs and remains liquid at atmospheric pressure; petroleum.

GAS OIL—liquid petroleum distillate having a viscosity between that of kerosene and lubricating oil, used to produce distillate-type fuel oils and diesel fuel.

NATURAL GAS—a mixture of hydrocarbon compounds existing in the gaseous phase or in solution with crude oil in natural underground reservoirs at reservoir conditions.

PERMEABILITY—the quality of porous rock and soil structures which allows crude oil and natural gas to migrate under pressure.

QUAD—1 quadrillion (10^{15}) Btu; the energy equivalent of 1 trillion cubic feet of natural gas or 170 million tons of coal.

REFORMING—the use of controlled heat and pressure with catalysts to effect the rearrangement of hydrocarbon molecules without altering their basic composition.

INTRODUCTION

The entire population of the world, with the exception of a very few isolated groups, is dependent on natural gas and the products of petroleum refining. We use these fuels to heat our homes, cook our food, light our buildings, power our means of transportation, and to serve as the primary source of energy for the production of electricity. In addition, these materials serve as feedstocks for a vast petrochemical industry which is the source of the enormous quantities of chemicals used for fertilizers, pharmaceuticals, and increased use of polymers (plastics).

Petroleum and natural gas are easily transportable by use of pipelines, tank cars, trucks, and even small fuel tanks such as those used in personal recreational vehicles and family cars. Of all fossil energy sources, petroleum is the one on which Americans are most dependent. The problem is that both resources are becoming scarce, potential sources are not as easily located or exploited (wells must be drilled to ever-increasing depths), and prices have been increasing dramatically. All the "easy" sources have been used.

PETROLEUM: ORIGINS AND COMPONENTS

Petroleum is the end result of decaying residues of microscopic organisms from ancient oceans of the Paleozoic Era. For approximately 570 million years, pressure and heat combined with chemical decomposition to change the organic components into the valuable but exhaustible resource we call "oil." Geologists do not agree on the exact processes that caused petroleum to be formed; however, certain general conditions can be assumed.

Single-celled aquatic plants (diatoms and blue-green algae) and animals (foraminifera) are known to have existed in abundance during the Paleozoic Era. It is believed that these free-floating organisms mixed with sediments when they died, and were buried deeper and deeper as other organic and inorganic materials from both land and ocean areas mixed together through the action of moving water. Fine-grained clay-type sediments were laid down on top of this residue along the margins of lakes and seas. Coarser silts were deposited, causing the first-formed materials to be compressed. The resulting pressure and heat solidified the sediments into layers. The pressure and decomposition helped change the organisms into compounds now known as petroleum and natural gas.

Crude oil varies greatly in its composition. This is not surprising given the wide range of conditions leading to its formation. This non-uniform composition is made up of highly complex mixtures of paraffins, naphthas, and aromatic hydrocarbons. Small, variable quantities of sulfur, nitrogen, and oxygen compounds are also present. Natural gas is frequently present in the same types of geological formations in which crude oil is found. Petroleum has been found at depths of less than 100 ft (30 m) to at least 25,000 ft (7500 m), indicating that a wide variation in pressures and temperatures existed to produce crude oil and gas.

Pressure also helped to distribute and disperse both materials after they had formed. Formations made of sandstone and limestone have greater permeability than layers made of shale, which is formed from clay particles under great pressure. Crude oil and gas will disperse under pressure until they reach some type of barrier.

For the gas and oil to collect in quantities that can be tapped with wells, it is necessary to have a "trap" which prevents further migration and dissipation in the formation (Figure 5-1). Two types of traps are common: structural, formed by folding or faulting of reservoir rock layers, and stratigraphic, formed when porous, permeable rocks are sealed off by superimposed impermeable rock layers. A seal of im-

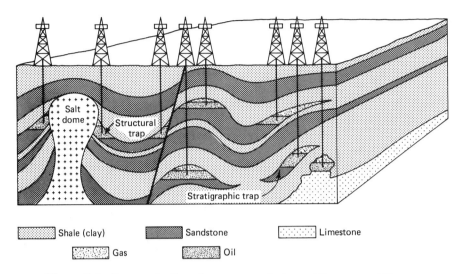

Figure 5-1 Types of oil and gas accumulations. (Courtesy of Exxon Corporation.)

permeable rock must be present over the trap to prevent gas and liquids from escaping.

All of these features must have been established during and after the Paleozoic Era in order for petroleum and natural gas concentrations or deposits to be formed, and they must have occurred in proper relation to each other in time. There are known fields of petroleum and natural gas on every major landmass in the world with the exception of Greenland. Some of the continental shelves are known to have reserves as well.

PETROLEUM AS A RESOURCE

Petroleum has been known to exist for centuries. People of many countries used oil seeping from the ground to seal seams in boats and to make torches as long as 5000 years ago, long before the fuel characteristics were known or needed.

The world's first oil well was drilled in Pennsylvania in 1859, partly in an effort to supplement the decreasing supplies of whale oil used for lighting. It took only until the 1940s for petroleum to replace coal as the primary fuel in the United States. Since then, nearly 700 billion barrels of crude oil have gushed or been pumped out of the earth. The United States is the world leader in cumulative production, but not currently the leader in annual yield. The USSR has held a substantial annual lead over both the United States and Saudi Arabia for more than 10 years. What began as a substitute fuel for dwindling supplies of whale oil over a century ago has become the apparent life blood of transportation systems and industries all over the world.

The demand for petroleum products during 1981 was approximately 50 million barrels per day in the free world. Consumption in the United States reached a high of 18.8 million barrels per day in 1978 but as a result of higher prices and both governmental and individual conservation measures, declined to 14.9 million barrels per day during the first half of 1983. The primary reduction in demand for petroleum products is in the use of distillate and residual fuel oils. One immediate result of the lowered demand was the reduction of the price of crude oil from $35 in the late 1970s to $29 per barrel in mid-1983, set by the Organization of Petroleum Exporting Countries (OPEC).

PETROLEUM RESERVES

Almost two-thirds of the current reserves of crude oil are in the lands of the 13 member nations of OPEC: Algeria, Iraq, Kuwait, Libya, Qatar, Saudi Arabia, United Arab Emirates, Indonesia, Iran, Nigeria, Venezuela, Ecuador, and Gabon. About 13% can be found in the centrally planned economies, primarily China, the USSR, and its satellites; about 10% in the principal industrialized nations, the major consumers; and the remaining 10 to 12% among all the remaining countries.

The large reserves found in the OPEC countries create a political and economic condition generally unfavorable to the industrialized and developing nations of the world. There are conditions of political unrest in many of the oil-producing countries, which threaten production and export supplies. There have been wars and strife since the beginning of recorded history. Governments and their contracted treaties last for only a few years and there is little hope that this situation will change in the near future. In some areas of the world climatic conditions are severe, making exploration, development, and production a risky and expensive proposition. All of these factors strongly support the development of energy resources not dependent on outside sources of supply.

The quantitative determination of "reserves" is a difficult process. It depends primarily on exploration which has taken place in geological formations that have been found most likely to provide conditions for the presence of petroleum (i.e., measured and indicated reserves). The second factor is an economic one.

The horizontal line shown in Figure 5-2 between economic and subeconomic resources is conditional, dependent primarily on crude oil prices in the marketplace. Those reserves labeled "inferred" and "undiscovered" are subject to conditions of exploration and may be highly

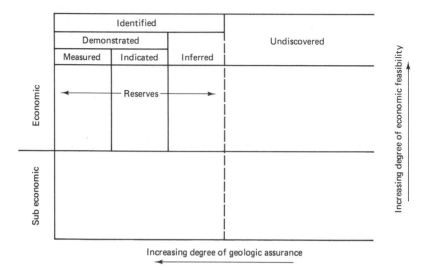

Figure 5-2 U.S. petroleum resources. (From the U.S. Geological Survey.)

variable both in quantity and in accessibility. Nearly all the easily available oil has been pumped, processed, sold, and consumed. Reserve conditions move inexorably closer to that horizontal "margin" line where development and extraction are ever more costly.

In a simplistic model [Bohi and Toman], the only force other than physical scarcity that limits extraction is a decline in price, caused by either the development of a cheaper substitute fuel or feedstock or by a decline in demand. The former is dependent on other technologies, while the latter is largely in the hands of the one who ultimately pays the bill—the consumer!

PRODUCTION

World crude oil production hit a peak of 62,535 million barrels per day (average) during 1979. Production in the United States was at 9208 million barrels per day (average) during 1973 and declined sharply for 4 years, then rose to and has maintained a level of between 8500 and 8700 million barrels per day (average) since then. Domestic production represented 46.3% of consumption in 1978, but represented 57.7% of consumption during the first half of 1983. The combined effects of increased price and conservation were the major factors in the reduced levels of consumption. Increased prices, although somewhat lower than the peak prices of the late 1970s, have made some additional national exploration economically feasible. Production of certain "marginally profitable" fields is more likely to occur when prices are higher.

EXPLORATION

The technology used to locate potential oil- and gas-bearing formations is well developed in contrast to early discoveries, which depended almost solely on guesswork or on locating outcroppings similar to those which had produced oil previously. "Wildcat" wells continue to be drilled, but the success rate for those drillers is low, on the order of 1 in 70 successful wells. Wildcatters can determine the presence of an oil or gas resource, but the determination of the quantity of the field or pool is still in question with these methods.

Current exploration technology uses several means to assess underground or undersea deposits. In addition to visual observation, several scientific measurement devices are used. The first of these is the seismograph. A small explosive charge is set off over a potential oil-bearing area. The seismograph measures the intensity of the returning shock waves and creates a strip chart showing the echos from sensors set at various distances from the charge. Permeable and impermeable rock layers send back different echo patterns, allowing a technician to estimate the potential for oil- and gas-bearing conditions and the approximate depth of the field.

A magnetometer measures the force of magnetism in rocks. The presence of iron helps to determine permeability, thereby helping to identify potential oil- and gas-bearing geological formations. The gravimeter uses a similar principle to measure the force of gravity. Since nonpermeable rock is more dense than semipermeable rock, it tends to show as a higher reading, again indicating potential formations suited to oil and gas accumulations.

In addition to these surface operations, aerial and satellite surveys are conducted over terrain that is difficult to reach on the surface. All the techniques listed above are only indicators of conditions. To make more accurate estimates of the presence of oil, core samples may be taken. Although this is a much more expensive procedure, it is an exceptionally strong indicator. Hollow drills cut through layers of rock to remove cores which are brought to the surface, where technicians analyze each segment of the core to determine whether oil is present. This technique does not indicate the full size of the field, however. Several techniques are generally used: combinations of aerial surveys and core samples plus magnetometer and gravimeter readings to confirm the existence and/or relative size of the pool.

The uncertainties present in each of the techniques lead to varia-

TABLE 5-1 Estimated International Crude Oil Proved Reserves

Country	Reserves[a]	Percent of Total
Saudi Arabia	165.3	24.7
Kuwait	67.2	10.0
USSR	63.0	9.4
Iran	55.3	8.3
Mexico	48.3	7.2
Iraq	41.0	6.1
United Arab Emirates	32.4	4.8
United States	29.8	4.4
Libya	21.5	3.2
Venezuela	21.5	3.2
China	19.5	2.9
Other	105.4	15.7
World total	670.2	100.0[b]

[a] Billion barrels.
[b] May not total 100% due to rounding.
Source: Energy Information Administration, *1982 Annual Energy Review*, p. 45.

tions in the estimates of discovered potential and undiscovered potential "reserves." A total of 160 oil basins worldwide have been found to be productive to date. Of these, only 25 contain as much as 10 billion barrels of oil (the equivalent of the Prudoe Bay field on Alaska's North Slope). Twenty-five basins contain more than 85% of the hydrocarbons discovered to date, but only six basins have reserves in excess of 50 billion barrels. This uneven distribution pattern is also expected to apply to the 440 basins discovered but as yet unexplored. The consensus of oil producers is that none of these basins will equal the vast reserves in the Arabian/Persian Gulf, providing further argument for the development of renewable energy resources.

Drilling technology has improved as the process of drilling has become more difficult. Artificial diamonds are bonded to the tungsten carbide drilling "studs" normally used in soft to medium-hard rocks (see Figure 5-3).

One of the newest developments is the Tround drill, shown in Figure 5-4. The 30-ft long unit is composed of a three-cone rotary unit with three small tubes (one is hidden) located between the cones. They are "guns" which fire up to four bursts of ceramic projectiles per minute at speeds of 4500 ft/sec (137,160 cm/sec). These high-speed projectiles disintegrate on impact, but they produce shock waves rather than penetrating the oil-bearing rock. The stress fractures the rock, making drilling much easier and faster.

Drilling Structures

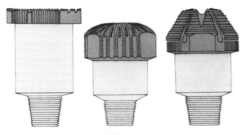

Types of diamond bits for drilling through formations of differing hardness.

Figure 5-3 Diamond bits. (Courtesy of Exxon Corporation.)

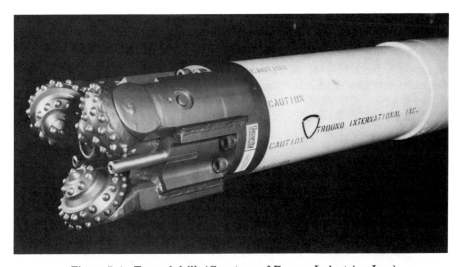

Figure 5-4 Tround drill. (Courtesy of Dresser Industries, Inc.)

The new process more than doubles drilling rates, especially in the hardest rock formations, where conventional drill bits will wear most rapidly. The drill can be used for drilling for oil and natural gas as well as making borings for geothermal wells, as described in Chapter 9.

DRILLING STRUCTURES

Standards for derricks and drilling platforms for both land-based and offshore use are established by the American Petroleum Institute (API). The standards specify factors of loading, including that of the derrick itself as well as loads produced by the drilling operation and outside factors such as wind, earthquake, and for offshore rigs, the forces of waves, tides, and currents.

Drilling for oil and natural gas is seldom an exact vertical boring operation. Subsurface structures are rarely uniform in density. Strata of varying types are the norm, and different drilling pressures as well as the flexibility of a long drill string may all affect the straightness of the bore.

Directional drilling makes use of the factors that cause a well bore to turn away from the vertical. It may be desirable to deviate around an obstacle, to drill under surface structures which are difficult or impossible to move, or it may be more economical to drill into a reservoir from an angle to avoid one or more layers of impermeable formations. It is also possible to use this technique to drill from nearby landmasses for reservoirs located under bodies of water—land-supported equipment is less expensive to construct and service.

When oil and gas formations are reached it is necessary to complete the well with one or more casings to allow free flow of the oil and/or gas and to seal out sand, water and other unwanted materials. This process also protects water supplies in subsurface aquifers.

Figure 5-5 Offshore drilling platform. (Courtesy of Texaco, Inc.)

EXTRACTION

Early discoveries of oil reservoirs were often accompanied by the eruption of "gushers," caused by the force of natural gas and geological pressures pushing the oil out and above the earth's surface. Once the initial pressure has been relieved, wells produce only by being pumped. This primary extraction step typically removes only one-fifth to one-third of the oil and half of the natural gas in the formation. Current environmental regulations prevent or restrict the uncontrolled release of oil and natural gas, so precautions are taken to keep this from happening.

Secondary recovery consists of injecting water or gas into the well to displace more oil or natural gas. This technique is widely used, but it too leaves large quantities of oil and natural gas in the reservoir.

Tertiary or "enhanced" recovery, a more expensive set of processes, is used to recover more of the remaining resource. Three principal techniques are employed. Thermal methods are used in relatively shallow and low-pressure reservoirs containing highly viscous oils. Steam is injected by itself or in combination with in-situ combustion (see Chapter 4) to heat the rock and oil (see Figure 5-9). This reduces viscosity and allows the oil to flow into the well for pumping.

Chemical methods are used with a wide range of oils and reservoir conditions (Figure 5-10). Detergentlike materials are injected to wash oil from the surfaces of flakes and particles of rock in the reservoir.

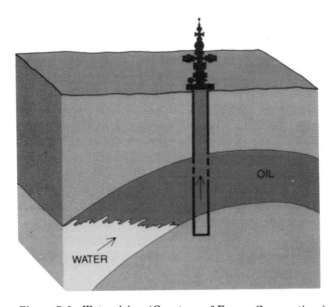

Figure 5-6 Water drive. (Courtesy of Exxon Corporation.)

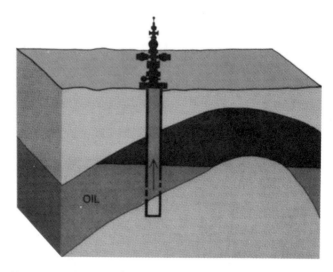

Figure 5-7 Gas cap drive. (Courtesy of Exxon Corporation.)

Polymers are then added to thicken the solution so that it may more easily be forced toward the wells.

Miscible methods are used for medium and light oils in fairly deep wells at medium to high pressures. This process injects chemical solvents into the reservoir together with carbon dioxide or nitrogen (Figure 5-11). This causes a better mixture of reservoir fluids and injected fluids and results in improved flow characteristics.

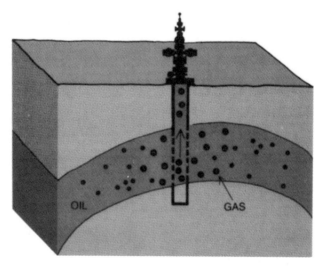

Figure 5-8 Solution gas drive. (Courtesy of Exxon Corporation.)

Extraction

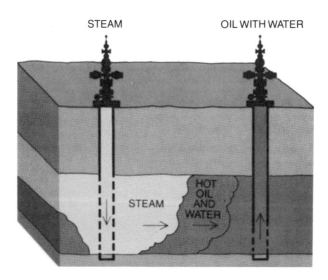

Figure 5-9 Steam flooding. (Courtesy of Exxon Corporation.)

Each of the three enhanced recovery techniques involves additional borings into the reservoir at some distance from the well. The drilling, chemicals, and related operating costs make tertiary recovery expensive. With the world oil and gas resource base estimated at 3000 to 5000 billion barrels, secondary and tertiary methods are expected to

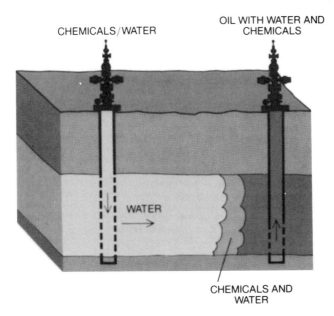

Figure 5-10 Chemical flooding. (Courtesy of Exxon Corporation.)

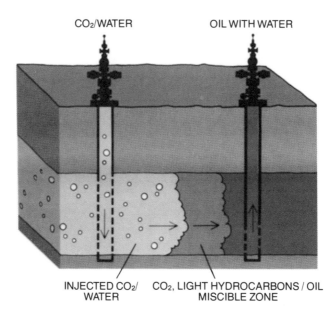

Figure 5-11 Carbon dioxide miscible flooding. (Courtesy of Exxon Corporation.)

recover up to 50% of the total available oil in a reservoir and up to 75% of the natural gas ["Improved Oil Recovery"].

REFINING

Crude oil is a complex mixture of hydrocarbons unsuitable for direct use. The primary separation of each component or fraction takes place in a process called fractional distillation. Modern units operate continuously over long periods of time and are of a size that makes it possible to process 100,000 barrels per day. Some units have been built to handle more than twice that amount.

A typical fractionating tower schematic diagram is shown in Figure 5-12. Crude oil is pumped into a furnace unit where it is heated to between 600 and 700°F (315 to 379°C). This vaporizes some of the oil and both vapors and liquid are passed to the fractionating column, a vertical cylindrical tower some 150 ft (45 m) high. The column is equipped with perforated trays which have caps over the perforations to aid in separating the different hydrocarbon fractions. The more complex, heavier fractions generally have higher boiling points and separate from the other hydrocarbons first due to condensation. The less complex, lighter fractions with lower boiling points rise through the column until their temperature has cooled so that they too condense. Temperature levels in the column are carefully controlled for each of the products desired. For example, light gasoline with a boiling point between 75 and 200°F (25 to 95°C) is withdrawn near the top. Naphtha, kerosene, and gas oil are taken off in turn moving down the column.

Other processes are used to break complex, long-chain molecules of hydrocarbons such as gas oil into simpler molecular structures. Thermal cracking and reforming are used to obtain the desired fuels. In the typical case, a catalyst such as finely divided zeolite or an alumina–silica compound is used.

Gas oil is heated to 850 to 950°F (455 to 510°C) in an atmosphere

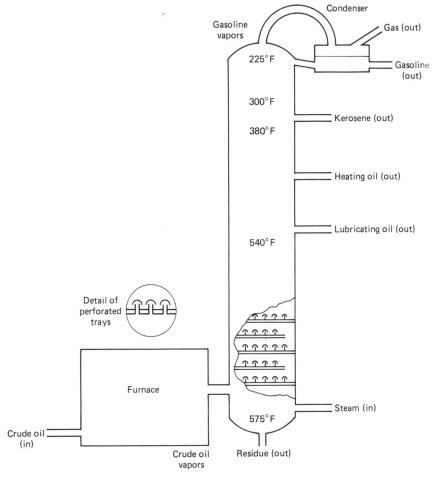

Figure 5-12 Fractionating tower. (Adapted from Energy Research Development Authority, *Fossil Energy*.)

of 10 to 20 lb/in.^{2}. The oil vaporizes, forcing the catalyst to the top of the chamber keeping it in suspension. Carbon molecules are collected by the catalyst, and the hydrogen, being more volatile, passes on to the fractionating column. Yields of 40 to 60% gasoline are possible with this process.

Hydrocarbons can also be reformed by use of a catalyst, usually platinum. Naphtha is used as the feedstock to form other hydrocarbons. The desirability of this process is due to the lack of a residue of unwanted heavy products and coke.

Polymerization is the process whereby hydrocarbon molecules are induced to combine (polymerize) into molecules two or more times the original molecular weight. Gaseous fractions can be made more useful by this process, forming a liquid fuel for internal combusion engines.

In addition to the molecular rearrangements that are accomplished, certain impurities occurring in the crude oil must be removed. Sulfur is the most common impurity, and aside from its obnoxious rotten-egg odor when in combination with hydrogen (H_2S), it is corrosive to many engine parts. Other undesirable effects are the reduction of antiknock qualities and the effects on the environment when burned. Since sulfur combines in many ways with components of crude oil, removal takes several forms. Sulfuric acid removes sulfur by placing it in solution, caustic soda removes hydrogen sulfide as well as heavier phenols, and clay treatment is used to remove gums and resins and aids in removing heavy sulfur compounds as well. Hydrosulfurization mixes hydrogen with oil, which is then vaporized before undergoing treatment with a catalyst. This process recovers hydrogen sulfide, which is then treated to produce elemental sulfur of high purity.

Air quality standards specify a maximum permissible sulfur content for fuel burning but are unfortunately not as clear on the amount of sulfur emissions. The resulting organic sulfur compounds, although minute by comparison with the volumes of fuel burned, are the major contributor to the acid rain that is devastating forests, lakes, and streams as well as adding to the discomfort of human and animal life in both North America and western Europe.

DISTRIBUTION AND STORAGE

Oil inventories constitute all the various types and levels of processed petroleum products from crude oil to finished products. Three categories make up the various levels of supply and storage: primary (producing and refining facilities, pipelines, tank cars, trucks, barges, and ships), secondary, and tertiary. The supply chain is estimated to contain a volume equal to about 6 months' supply or the equivalent of about 10 billion barrels of oil.

A part of the primary distribution and storage system includes large quantities of crude oil that are stored near seaports for on-loading of tankers. Seagoing tankers of the very large crude carrier size also require large off-loading storage tanks, since those in the 200,000 deadweight-ton class may carry as much as 1.5 million barrels of crude oil. Since these ships cost an average of $15,000 per day while sitting idle, it is important to load and off-load them as rapidly as possible. With large volumes of crude oil in transit on the high seas, as much as 20% of the inventories in the primary sector may be on board at any one time.

Crude carriers of twice the size mentioned above are also used. Since they have a draft of approximately 100 ft and no U.S. ports are equipped to handle these monsters, they must stand offshore and off-load into smaller vessels, which in turn must move into port to off-load. The double handling increases costs and the risk of crude oil spills.

These ultralarge crude carriers are much more difficult to maneuver, and when disaster strikes in the form of hurricanes, typhoons, or collisions, the volume of the potential oil spill is enormous. Techniques are being developed to contain spills, to break up the resulting globs of oil and tar, and to reduce the environmental impact. No one is certain of the long-range effects on wildlife and coastlines subjected to this abuse.

Figure 5-13 Baytown, Texas, refinery and tank farm. (Courtesy of Exxon Corporation.)

Secondary distribution begins with the transport of products from large distribution terminals via tank truck, barge, or rail, bound for smaller distributers in regional stations. Also included in this sector are the smaller distributors, wholesalers, and retailers who sell directly to consumers.

Tertiary inventories are those held by consumers, found in tanks of autos and trucks, heating fuel tanks in homes, and in commercial and industrial plants.

ENVIRONMENTAL FACTORS

Environmental concerns are of growing importance to the oil and natural gas industry primarily because of governmental legislation. This legislation has been established because of growing concern about the potentially harmful effects of development, production, and transport of oil and natural gas. Almost every regulation increases the initial cost of all energy resources. What is often overlooked in such cost determinations is the long-term benefits to people—employees, consumers, and the public in general—as well as to the quality of the land, water, and air, the vital parts of the environment on which we depend.

Environmental regulations may increase both the time and cost of development by two or three times that required if such controls were not in place. However, when costs of noncontrol of similar factors from past developments are considered (asbestos poisoning, black lung disease, allergic reactions to formaldehyde gases—to mention just a few), the "burden" seems less difficult to bear. The consumer must ultimately bear the burden of all the added expenses.

Nearly every step of the petroleum industry has the potential for environmental damage. Oil-well fires during the drilling process, while fairly uncommon, consume vast quantities of crude oil and gas. When fires occur on offshore rigs, the potential for oil spills and contamination of beaches and the habitat for both marine wildlife and shoreline fauna is great. Each year brings reports of lost resources, danger to personnel, and damage to tankers as a result of hurricanes, collision, or fire. As the cargo capacity of tankers increases, there is the potential for greater loss from each individual incident.

Oil spills from tankers where refineries or storage facilities are located near the shore contaminate the sea and seacoast each year, causing temporary (and sometimes permanent) ecological damage. Pipelines are equipped with automatic shutoff valves, but because of their spacing (about every 10 miles) as much as several million gallons of oil could escape during a leak.

The consumption of petroleum products can be especially harmful to the air we breathe as well as to soil and water conditions. The wide variety of end products of petroleum consumption produces more than 13.22 million tons (12 billion kilograms) of carbon monoxide each year, about 3.86 million tons (3.5 billion kilograms) of sulfur oxides, and nearly 12.12 million tons (11 billion kilograms) of gaseous hydrocarbons. Much of this remains suspended in the atmosphere to form "smog," but it eventually descends onto the land and into water supplies in the form of acid rain. Other effluents from refineries include ammonia, phenols, naphthalene, cresols, sulfides, and other noxious chemicals.

Oil-fired power plants often use "residual" oil, the dregs of the refining process. Many of these plants do not operate at top efficiency and thus release noxious emmissions into the atmosphere. Stack scrubbers, in the form of electrostatic plates, have been installed in some plants and are reducing the amount of emission. New plants have built smokestacks which are several hundred feet high, thereby carrying the pollutants higher into the atmosphere, where they have the opportunity to spread over a wider area. It has been believed that this will disperse these materials sufficiently to reduce their impact, but we have seen increasing evidence that the detrimental cumulative effects are being noted on human beings and on plant and animal life many hundreds of miles down wind from the source.

Oil shales are another source of petroleum. Experiments to coax out a rubbery hydrocarbon called kerogen have proved to be successful but expensive. Nearly 1.3 tons (1180 kilograms) of rock must be processed to extract 1 barrel of oil. In the process, the rock debris expands several times in volume over that of the unprocessed shale. A shale oil plant will need thousands of gallons of water each day to operate, and most of the oil-bearing shales are in a small area of southwestern Wyoming, northeastern Utah, and northwestern Colorado, which is one of the most arid parts of the United States. Tar sand deposits are known to exist in many parts of the world, the largest being in northeastern Alberta Province in Canada. The very heavy, viscous crude oil in these deposits cannot be recovered by conventional means. Both of these resources are discussed further in Chapter 15.

The "love affair" that Americans have with the automobile has created a life style of suburbs, two- and three-car families, and so little reliance on mass transit that many public transportation systems could not survive without heavy subsidies. Efforts to reduce private auto use take many forms, including car pooling and express lanes on superhighways for vehicles with multiple passengers. The need for this reduction is evident from the high levels of pollution in cities and heavily populated regions. The inefficient automobile engine produces nearly 2.5 times as much carbon monoxide at idle as at cruising speed. The CO level on some city streets may reach 50 parts per million. It takes only 1 part per million to produce measurable physiological effects.

NATURAL GAS

Natural gas is a mixture of hydrocarbon compounds and small quantities of nonhydrocarbons existing as "nonassociated" or free gas or as "associated" or dissolved gas in solution with crude oil in natural underground reservoirs. Natural gas frequently occurs with crude oil, but this is not always the case.

Since both crude oil and natural gas are hydrocarbons, they are believed to have originated from the same components and natural processes. The conditions that produce a gas in one instance and a liquid in another are not well understood. The differences are primarily that gaseous hydrocarbons are of lighter molecular weights than are the liquid hydrocarbons.

The components of natural gas are methane (CH_4)—making up about 75 to 95% of the mix—ethane (C_2H_6), propane (C_3H_8), and butane (C_4H_{10}). All these components are paraffin isomers with the general formula C_nH_{2n+2}.

BACKGROUND

When the first gusher oil well "blew in" there was recognition of the presence of high pressures existing beneath the earth's surface. The knowledge of the existence of natural gas in formations that also held crude oil was not a benefit but rather a nuisance to the driller. Only a small amount of this "free fuel" was used at the site of the well or for local use; the rest was "flared" or just vented into the atmosphere.

Two problems existed—there was no use for the gas (other fuels served just as well), and there was no distribution system established to take it to locations where it could be used. Not until the 1920s were techniques for the development of pipelines of welded, high-pressure steel pipe perfected and pipelines installed to provide gas to industrial and residential consumers.

The advantages of natural gas rapidly became apparent. It does not require storage tanks at the point of use, it can be piped like water, has a high Btu content, and burns cleanly in inexpensive, easily maintained furnaces and stoves. Natural gas currently makes up about 25% of the total energy supply of the United States.

The price of this fuel has been kept relatively low by governmental regulation. Recent legislation (the Natural Gas Policy Act of 1978) will remove the controls in several stages and allow the price of the fuel to react to market conditions. Large industrial users are expected to bear the burden of potentially increased prices until controls are totally removed in 1989. Small consumers and individual homeowners may have some difficult choices to make if the price increase is higher than income levels will support. The impetus for deregulation is purported to be to increase the development of potential reserves, to foster the search for new discoveries, and to conserve the resource.

SUPPLIES OF NATURAL GAS AND RESERVES

Nonassociated and associated/dissolved sources of natural gas as well as temporary underground storage are all considered in the tabulation of "proved reserves" of this resource. Natural gas is often transported by pipeline to storage areas near the point of consumption. The storage facilities often take the form of gas wells which have become unproductive. Other suitable formations are salt domes, aquifers, and coal mines. Steel storage tanks with floating covers are also used to maintain relatively large quantities of natural gas for local use.

Three states provide nearly three-fourths of the proven U.S. reserves of natural gas: Texas, nearly one-third; Louisiana, at least one-fourth; and Alaska, more than one-tenth. When Oklahoma, Kansas, and New Mexico are added, the total is nearly 90% of the total reserves.

Several potential gas supply categories exist: (1) supplies from known fields may be expanded by extensions of existing pools, by discovery of new pools either in those fields or on other levels; (2) discovery of formations under similar geologic conditions or from different geologic formations; and (3) from new pools from formations not previously found to be productive. The first levels are said to be in the "probable" category, the second are "possible," and the latter are highly "speculative" in nature.

The well-known predictions of M. K. Hubbert describe the cycle of discovery, development, production, and declining reserves for both oil and gas (Figure 5-14). The cycle in both instances takes the form of the bell curve so often used by statisticians. Oil production in the United States was seen to peak about 1973 and natural gas about 1980. More than 80% of the known oil will have been produced and consumed during the years 1930 to 1995 (or 2020 if a more optimistic "ultimate" reserve level is used).

The years between 1950 and 2040 will see the production and consumption of 80% of known natural gas. Based on Hubbert's calculations, Gibbons and Chandler predict oil production in the United States at the beginning of the twenty-first century to be 1.5 billion barrels per

TABLE 5-2 Estimated International Natural Gas Proved Reserves

Country	Reserves[a]	Percent of Total
USSR	1240[b]	40.9
Bahrain	483	15.9
United States	204	6.7
Saudi Arabia	121	3.9
Algeria	111	3.7
Canada	97[c]	3.2
Mexico	79	2.6
Qatar	62	2.0
Norway	58	1.9
Venezuela	54	1.8
Netherlands	52	1.7
Kuwait	34	1.1
Malaysia	34	1.1
Nigeria	32	1.0
China	30	1.0
Indonesia	30	1.0
Iraq	29	0.96
United Arab Emirates	29	0.96
Argentina	25	0.82
United Kingdom	25	0.82
Libya	22	0.73
Pakistan	19	0.63
Australia	18	0.59
Other	146	4.8
Total	3034	100.00

[a] Trillion cubic feet.
[b] Explored reserves.
[c] Proved and probable reserves.
Source: Energy Information Administration, *1982 Annual Energy Review*, p. 45.

year by conventional means plus an additional 500 million barrels by tertiary recovery. The two sources combined would equal about 13 quads of energy per year. This is about one-third of the oil consumed in the United States in 1981.

The unknown long-term effects of deregulation cause Gibbons and Chandler to make tentative projections for natural gas production of about 20 quads per year, close to that of current consumption rates and about one-fourth of the total annual energy consumption in the United States. Future production will depend on secondary and tertiary production methods and thus result in greater cost per foot of drilled well and higher consumer cost.

Offshore areas along the continental shelf, particularly the Gulf coast and the area off the East coast known as the Baltimore Canyon,

Supplies of Natural Gas and Reserves

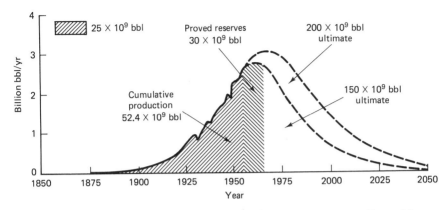

Figure 5-14 Hubbert curve, peak of U.S. oil production. (From M. King Hubbert, "Nuclear Energy and the Fossil Fuels," in *Drilling and Production Practice*, Dallas, Tex.: American Petroleum Institute, p. 17, Fig. 21.)

continue to be explored for gas deposits. The Gulf of Mexico is the major source of natural gas outside state-controlled areas. Gas dissolved in seawater, lying under huge caps of impervious rock at great depths, has been estimated to exist in quantities of up to 100,000 trillion cubic feet—enough gas to supply the needs of the United States for 1000 years! But there are problems!

The trick is to get the gas out and to do so without massive negative effects on the environment. An enormous amount of water must be brought to the surface to remove this trapped gas. One barrel of water at a temperature of 240°F (115°C) might contain, for example, 50 to 100 ft^3 of gas. A well might produce as "little" as 40,000 or as much as 100,000 barrels of brine per day. Brine from wells with a depth of less than 13,000 ft (4000 m) could have a pressure as high as 11,000 psi, while that from deeper wells, down to 30,000 ft (9200 m), might be pressurized to three times that amount. The potential for gas is increased by a factor of 10 at that depth. The cost of drilling and recovery will also increase dramatically. No reliable estimates exist for this type of drilling, but it is expected that drilling costs alone could exceed current costs by a level of 200 or 300%.

Where this process is used on land, the geothermally heated high-temperature brine could be used for process heat or for cogeneration. The major negative consideration is the disposal of vast quantities of water with concentrations of salt as high as 10 times that of seawater near the surface of the ocean. In addition, the potential for land subsidence is greatly increased. Offshore wells could be of serious consequence with regard to lowering the ocean floor, both in terms of short-term damage to fragile aquatic environments and long term with

98 Petroleum and Natural Gas: Our Most Popular Fuels Chap. 5

Figure 5-15 Offshore drilling vessel. (Courtesy of Global Marine Drilling Company.)

regard to changing the temperature of ocean currents. Brine disposal is just as serious at sea as on land, posing serious ecological problems for fish and plant life.

There are three other unconventional sources for gas production. The first are the gas-bearing shale formations of the eastern United States in the Devonian shales found on the western side of the Appalachian Mountains, running from New York to northeastern Tennessee. An estimated 10 to 520 trillion cubic feet (up to 522 quads) may be available by means of water fracturing (injecting water at high pressures into gas-bearing strata), but the wellhead price would be about 25% higher than existing gas prices.

Another potential source area is the low-permeability (tight) sands of the Rocky Mountain region (Figure 5-16). A potential 50 to 320 trillion cubic feet may be available, but a shortage of water needed to

Tight Gas Sand Basins

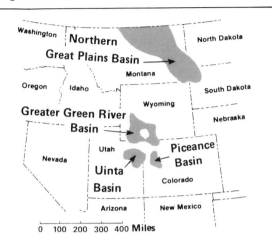

Figure 5-16 Tight gas sands basins. (From the U.S. Department of Energy.)

aid in the recovery process may reduce the feasibility of this potential source.

A third proposal is the recovery of methane from coal seams which have not yet been opened. Methane is the cause of mine explosions. Its removal would be a safety measure, but the process would have to occur several years prior to the development of the coal mine to allow the gas to migrate through the seam. The gas is currently vented to prevent explosion and is therefore "wasted." Use of this gas would be a conservation measure as well as a safety precaution. An estimated 800 trillion cubic feet may be available from coal mines. Nonmineable coal seams can also be a source for the extraction of methane gas.

SUMMARY

The formation of oil and natural gas is known to be a process of many centuries. Petroleum products to be used in the foreseeable future will depend on known measured reserves and on indicated and inferred sources based primarily on geological information. Estimated but undiscovered reserves may yet be found in locations that are even more difficult to explore because of geographic, climatic, and political conditions. Oil reserves have been estimated to be sufficient to last to the year 2000 at a minimum, and to somewhere between 2020 and 2075 at maximum. Several factors control the longer-range estimates: the price of energy, new discoveries, cost of recovery, rate of consumption and conservation, and introduction of synthetic fuels, all unknowns over the range of time expressed above.

Production levels reflect consumption rates and the latter have been falling or remaining stable for the past few years. Lower prices also mean that fewer marginal reserves are being developed. There is a growing concern over costs of energy and a recognition of the dwindling resources. The further development of technologies for increased use of renewable energy sources may mean an extension of the time when nonrenewable sources are the major source of energy for commerce, industry, and personal comfort.

The use of natural gas, oil, and coal for heating and transportation may become more restricted due to higher cost. It is possible that higher cost may also restrict the use of these nonrenewables to the preparation of fertilizers, pharmaceuticals, plastics, and other products prepared from the carbon-based components of these materials.

ACTIVITIES

1. How are each of the various petroleum products delivered to your community? How are they distributed to the point of use?
2. What uses are made of petroleum and natural gas products: heating, fuel source for production of electricity, and what others?
3. Determine the cost of installation of equipment for the use of one oil or natural gas appliance. Compare the cost of installation for a renewable system to perform the same task. Estimate the operational cost of each system over periods of 1, 5, and 10 years. Which system is the best buy?
4. Develop an outline and moderate a discussion on how to reduce our dependence on imported petroleum products.
5. Investigate and discuss the environmental problems related to the discovery of oil and natural gas, their extraction, refining, and transportation.
6. Discuss in class how declining supplies of natural gas and oil will affect life in the next 5, 10, and 20 years. Include such aspects as costs, availability, and new developments such as heating and transportation. How will your life style be affected? How might these factors be affected by developing countries, industrialization, and use of renewable energy resources?

BIBLIOGRAPHY

Bohi, Douglas R., and Michael A. Toman, "Understanding Nonrenewable Resource Supply Behavior," *Science*, Vol. 219, February 25, 1983, pp. 927-932.

Considine, Douglas M., ed., *Energy Technology Handbook*. New York: McGraw-Hill, 1977.

Energy Information Administration, *Natural Gas Monthly* (various issues), Washington, D.C.

Energy Information Administration, *Petroleum Supply Monthly* (various issues), Washington, D.C.

Federal Energy Administration, "Geological Estimate of Undiscovered Oil and Gas Reserves in the United States," Geological Survey Circular 725. Washington, D.C.: FEA, 1975.

Gibbons, John H., and **William U. Chandler,** *Energy—The Conservation Revolution.* New York: Plenum Press, 1981.

Hubbert, M. King, "Nuclear Energy and the Fossil Fuels," in *Drilling and Production Practice.* Dallas: American Petroleum Institute, 1956.

Hunt, V. Daniel, *Handbook of Energy Technology.* New York: Van Nostrand Reinhold, 1982.

"Improved Oil Recovery," Exxon Background Series. New York: Exxon Corporation, December 1982.

"Petroleum Refining," *Encyclopaedia Britannica*, Vol. 14. Chicago: Encyclopaedia Britannica, 1981, pp. 180-188.

Steinhart, Carol E., and **John S. Steinhart,** *Energy, Sources, Uses, and Role in Human Affairs.* North Scituate, Mass.: Duxbury Press, 1974.

"The Upstream." New York: Exxon Corporation, December 1982.

Chapter 6
Nuclear Fission
The Tarnished Star of Our Energy Future

CONCEPTS

1. Nuclear energy provides 12.6% of the nation's electrical energy (1982).
2. The nuclear fission electrical power industry has serious fiscal and social acceptance problems.
3. The general public does not understand nuclear fission technology.
4. Nuclear-fission-fueled electricity-generating plants are pricing themselves out of the market.
5. Uranium is a nonrenewable resource.
6. Einstein's principle: any loss of mass during a reaction is accompanied by a release of energy.
7. Overpopulation, air and water pollution, depletion of resources and agricultural lands, and released toxic chemicals are a greater problem for the future than radiation.

GLOSSARY

CARCINOGEN—any substance that tends to produce cancer.
FISSION—splitting an atomic nucleus with the subsequent release of energy.
FUSION—combining atomic nuclei with the subsequent release of energy.
MELTDOWN—melting of the reactor core and containment vessel with the resulting release of radioactive materials.
RADIOACTIVE—having the property to emit particles or radiation from an atomic nucleus.

INTRODUCTION

In the early years of nuclear fission development, proponents of nuclear energy heralded it as the ultimate answer to our ever-increasing energy requirements. The abundant clean thermal energy produced by nuclear reaction was to be used to produce electricity in new plants throughout the nation. Electricity rates were projected to drop to levels where metering of use would be almost unnecessary.

A majority of the initial development of nuclear energy was done by the U.S. government under the nuclear weapons program. After World War II, the Atoms for Peace effort began to involve private industry. There are six manufacturers of nuclear reactors in the United States, each with its own designs: Westinghouse, General Electric, Babcock and Wilcox, Combustion Engineering, General Atomic, and Allis-Chalmers.

Several system designs have been developed to create and control nuclear fission reactions since the first successful occurrence in 1942 at the University of Chicago. During the 1940s, the Atomic Energy Commission (AEC), forerunner of the Nuclear Regulatory Commission (NRC), began a program to build several experimental reactors for operation beginning in the 1950s. The first civilian unit was completed and began producing electricity in 1958 at Shippingport, Pennsylvania.

As development continued and larger electrical power plants were built and placed in service, problems began to appear that had not been anticipated during the earlier years. The technology once thought to be the "answer" became a technology fraught with problems of poorly designed and engineered systems: faulty equipment, materials, and workmanship; large cost overruns; construction delays; constantly changing governmental regulations; concerns for very long-lasting environmental impacts; carcinogen exposure; and gross misunderstanding and mistrust by the general public. Because of these many problems, some energy leaders see fission nuclear energy only as a technology to "bridge the gap" between the use of fossil fuels and the sought-after nonpolluting, unlimited fuels of the future.

NUCLEAR OVERVIEW AND CURRENT STATUS

Continually increasing electrical energy costs, cogeneration, the use of renewable energy sources, conservation, and load management have reduced the former rapid rate of increasing demand for electricity and in some states have actually created an overall demand reduction. Also contributing to the slowing of growth in electricity demand has been the recent economic recession. These influencing factors have resulted in a sizable overcapacity in the nation's electricity supply industry. In fact, 1982 saw a 0% growth in electrical demand and New York State had a 46% overage of generating capacity as of January 1984 [Borfitz].

This current abundance of available electrical energy, exponentially increasing construction costs for new nuclear power plants, and the related concerns about the technology have resulted in a dramatic slowdown in the nuclear power industry. There have been no new proposals approved for nuclear-powered electricity generating plants in the United States since 1973, and none are expected in the near future. Between 1974 and 1984, utility companies actually canceled 100 nuclear-fueled plants which were proposed or under construction. In the first 3 months of 1984, three more were canceled. As of April 1984, 29 utilities had 33 plants under construction. Most were expected to be completed within 10 years. Five more were on firm order but holding. Adequate electricity generating capacity is expected in the United States through the year 2000 even if some of the units under construction are terminated and the nation's economic situation improves. Worldwide, excluding the United States, there are 40 countries with nuclear energy programs but only 22 had operating reactors as of 1983. There were 199 operable units, 151 under construction, 29 on order, and 197 in firm planning stages as of December 1981.

The United States is the current world leader in the manufacture of nuclear-fueled electricity generating units and with 68,000 MW of generating capacity has more facilities in service than any other country. France, the USSR, Japan, Belgium, Sweden, Switzerland, Finland, Taiwan, and West Germany are prominent leaders, with France first in growth rate, followed by the USSR. It is projected that by 1985 France will produce 70% of its electricity via nuclear power. France, the USSR, West Germany, India, Japan, and Great Britain are continuing to build and operate fast breeder reactors, and Canada, France, West Germany, India, Japan, Great Britain, and the USSR are reprocessing nuclear fuels. Interest in the expansion of nuclear fission has slowed in the United States, Austria, and Sweden.

Overseas, there have been no new orders for plants in the last 3 years. Spain has indefinitely deferred construction of five new plants, including three which have been partially constructed. France has placed six planned facilities on "hold."

Nuclear electricity generating plants take a long time to build. In foreign countries, construction generally is completed within 6 years, while in the United States 14 years is the average. This considerable time differential is due primarily to the extensive safety precautions that we require for the use of this technology. Also contributing to the long construction time is the fact that each plant has been custom-designed and is site-specific, while those sold to foreign countries have been "packaged" facilities. There were 56 operating nuclear-fueled electricity generating plants in the United States in 1975. This number grew to 74 in 1980. As of July 1, 1983, there were 84 licensed to operate, 56 with permits for construction, and 5 plants on order (see Figure 6-1). Only 74 were in commercial operation.

Key
● Reactor With Operating License
○ Reactor With Construction Permit
△ Reactor On Order

84 Reactors with operating licenses .... 67,665 MWe
56 Reactors with construction permits .... 61,712 MWe
  5 Reactors on order .... 5,140 MWe
145 Total .... 134,517 MWe

July 1, 1983

Figure 6-1 Nuclear power plants in the United States. (Courtesy of Atomic Industrial Forum, Inc.)

Nuclear-fueled plants provided 12.6% of the nation's electrical energy in 1982, up from 11.8% in 1981. Thirty-four percent (34%) of New England's electricity and 83% of Vermont's is generated using nuclear energy [Atomic Industrial Forum, January 12, 1983]. If construction of the units being built is completed, an estimated 20% of all U.S. electricity will be from nuclear facilities. Most of these units are expected to be completed, because delay or cancellation would add costs and create unemployment, and considerable investment has already been made.

HOW NUCLEAR PLANTS WORK

All matter is made up of tiny particles called atoms and each atom has a dense center called a nucleus. Nuclear energy begins in the nucleus of an atom.

Each nucleus contains protons (plus charge) and neutrons (zero charge). The protons should repel each other because they are like charges; instead they are bound tightly together by nuclear force. The greatest energies of all are those that bind particles together in the atomic nucleus. If the nucleus can be forced to split, energy in the form of electromagnetic radiation, energetic particles, or both will be released. The process of splitting an atom is called nuclear fission and the system used to contain and control fission is a nuclear reactor.

The atomic structure of most materials does not split or fission easily. Uranium-235 is used as the fuel in reactors because it possesses a large, heavy nucleus that will split relatively easily. Free neutrons are fired at the U-235 atoms, and when a nucleus is struck, it splits into two pieces (called fission fragments) and two or three neutrons, which then may go on to split additional U-235 nuclei (see Figure 6-2). These nuclei in turn release more neutrons. If the conditions are suitable, the reaction will continue as more and more nuclear fissions occur. This is called a chain reaction.

During a chain reaction, thermal energy and radiation are produced in the reactor, and the heat is used to boil water to make steam to drive the turbine connected to an electric generator. Basically the plant is the same as other electricity-generating plants except that nuclear energy, rather than coal or oil, provides the heat-driving force for the turbine. Nothing is burned or exploded in the nuclear reactor. The energy in 1 lb of nuclear fuel contains 3 million times the energy in 1 lb of coal.

The U-235, fabricated into small pellets that are placed end to end in a fuel rod assembly, is located inside the nuclear reactor vessel. Control rods, which soak up neutrons, are inserted between the fuel assem-

How Nuclear Plants Work 107

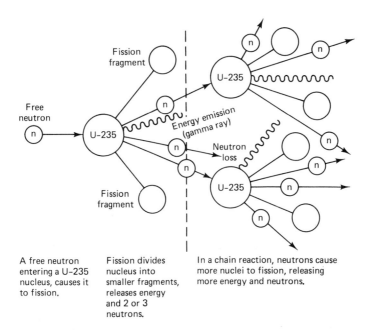

| A free neutron entering a U-235 nucleus, causes it to fission. | Fission divides nucleus into smaller fragments, releases energy and 2 or 3 neutrons. | In a chain reaction, neutrons cause more nuclei to fission, releasing more energy and neutrons. |

Figure 6-2 Nuclear fission reaction. (Courtesy of Atomic Industrial Forum, Inc.)

blies to control the chain reaction and the resulting release of thermal energy.

Approximately two-thirds of the energy from a nuclear reaction comes from the fission of the uranium in the fuel. During reactor operation, plutonium-239 and -241 are created, providing the balance. Plutonium fissions as soon as it is formed, and near the end of the fuel life cycle it contributes as much energy as does the uranium. Approximately one-half of the plutonium remains in the spent fuel rods together with the residual U-235. Plutonium, a newly created element, is actually a by-product and is carcinogenic as well as very toxic.

Several different nuclear reactor designs are being used or are proposed for future development:

Boiling-water reactor (BWR)
Pressurized-water reactor (PWR)
High-temperature gas-cooled reactor (HTGR)
Liquid-metal fast breeder reactor (LMFBR)
Light-water breeder reactor (LWBR)
Graphite-controlled breeder reactor (GCBR)

There are two types of commercial reactors widely used in the United States: BWRs and PWRs. Both are called "light-water reactors" because their coolant or heat-transfer medium is ordinary water.

The BWR was the first design to be built commercially and the first unit was placed on-line in California in 1957. In this unit, steam is produced by pumping water through tubes in the reactor core. The steam is then directed to drive the turbine (see Figure 6-3). These units are less efficient than other designs and there is concern for safety since the radioactive steam created in the reactor circulates throughout the entire system.

Pressurized-water reactors are the most common. From Figure 6-4 you can see that the electricity-generating portion is the same as in the BWR but the system inside the containment structure varies. Water is circulated through the reactor core, as in the BWR, but is kept under pressure so that it will not boil. This superheated water is circulated through a heat exchanger called a steam generator, where a separate supply of water is heated to produce steam. The steam is then dried and piped to drive the turbine. The radioactive water flowing through the primary loop of the reactor core is contained within a closed system, thereby providing a much greater margin of safe operation.

Breeder reactors are considered to be the next generation of energy producers because light-water reactors produce large amounts of radio-

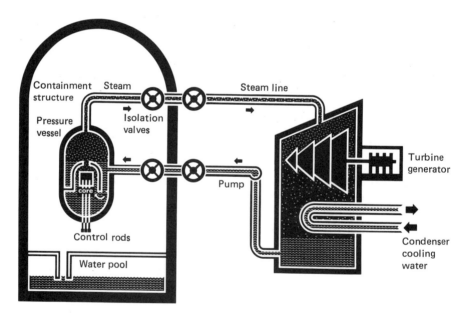

Figure 6-3 Boiling-water reactor. (Courtesy of Atomic Industrial Forum, Inc.)

active wastes, and U-235, the fuel for light-water units, is a limited resource expected to last only until around 2020. Breeders are also more efficient in that they extract 60 to 70% of the energy potential from the uranium, compared to 1 to 2% for light-water units. With this efficient use of uranium and the production of plutonium, breeder reactors could extend our supply of uranium until nearly 4500, meeting all U.S. electricity needs without importing fuel, and providing us with energy independence.

The LMFBR works differently from the other units. The breeder reactor uses U-235 for fuel but the reactor is lined with U-238, the nonfuel, very plentiful type of uranium. While the U-235 is undergoing fission, some of the free neutrons that are released bombard the U-238 liner, changing it to fissionable plutonium. This plutonium can be processed and used as fuel, and since more is produced than is consumed in the core, there is a net fuel gain of 10 to 20% per year. Eventually, the reactor creates enough plutonium to replenish its own fuel supply and operate others. They are called "fast" breeders because of the fast or high-energy neutrons.

The cooling and heat-transfer medium in the LMFBR is liquid metallic sodium, which has a higher boiling point than water—1621°F (883°C)—and conducts about 100 times better than water. Water also slows down the nuclear reaction, while sodium allows a faster flow of

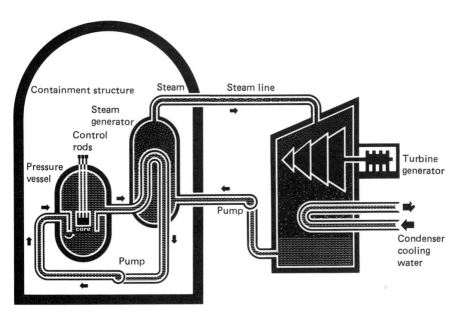

Figure 6-4 Pressurized-water reactor. (Courtesy of Atomic Industrial Forum, Inc.)

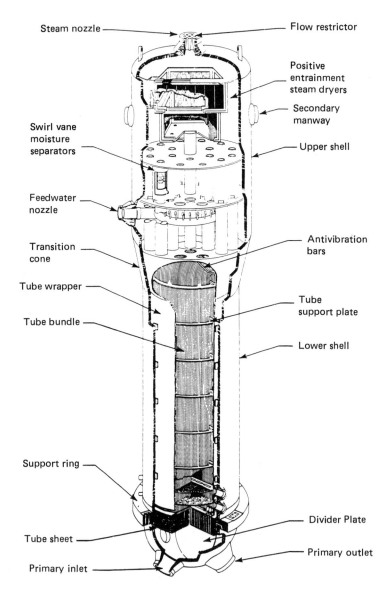

Figure 6-5 Westinghouse model F steam generator. (Courtesy of Westinghouse Electric Corporation.)

Figure 6-6 Indian Point 2, 873-MW PWR. (Courtesy of Atomic Industrial Forum, Inc.)

neutrons. The problem of using sodium is that it is corrosive, burns violently, and can explode if it comes in contact with air or water. This may have happened in 1973 in the USSR at the now-abandoned unit on the Caspian Sea.

Historically in the United States, the first experimental breeder reactor, called Clementine, was demonstrated between 1946 and 1953 at Los Alamos, New Mexico. The first unit to produce electricity was built by Argonne National Laboratory and operated from 1951 to 1963.

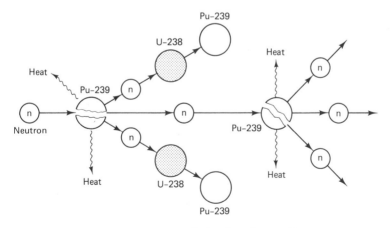

Figure 6-7 Nuclear fission breeder reaction.

Two other experimental breeder reactors were built by the U.S. government but both reportedly had partial core meltdowns: Idaho (1961) and Enrico Fermi (1966).

The most recent efforts toward U.S. breeder reactor development have been at the Clinch River site near Oak Ridge, Tennessee. This demonstration unit, approximately one-fourth the size of projected future commercial units, was to provide the technical basis for extending breeder units and allow environmental impact studies. The $800 million (1970) estimate with a completion date of 1980 grew to a projected $4.2 billion with startup in 1990. Congress withdrew additional funding for fiscal year 1984 with the facility about 90% complete. The decision was due to the massive cost overruns. Opponents of the unit projected a nearly $10 billion total cost and expressed serious concern about plutonium transportation and reprocessing, potential radioactive material releases, and nuclear weapons proliferation. The DOE is expected to continue breeder development with international cooperation.

Although the United States has terminated breeder reactor development at this time, France continues its leadership, although it too has trimmed its programs due to heavy financial costs. Following testing in demonstration plants, France's first commercial-scale unit was projected

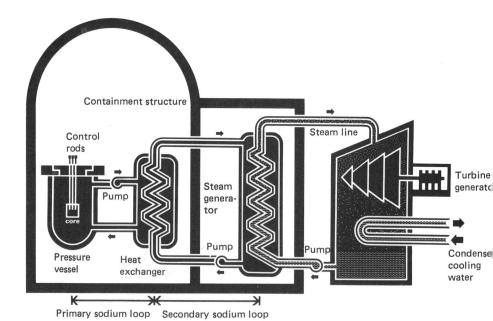

Figure 6-8 Liquid-metal fast breeder reactor. (Courtesy of Atomic Industrial Forum, Inc.)

to be on-line in 1983, but no announcement had been issued as of late 1984. The breeder leads them toward the promise of a limitless supply of energy.

THE FUEL CYCLE

Uranium ore, as found in its natural state, has only a 0.7% concentration of fissionable U-235. Through gaseous diffusion and centrifuge technology, this concentration is increased to 3 to 4% for use in reactors. The U.S. government is the sole domestic source of enriched uranium. Each reactor will require the processing of 5500 tons of uranium ore during its lifetime.

The enriched uranium is pelletized and placed in sealed 12- to 14-ft (3.6- to 4.3-m) tubes of zirconium alloy. These tubes are then grouped together in bundles to form fuel assemblies (Figure 6-9). The fuel assemblies and control rods comprise the reactor core (see Figure 6-10).

After 3 years in the reactor, spent fuel can be reprocessed or undergo proper disposal. Reprocessing involves solvent (nitric acid) extraction of the unused U-235 and plutonium that was produced during the fission process. The balance of the fuel is still a highly radioactive waste product which requires disposal. Figure 6-11 shows the graphic sequence of the uranium fuel cycle.

The United States no longer operates a nuclear fuel reprocessing plant and has no permanent storage facilities. Since the beginning of the nuclear fuel program, spent-fuel assemblies have been kept at each of the respective plants in large water storage pools (Figure 6-12). The water retains the radioactivity and keeps the assemblies cool. A problem to be faced in the very near future is what to do with these spent-fuel assemblies, because several plant storage pools are nearly full.

Many proposals have been made for handling radioactive nuclear waste that is not reprocessed. These range from direct burial to loading the material into rockets and shooting them into the sun. Others suggest that it be mixed with cement or dissolved in metals. Since the major concern is to keep the waste from entering the environment, the present thinking is to bond the material chemically at the atomic level, thereby reducing its ability to be dissolved by the environment. Vitrification with a borosilicate glass composition is currently the most popular proposal and is now in use in France.

Following vitrification, the glasslike material will be sealed in titanium alloy or stainless steel cylinders and buried deep underground in geologically stable formations. What is needed now are highly productive, simple, economical processes.

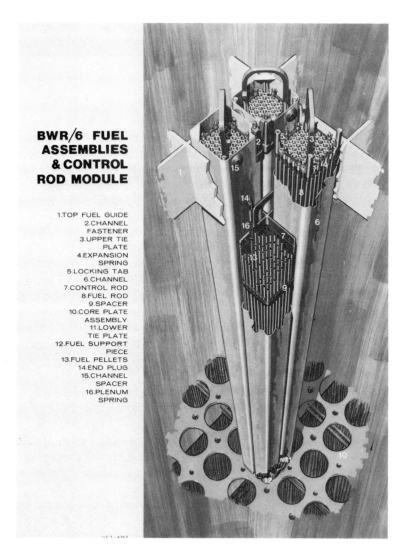

BWR/6 FUEL ASSEMBLIES & CONTROL ROD MODULE

1. TOP FUEL GUIDE
2. CHANNEL FASTENER
3. UPPER TIE PLATE
4. EXPANSION SPRING
5. LOCKING TAB
6. CHANNEL
7. CONTROL ROD
8. FUEL ROD
9. SPACER
10. CORE PLATE ASSEMBLY
11. LOWER TIE PLATE
12. FUEL SUPPORT PIECE
13. FUEL PELLETS
14. END PLUG
15. CHANNEL SPACER
16. PLENUM SPRING

Figure 6-9 BWR fuel assembly and control rod module. (Courtesy of General Electric Company.)

FINANCIAL COSTS

The multitude of factors affecting the financial aspect of the nuclear industry appear never-ending. They include:

Government support for research and development
Inflation

Figure 6-10 Fuel assembly being lowered into the reactor core. (Courtesy of Atomic Industrial Forum, Inc.)

Construction costs
Changing technology
Changing regulations
Fuel costs
Material problems: inappropriate; degradation
Fines for violations
Off-line time for repairs and refueling
Storage of low- and high-level wastes
Evacuation plans
Fuel reprocessing
Decommissioning of plants

The actual costs related to the nuclear-fueled electricity-generating industry are impossible to determine. Compounded in the estimate of costs for any energy source should be all the related expenses from beginning to end—resource recovery through environmental impacts. Estimates of costs for nuclear energy, therefore, cannot be made at this time because some data are not released by the government, no commercial-size plants have been decommissioned, nor have the costs for long-term waste storage been encountered.

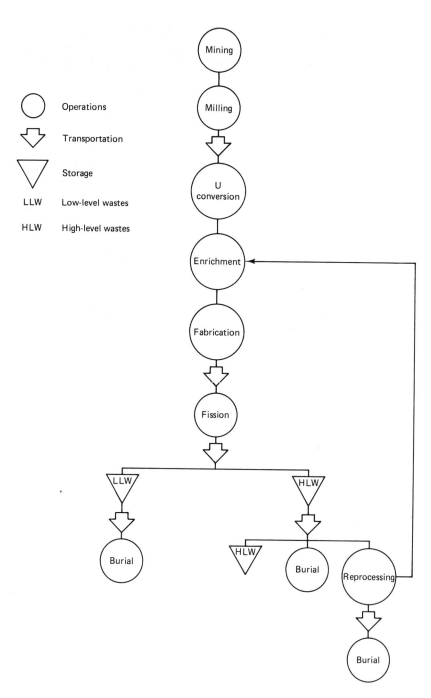

Figure 6-11 Uranium fuel cycle.

Financial Costs

Figure 6-12 Spent-fuel storage pool. (Courtesy of Atomic Industrial Forum, Inc.)

Construction costs of new plants have been skyrocketing beyond even the best-prepared estimates. One reason is that contracts are made on a cost-plus basis rather than by lowest bidder. This contractual arrangement is necessary due to the constantly changing NRC regulations, design alterations, and rates of inflation during the time of construction. Other variables include labor union demands, delays, and problems with material and workmanship quality. It is estimated that nuclear-powered plants cost 60% more to build than comparable coal-fueled plants with the latest pollution control equipment [Parisi]. Experts on both sides of the issue project costs of 25% more to 8% less than those for comparable coal-fired units. A 1982 study by the Energy Information Agency (EIA) reported that the cost of coal- and nuclear-fueled units was nearly equal. This great disparity in reports is disconcerting and adds confusion for those who study the nuclear industry.

Illustrative of problems encountered during the construction of nuclear electricity generation plants were those encountered at the Shoreham plant of the Long Island Lighting Company. The permit to build this 820-MW plant was issued in 1973 with a completion date of 1977 at a cost of $350 million. The unit has been rescheduled to be on-line in 1985, and the reestimated finished cost is $3.7 billion. The Seabrook, New Hampshire, 2300-MW plant, grossly behind schedule and $3,000,000,000 over budget, was terminated March 30, 1984.

The Washington Public Power Supply System (WPPSS) suspended its nuclear energy program in July 1983. Five units were to be built, and Unit 1 and Unit 3 were 60% and 70% completed, respectively, when the $2.25 billion financial burden became too great and the company defaulted on its bonds. In addition, electricity rates were increasing dramatically and rate payers were rebelling with law suits and regulatory intervention.

On January 16, 1984, Public Service Indiana announced that they were "financially unable to proceed" with the construction of their Marble Hill plants on the Ohio River near Madison. Originally projected to be on-line in 1983, one unit was 60% completed and the other about 30%. The most recent estimate for completion was 1988, with the cost jumping from $1.4 billion to $7 billion. Within a week after this announcement, on January 22, 1984, the Zimmer plant in Moscow, Ohio, was terminated. Although it was about 97% complete, the owners decided to convert the unit to a coal-fired facility after costs, originally $240 million, ballooned to an estimated $3.5 billion. The conversion is expected to cost $350 million and be completed in May 1986.

The most expensive nuclear-fueled plant ever built in the nation by investor-owned utilities is Nine Mile 2, located on Lake Ontario in New York State. This 1080-MW unit was approved for construction in 1972 with a cost estimate of $356 million and a completion date of December 1977. The project has been plagued with new and changing NRC requirements, cost overruns, work stoppages, management problems, faulty workmanship, improper materials, and low worker productivity. The 1983 cost estimate by the utilities was $4.2 billion with a completion date of October 1986. This estimate was $3,844,000,000 and 9 years over the original plans when the March 30, 1984, revised utility estimate jumped to $5.1 billion, an increase of $900,000,000 over 1983's estimate. Some nonutility reports have projected a total cost of nearly $7 billion and a 1988 completion date. The Long Island Lighting Company, one of the five utility partners, withdrew its financial support in April 1984. Studies concerning continuance of construction were being conducted by several groups. The Public Service Commission's May 1, 1984, estimate of $5.4 billion and completion date of late 1987 was followed by a May 3 estimate by the New York State Energy Office of $5.9 billion. Which study is right? Is either correct?

The total cost of enriched uranium fuel has yet to be determined.

The three uranium enrichment gaseous diffusion plants at Oak Ridge, Tennessee; Paducah, Kentucky; and Portsmouth, Ohio, are under direct government control for safety and security reasons. Although uranium ore processed at these facilities is sold to utility companies for fueling reactors, actual total costs for producing the fuel are not available. One ton of ore yields about 3 lb of uranium oxide.

Japan was successful in removing uranium from seawater in early 1984. A filter trap concentrates the extremely dilute uranium. This development allows tapping an unlimited uranium source but at approximately 10 times the cost of processing uranium ore.

Whenever a nuclear plant must be shut down for refueling or repairs, electricity must be acquired from other sources. Since these plants produce large quantities of electricity, several different sources are often required, including purchasing expensive electricity from oil- and coal-fueled plants, some located in other countries. A complete refueling cycle takes about 3 years, with one-fourth to one-third of the fuel assemblies being replaced annually, and repairs may be necessary at any time. Refueling takes about 2½ months. Some units have been down (off-line) for repairs for nearly a year, with costs up to $5 million. In 1981 the national average for on-line time was 77%. New York's nuclear-powered units were on-line only 46% of the time in 1982. The national average dropped to 65% in 1983, with two-thirds of the plants on-line less than 60% of the time.

NRC safety precautions require that an evacuation plan be established and tested for each plant. This plan is necessary so that all people within a 10-mile radius of a plant can be evacuated in an orderly manner should an uncontrollable problem occur. These plans require sophisticated warning systems as well as the coordination of many groups, including police, fire companies, civil defense, utility workers, and others. In some areas, local officials are resisting participation in both the planning and testing of evacuation systems, which must be performed on a periodic basis. A portion of the financial burden for these precautions is taxpayer supported.

Historically, federal government subsidy for nuclear energy research and development has been substantially higher than support for other energy forms. The basic research and development for the technology was funded and conducted by the federal government as a part of the military development program. Nearly three-fourths of the federal energy research monies go to continued nuclear-related research and development.

Some states have established nonprofit utility corporations to ease the financial burden on utility companies and energy users. Since these government-owned nonprofit corporations are tax exempt and have no stockholders to satisfy, they can charge lower rates for the electrical power they produce. Some of these corporations have already purchased nuclear plants from utilities to help them survive financial crisis.

Most utility companies have no competition and are consequently legalized monopolies. For this reason, each state has its own public utilities agency. This group of people is saddled with the responsibility of setting the rates paid by consumers for the energy they use. Additionally they have the responsibility of seeing that these rates are sufficiently high to ensure the financial health of the utilities and that service is maintained. Basically they are in the difficult position of helping to ensure that the utilities stay in business by making a profit and that consumers are getting the best rates possible. Public utility agencies are authorized to allow up to a 15% return on investments made in utility companies.

By law in many states, utility companies cannot charge customers for electricity-generating facilities until they are operational. During construction, loans or bond issues are made by the utilities and repayment is initiated when the plant goes on-line. According to a study by Cambridge Energy Research Associates of Massachusetts, new nuclear plants currently under construction will produce electricity at rates equivalent to double or triple the price of OPEC oil, thus requiring rate increases ranging from 30 to 80%. In each case, the public utilities agency will be in the middle between the financially burdened utility and the consumers, who have no desire for rate increases. Nuclear-powered plant rates are expected to increase about 7% above inflation each year.

Utilities that terminate nuclear projects unfortunately still must repay loans and interest incurred. These planning and construction costs may be passed on to the consumer even though no energy benefit will ever be obtained. This is the position in which WPPSS, Rochester Gas and Electric, and others are finding themselves.

Enforcement of NRC safety and operation regulations is conducted by inspectors, and if necessary, hearings are held and fines assessed for violations of those regulations. Fines have been levied for several occurrences—some beyond utility control. Some violations that have been reported range from operating with faulty radiation monitoring units to employing unsupervised trainees to inspect work that required certified inspectors. In addition to reducing public confidence in the utility, the violations create a financial burden that is passed along to the consumer.

RADIOACTIVE WASTES

The radioactive wastes associated with the use of nuclear energy are of great concern and considerable expense. Wastes are created at every step in the use of this energy source. Uranium mining, refinement, use in a reactor, reprocessing (if done), and plant decommissioning each must use procedures that ensure the containment of radioactive materials. Many different wastes are created, but the ones causing the greatest concern are krypton, plutonium, strontium 90, iodine 131, and cesium 137.

A typical 1000-MW plant produces 30 tons (27,216 kg) of spent fuel annually, including 435 lb (197.3 kg) of highly toxic plutonium. There are no known methods for neutralizing most of these wastes, and the methods for containment and permanent storage are questionable because of the many thousands of years that these materials remain active.

A 1983 U.S. Supreme Court decision has required that each state accept the responsibility for deciding what to do with low-level (non-fuel) nuclear wastes produced in that state until safe ways are found for disposal. It has been determined that utility companies are not to be held responsible for the future disposition or long-term storage of the wastes they create. Under current law, individual states or regional groups of states are working toward solving the waste problem. Potential burial sites are being studied and when a repository is constructed, utilities will probably pay by the cubic foot of waste to be buried. Each nuclear plant customer is currently paying a tax of 0.1 cent per kilowatt-hour to help support research and development of these facilities.

High-level (spent-fuel) waste disposal has been made a national responsibility, and therefore final disposition rests with the federal government. The first geologic burial site for nuclear waste has been started near Carlsbad, New Mexico, 2150 ft (655 m) underground in a massive salt formation. Called the Waste Isolation Pilot Plant, the DOE $2.1 billion project is intended for experimentation and for high-level radioactive wastes created primarily by the nation's nuclear weapons program and is expected to be ready in 1988. Although the project does not specifically address the storage of wastes from civilian atomic-powered plants, the state is greatly concerned that it will become a repository.

The New Mexico site is considered a pilot repository, and two additional sites, one in the West and one in the East, are projected to be established. By June 1985, three to five potential sites are to be selected for further testing, with one to be selected by the President as the first permanent geological repository by May 1987. Sites are now being considered in Texas, Washington, Utah, Nevada, Mississippi, Louisiana, and New Mexico. Construction is projected to begin in 1989. This facility should be operational by 1998 and should begin accepting waste in 2002. At this time the waste is planned to be converted to a stable glasslike material prior to burial inside canisters of highly corrosion-resistant alloys. Because they are still experimental in the United States, projected costs of these facilities are not available.

A new alternative to vitrifying radioactive wastes with molten borosilicate glass appears promising. Under development in Australia, the Synroc process, which blends nuclear waste products with synthetic rock, appears to have the advantages of safety, high leach resistance, and equal cost compared to vitrification. Synroc binds radioactive components into atom-sized spaces within rock crystals. It is also a tested process that would be easier and cheaper to bury and would not require large, expensive national waste repositories. Burial is said to require only a drill hole, which could be close to the plant, thereby reducing transportation. Australia plans to build a demonstration plant, but because Synroc must be tailored for each kind of waste, DOE was expected to begin its $970 million glass plant in 1984.

Peterson reports that should a permanent storage facility not be available when needed, monitored retrievable storage could be used. This proposal involves sealing radioactive waste in special casks that are then stored in a field of dry wells or in large surface vaults. Continuous monitoring would be possible and the system permits recovery for future use.

The nation's only reprocessing plant for spent nuclear fuel was operational at West Valley, New York, from 1966 until 1972. This $32.5 million facility separated unused uranium and plutonium from the spent fuel. The reclaimed materials were to be used again in the fuel cycle. When radioactive material escaped into places where it was not supposed to be and expenses became excessive, Nuclear Fuel Services withdrew, leaving the facility to the DOE and the state of New York (as per the original contract). West Valley Nuclear Services Company, a subsidiary of Westinghouse, has been hired, and cleanup of the estimated 520,000 gal (1,968,000 liters) of highly radioactive liquid wastes, stored in two carbon steel and two stainless steel underground tanks is underway. There are also 2 million cubic feet (56,640 m^3) of buried low-level radioactive wastes on the site. Westinghouse scientists estimate that they can prepare the high-level liquid wastes for burial for less than $500 million. The project uses federal, state, and private funds [Coy].

Since the closing of West Valley, the United States appears to have selected storage for high-level nuclear wastes, while France, West Germany, and others continue their efforts in reprocessing. All U.S. efforts have been slow, conservative, meticulous, methodical, and expensive. The storage of both low- and high-level nuclear wastes does not appear to be a technical problem to the industry, but the costs are nearly impossible to project.

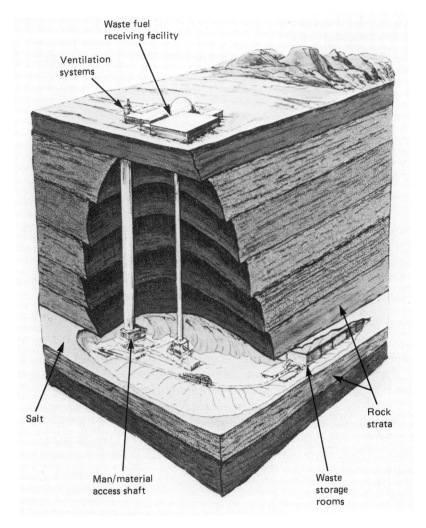

Figure 6-13 Geological radioactive waste disposal site. (Courtesy of Dames and Moore; Atomic Industrial Forum, Inc.)

Nuclear electricity-generating plants are anticipated to have a useful life of 30 to 40 years. At that time, maximum safe levels of radiation will be reached in the reactor materials and surrounding structure. Decommissioning of a commercial-size plant has never been done, and no decision has been reached as to the best method for handling the situation. The two present plans call for dismantlement or entombment and guarding for at least 200 years. Dismantlement, in 1983 dollars, is projected at $1.07 billion and entombment at $266 million per plant due to the thousands of tons of steel and concrete permeated with radiation. More than 15 U.S. plants have already closed and their disposal is a major problem. It is projected that future generations will be paying for the security of nonproducing plants which are now providing our electrical energy.

PUBLIC OPINION

The nuclear industry has more organized groups supporting and opposing it than any other energy conversion technology. These groups range from small local groups to large, well-organized, highly financed, nationwide organizations. These groups and their members represent all sectors of society, including the political, scientific, technical, educational, structural, and industrial communities. The activities of these groups range from providing literature, films, and speakers for various gatherings to political lobbying at state and national levels.

Some newspapers have taken a position on the nuclear issue, and the groups opposing that position frequently claim irresponsible journalism. As you attempt to keep up to date, close examination of the wording in articles is necessary. For example, there may have been a release of radioactive gas at a plant, but further study may reveal that you actually could receive more exposure from an airline flight to visit your grandma. Sensationalism rather than objective reporting to the public is believed to have caused considerable financial impact on the commercial nuclear-fueled electric industry through excessive regulation.

Opposition groups are concerned principally with the safety, health, environmental, and financial characteristics of the nuclear energy industry. Proponents often cite the need for additional electricity-generating capacity, the declining availability and increasing costs of fossil fuels, and the volume of fossil fuels which are or can be replaced by nuclear fuels. Additionally, a beneficial economic characteristic of the nuclear-powered electricity-generating plants is that they require an abundance of materials and employ many people for 12 to 15 years during construction. This has been especially beneficial during the recent period of great fiscal distress when the construction industry was releasing employees because there was no work. Unions and their members have been strong proponents for nuclear energy.

Some state governors have taken a position of opposition to the development of nuclear power plants in their states. In some cases, the utilities have plants in operation and/or under construction and the financial burden is becoming too great for residents. In others, plants located in urban areas cannot assure an acceptable public evacuation plan. State responsibility for low-level waste storage is also an important factor.

The uncertainty in the nuclear fission financial picture is being reflected in the purchase of utility investments in the stock market. The construction of several electricity-generating plants is long overdue and far over budget, threatening to defeat some utilities that have invested billions of dollars. The WPPSS plants' termination and bond default were the first pronounced indicators.

The accident at the Three Mile Island Nuclear Power Station at Harrisburg, Pennsylvania, on March 28, 1979, initiated a sudden and dramatic change in public opinion. If it is to survive, the fission nuclear industry must restore public acceptance of plant safety, systems for disposing of wastes, and public awareness of the social and economic consequences of not having nuclear power, in addition to solving the related financial problems. An extended period of accident-free operation would help to achieve restoration of public confidence in safety. According to the Atomic Industrial Forum (1/12/83), the utility industry would spend approximately $25 million in 1983 to counteract publicity by nuclear critics.

Reports about the nuclear fission industry seldom include mention of the economically competitive units which are successfully providing consistent electrical energy. The St. Lucie Unit 2 (Florida), Dresden 2 (Illinois), and Connecticut Yankee have set and maintained credible records which no longer appear attainable with the new units under present NRC regulations.

ENVIRONMENTAL IMPACT AND SAFETY

As with the costs related to the use of nuclear energy, human safety and the environmental impact issues are fraught with controversy. Opponents argue that there are serious problems with radiation, accidents, waste handling, and potential for nuclear weapon proliferation. Proponents contend that the technology is available to handle these problems through upgraded equipment and methods, engineering, proper procedures, and security systems.

Radiation and its potential effect on human life is of greatest concern to all factions of nuclear energy. Its potential for causing death, cancer, leukemia, eye cataracts, and genetic diseases and defects are well known.

Radioactivity is constantly being encountered in our daily lives through natural sources. It is released when minute quantities of matter disintegrate and is found in the air, water, food, earth, and our homes. Its intensity varies according to geographic location and altitude. Radiation is measured in rads, which are units of energy that ionizing radiation deposits in tissues. Scientifically, a rad is the amount of nuclear radiation which, when absorbed in 1 g of material, liberates 10 joules of energy. Ionization produced in individual cells of living tissue causes breaking of the chemical bonds, destroying some of the cells and forming new substances.

The rem, an abbreviation of "roentgen equivalent mammal," is a measurement of the biological effect of different types of radiation. A millirem (mrem) is 1/1000 of a rem and is used when discussing low levels of radiation. Hulme describes the interrelationship of rems to rads by the formula

$$\text{dose in rems} = QF \times \text{dose in rads}$$

QF is the quality factor of radiation that is emitted. Bombs and x-rays have been assigned a QF of 1.

The average citizen is exposed to 130 mrems per year from natural sources. Cosmic exposure varies from 75 mrems per year for Wyoming residents to 38 for those who live in Florida. People in Colorado receive 140 mrems per year from their external environment, while those on the East coast receive only 15 mrems per year. Additional annual radiation comes from medical and dental x-rays, 90 mrems; fallout from nuclear weapons testing, 5 mrems; round-trip cross-country jet flights, 5 mrems; consumer products, 4 mrems; and smaller amounts from luminous watch dials and color television, which total about 2 mrems per year. In contrast, a properly operating nuclear-powered electricity-generating plant will expose area residents to less than 1 mrem per year. Significant levels of radiation are emitted from gasoline lanterns, fertilizers, fluorescent lights, some nineteenth-century pottery, and even the Vatican paving stones. Radiation is cumulative in the body, and therefore any additional unnecessary exposure should be avoided.

The EPA sets radiation control guidelines to protect the environment, public, and workers in the nuclear industry. Revisions have been made over the years which reflect a growing understanding of the effects of radiation on the human body. In 1910, safe levels of radiation were set at 100 rads per year for the general public. This was cut to 30 rads per year in 1934, to 15 rads per year in 1948, and in 1958 drastically cut to 5 rads per 30 years. Industry workers are permitted 5 rads per year. The NRC is responsible for the regulatory application of the EPA guidelines and for the implementation and enforcement of radiation exposure controls in nuclear-powered plants.

According to the Edison Electric Institute, symptoms of radiation sickness typically appear with doses of 200,000 to 600,000 mrems. Generally, 600,000 mrems is fatal to 50% of the human beings exposed. Below 10,000 mrems, experience with adults provides no significant link between exposure and illnesses. Although genetic abnormalities directly attributed to radiation in laboratory animals have been observed, it can only be assumed that similar damage may take place in human beings. The effects of radiation are generally not immediate and may take 15 to 20 years for cancer to develop or a generation or more for congenital defects to appear. After these lengths of time, it is impossible to attribute problems to any one specific exposure—from space walks to drinking a glass of milk.

Although the technology exists which can safely contain the nuclear reaction, equipment malfunctions, operator errors, design defects, poor construction practices, improper materials, and materials degeneration have contributed to many occurrences of potentially unsafe operation, ranging from radioactive water leaks to unit shutdown. Of the more than 20,000 "incidents" reported, most have not been serious, and excessive radiation has not been released to the exterior of the plants because the plants are designed to contain radioactive fission products in the event of an accident. Safety systems are backed up by safety systems, each separately controlled and operated.

The most publicized occurrences of facility problems have been at Brown's Ferry in Alabama, Enrico Fermi in Michigan, Three Mile Island (TMI) Unit 2 in Pennsylvania, and Diablo Canyon in California. The TMI incident of March 28, 1979, is considered to be the most serious to date. Owing to a combination of inappropriate operator actions and procedures, equipment malfunction, questionable instrument readings, deficiencies in control room design, and failure to learn from previous incidents, operators stopped the flow of cooling water to the reactor core. The residual radioactivity inside the reactor, produced by the fission, heated the reactor core to a temperature of several thousand degrees. Although a partial meltdown occurred, plant operators and NRC representatives were unable to identify the problem quickly. There was a small release of radioactive material, but most was contained within the plant. A theoretical maximum radiation exposure to the local public staying indoors for the 10 days was 90 mrems (average estimated exposure, 1 to 2 mrems). No injury to the public or clinical evidence of serious injury to employees due to radiation has been observed. Residents of the TMI area in January 1985 became concerned when a report prepared for the NRC indicated that the occurrence of cancer in the area is 700% higher than the national average. Study of the problem continues. The cleanup of TMI is estimated to take several years and cost nearly $1 billion.

Since the problem at TMI, several steps have been taken to reduce the possibility of a reoccurrence. Two new organizations have been established. The Nuclear Safety Analysis Center links the utility industry with experts and updated technical information, while the Institute of Nuclear Power Operations establishes industry standards, trains operators, certifies instructors, and conducts evaluations. The NRC now requires that every plant have emergency plans, including plans to evacuate all people within a 10-mile radius. Local agencies, the utility, and the NRC must all support this plan. This requirement is posing a serious problem for Indian Point, Shoreham, and other units which have been constructed in or near densely populated metropolitan areas.

The NRC estimates that there will be 10 to 15 years between serious nuclear-powered plant accidents. Statistically, the chance of a meltdown (called an "energetic excursion" by the industry) and release of radiation is estimated to be about 1:10,000. The Rasmussen Report estimates the occurrence of 1 time in 20,000 per reactor-year. Recent concerns published in newspapers indicate that there are several plants in operation with cracked pipes in the primary system for circulating water through the reactors to prevent them from overheating. There is controversy about the levels of safety and debate about shutdowns. Recent research for the NRC at the Naval Research Laboratory has shown that copper has an adverse effect on the radiation resistance of reactor-vessel steels and welds. Copper alloying has been used in the past to prevent rust. Abrupt changes in pressure and temperature in the reactor could cause the embrittled vessels and welds to crack, allowing emergency cooling water to escape [Edelson].

Accidents are averted by high-quality materials, conscientious construction, and intelligent plant operation—all under the watchful eye of the NRC. Minor incidents not considered dangerous are reported and the situations are improved, but serious regulation violations are accompanied with fines. Some utilities (customers) are paying hundreds of thousands of dollars for violations ranging from fraudulent safety records and operation with defective instrumentation to lack of attentive security personnel and operating with containment valves open, which could permit radioactive water to be discharged.

Contrary to the belief of many of the uninformed citizenry, nuclear-powered electricity-generating plants cannot blow up. They are not atomic bombs! Uranium-235 concentration of 90% is required for a bomb, and utilities use a 3 to 4% concentration. The relationship of nuclear-powered electricity-generating plants to weapons is that plutonium, an element produced by the nuclear reactor in the reactor vessel, is the essential ingredient for nuclear weapons manufacture. Plutonium is also very highly toxic and has an effective life of about 250,000 years. For these reasons, strong security measures are taken regarding nuclear fuels, fuel transportation, reprocessing, and worldwide nuclear plant proliferation.

The United States has been very careful in its development and use of nuclear energy, and this has resulted in zero attributable fatalities and limited occurrences of human exposure. Other nations, in their race to develop the technology, have had more serious problems. A report of Great Britain's worst nuclear accident attributes 33 cancer deaths to radioactive fallout from a 1957 fire at the Windscale power station.

The atmospheric impact of cooling towers must also be considered. Since new plants cannot return cooling water to its source at more than $5°F$ ($2°C$) above that at which it was removed, huge cooling towers are now being built to expel the "extra" heat into the atmosphere. This extra heat is in the form of latent heat of evaporation. Denton and Afgan state that a 1000-MW nuclear-powered plant discharges up to 35.13 ft^3 (1 m^3) of water per second into the atmosphere and that this corresponds to the clean water consumption of a sizable city. The potential for weather modification through fog, wetting, icing, and shading is high, especially if there are multiple units in a small geographic area.

THE FUTURE OF NUCLEAR FISSION

Nuclear fission is proving to be a poor technology selection for the solution to our energy needs. The near future for fission energy appears dim owing to the inability to predict and control construction and operating costs. The political pressures, technical problems, and governmental regulations are becoming so strong and complex that many utilities currently using the technology would like to abandon it in favor of other energy forms. The problem is that such extensive financial investment has been made that it is considered too late to withdraw.

The entire industry is looking at construction and quality assurance practices in the attempt to maintain a marketable position. It is projected that standardization of plants is needed for a safe and cost-competitive position in the market. Several different companies providing custom-designed and custom-engineered plants under rapidly changing regulatory demands, utility desires, and industrial standards cannot financially continue. Licensing time and procedures are expected to be streamlined from an average of 14 years down to 6 to 8 years once standardization occurs.

A look toward the near future with declining petroleum and natural gas resources leaves us with only two conventional options for the continuance of large, centralized electricity-generating plants: coal combustion or nuclear reaction. A growing population together with its increasing energy demands dictates that fission nuclear energy continue to play a role until appropriate sources are developed and/or implemented or people begin using energy appropriately. Even though fission nuclear energy is pricing itself out of the commercial electricity-generating market, a few new plants will probably be built.

Expansion of the use of fission nuclear energy will require recycling and reuse of fuel and permanent disposal of wastes. General Electric and Westinghouse are working on advanced light-water reactors which are to be simpler in design, more fuel efficient, and easier to maintain. Work on these units was shifted to Japan in 1982 due to higher labor costs in the United States.

There are several kinds of breeder reactors in the development stage. Although the LMFBR is the most prominent, the GCBR and the LWBR have distinct possibilities. The graphite-controlled unit uses molten salt in the reactor and features on-line fuel reprocessing, thereby avoiding storage and transportation before reprocessing at a separate plant. In the LWBR, thorium, more plentiful than U-238, is converted into fissionable U-233. Thorium is not as toxic as plutonium and is too low a grade for weapons application. The DOE is now testing one of these units near Shippingport, Pennsylvania.

Even though the growth rate has slowed considerably in the United States, the use of nuclear fission is expected to grow 7% annually through 2000 in Europe and the newly industrialized countries. At home, between now and 1990, $75 billion in business can be expected in plant construction expenditures. The "care and feeding" of plants is projected to have a service market of $24 billion per year for equipment, services, and fuel for the next 25 to 30 years. An additional $25 to $35 billion annual market is expected from other countries.

When developed, nuclear fusion—currently in experimental stages—is expected to become the ultimate energy source for all sizes of electricity-generation systems. Although there will still be some risks, they are expected to be much smaller and more acceptable compared to nuclear fission. The technology is complex and the costs are projected to be high. The first demonstration nuclear fusion electricity-generating plant may be in operation shortly after the year 2000.

As an informed citizenry and decision makers, we must be aware that all energy conversion technologies have safety and environmental hazards. It is our responsibility to ensure that the best technologies are selected to benefit our society.

SUMMARY

The use of fission nuclear energy to fuel commercial electricity-generating facilities, once considered the answer to the world's energy dilemma, is facing serious financial problems and reduced demand. Although the technology to handle the necessary functions exists, construction errors and delays, extensive cost overruns, equipment malfunctions, operator errors, continual regulation and design changes, and concern about long-term wastes and their effects on the environment and human health are plaguing the industry.

Research and development efforts continue worldwide to improve the technology in the areas of use and waste handling. Many people express the belief that nuclear fission is necessary to fill the void being created by the decline in availability of fossil fuels. This may be true without extensive efforts in conservation and the use of other energy forms. As with any technology, we must ask ourselves: Are the costs, risks, and benefits of nuclear power greater or less than those of the practical alternatives? Considering the current energy-use attitude of many Americans, is there another real choice?

ACTIVITIES

1. Read about and discuss the use of nuclear energy for the production of electrical power. Be sure to cover its advantages, disadvantages, and potential impact on the future.
2. Tour a nuclear-powered plant and discuss your experiences.
3. Examine newspaper accounts of current events relating the nuclear industry for a given period of time (e.g., 1 month, 2 months, etc.). Read carefully to understand exactly what is being reported. Attempt to detect journalistic bias. Discuss the topics, concerns, and emphases of these reports.
4. Discuss the use of nuclear energy with members of the general public (e.g., family, neighbors, etc.). Determine their opinions, concerns, biases, and general understandings of the technology and related issues.
5. Determine the energy sources used to generate electricity in your area. What portion of the total electrical energy needs does each supply?
6. Using data from major energy resources, determine what portion of our nation's electrical energy is being provided by various sources (i.e., solar, hydropower, coal, nuclear, etc.). Project each for the near-term and long-term future.
7. Describe possible social effects of having a greatly increased reliance on fission nuclear energy.
8. Explore information about nuclear wastes. Determine annual production of types, quantities, and hazards, if any, they present.

9. Determine which countries currently use and/or supply nuclear energy technology.
10. Using the form in Appendix F, estimate your annual exposure to radiation.

BIBLIOGRAPHY

Atomic Industrial Forum, "AIF Releases International Survey." Info News Release, Atomic Industrial Forum, Washington, D.C., March 29, 1982.

Atomic Industrial Forum, "Electricity from Nuclear Power." Atomic Industrial Forum, Washington, D.C., 1983.

Atomic Industrial Forum, "The Nuclear Industry in 1982." Info News Release, Atomic Industrial Forum, Washington, D.C., January 12, 1983.

Atomic Industrial Forum, "Plutonium in Perspective." Atomic Industrial Forum, Washington, D.C., 1977.

Borfitz, Debbie, "Public vs. Power Authority," *Syracuse Post-Standard*, January 16, 1984.

Cohen, Bernard L., *Before It's Too Late—A Scientist's Case for Nuclear Energy*. New York: Plenum Press, 1983.

Coy, Peter, "West Valley: A Nuclear Success Story," *Syracuse Post-Standard*, January, 1984.

Denton, J. C., and N. H. Afgan, *Future Energy Production Systems*, Vol. 2, New York: Academic Press, 1976.

Edelson, Edward, "Thermal Shock," *Popular Science*, June 1983.

Edison Electric Institute, "Nuclear Power: Answers to Your Questions." Washington, D.C.: Edison Electric Institute, 1983.

Hulme, H. R., *Nuclear Fusion*. London: Wykeham, 1969.

Hunt, V. Daniel, *Handbook of Energy Technology*. New York: Van Nostrand Reinhold, 1982.

Johnson, Tom, "Nuclear Waste—A Better Way to Bury It?" *Popular Science*, May 1983.

Krenz, Jerrold H., *Energy: Conversion and Utilization*. Boston: Allyn and Bacon, 1976.

Meador, Roy, *Future Energy Alternatives*. Ann Arbor, Mich.: Ann Arbor Science, 1978.

Moore, James S., "Nuclear Power Industry Scrambles to Survive," *High Technology*, July 1983.

Nero, Anthony V., *A Guidebook to Nuclear Reactors*. Berkley, Calif.: University of California Press, 1979.

Parisi, Anthony, "Future Looks Dim for Nuclear Power," *Syracuse Herald American*, April 12, 1981.

Peterson, I., "Radwaste Program: A Delay in Plans," *Science News*, January 7, 1984.

Severo, Richard, "Too Hot to Handle," *New York Times Magazine*, April 10, 1977.

Syracuse Post-Standard, "Once-Solid Electric Utility Stocks Now Feared as Volatile," *Syracuse Post-Standard*, January 5, 1984.

Syracuse Post-Standard, "Power Brokers—State Weighs Purchase of NM 2 and Shoreham," *Syracuse Post-Standard*, July 27, 1983.

Syracuse Post-Standard, "Senate Kills Clinch River Nuke Plant," *Syracuse Post-Standard*, October 27, 1983.

Syracuse Post-Standard, "Stop or Go, Nine Mine Point 2 Costs Keep Mounting," *Syracuse Post-Standard*, March 24, 1983.

Syracuse Post-Standard, "Study: Many N-Plants Won't Save Money," *Syracuse Post-Standard*, November 7, 1983.

"U.S. Energy Review," Energy Fact Sheet No. 25. U.S. Department of Energy. Washington, D.C.: U.S. Government Printing Office, 1981.

Chapter 7
Solar Energy
The Primary Energy Source

CONCEPTS

1. The sun is the primary source for all the food and fuel we use.
2. Solar radiation is of low intensity and is intermittent in availability (i.e., daytime, seasonal, and climatic variations).
3. Solar radiation is made up of energy in a wide range of wavelengths, but most is radiated in the shorter wavelengths (higher frequencies).
4. Solar radiation can easily be used to produce useful heat in low- and medium-temperature applications and for high-temperature applications with high-tech support systems such as computer-controlled heliostats.
5. Solar energy, striking certain materials, will produce electricity directly.

GLOSSARY

AMBIENT TEMPERATURE—the temperature, usually exterior, around a system.
ANGLE OF INCIDENCE—the angle from horizontal the sun's rays make when they strike a surface.

AZIMUTH—the angle between true south and the point on the horizon directly below the sun.

COLLECTOR ANGLE—the angle between the surface plane of a collector and the horizontal plane.

CONDUCTANCE (C)—flow of heat through a material; specifically, Btu per hour through 1 square foot with a 1°F delta T (ΔT).

DELTA T (ΔT)—any temperature differential.

FLAT-PLATE COLLECTOR—a solar absorber consisting of a flat plate, usually of metal and black in color, placed in an insulated box covered with clear glass or plastic, and used to convert solar radiation to heat.

FOCUSING COLLECTOR—a solar absorber with a reflecting surface of a semicircular or parabolic shape which concentrates solar radiation on a small area to develop high-temperature heat.

INSOLATION—the amount of all solar radiation striking a surface.

SELECTIVE SURFACE—a surface or coating that absorbs a high percentage of incoming solar radiation but emits a low amount of heat radiation.

SOLAR CONSTANT—the amount of radiation reaching the outer limits of the earth's atmosphere, 429.2 Btu per square foot per hour (1.94 calories per square centimeter per hour).

TROMBÉ WALL—a masonry wall that collects and releases stored solar energy into a building by both radiant and convective means.

INTRODUCTION AND HISTORICAL BACKGROUND

The use of solar energy in a planned way can be traced back at least 2500 years to the manner in which the Greeks built their homes and public buildings, orienting them toward the sun. The Greeks' veneration of the sun seems to have grown less from a scientific basis than from the innate sense that sunlight nurtures good health.

There is ample evidence in both Greek and Roman architecture of the knowledge of the advantages of south-facing structures for heating and lighting. Greek literature and theater of the time identified sun-heated rooms as "civilized" and in harmony with the "natural order of things," suggesting an attitude of prestige in their use.

Another reason for the apparent interest in and use of the sun to provide heat was the shortage of fuel for use in heating, either because there was concern about depletion of forests or because certain people controlled the fuel supply and another means of providing heat was needed (shades of modern monopolies). Heating with wood had become so popular with the wealthy in ancient Rome that the once heavily forested regions of Italy became depleted, and wood had to be imported. Wood or charcoal was burned in furnaces, and hot air circulated through hollow tiles in the floors of public baths and private villas. This "hypocaust" system could consume as much as two cords of wood per building per day.

The Romans of the first century A.D. went further than the Greeks in designing structures for different climatic conditions. They used their knowledge of the sun's path to construct homes that would let in direct sunlight during the colder months yet keep it out during much of the summer. Vitruvius recommended that certain rooms be placed appropriate to their functions for optimum comfort. Pliny followed this advice and built his dining room facing southwest to catch the evening sun at the dinner hour. His study was semicircular, with a large bay window to let in the sun from morning to evening, allowing him the advantage of natural daylight and warming during his working hours.

Another development during this era was the inclusion of clear glass in window openings to transmit light and help trap the heat that was generated. By the second century this practice was increasingly common.

Romans considered the sun a source of energy so important that they passed the first recorded "sun rights" legislation. As late as the sixth century the Justinian code called for care in the construction of buildings so that the shadow did not obstruct the sunshine where it was an absolute necessity. The Romans also made use of the tempering effect of a heated mass when they made floors into absorbers. Broken earthenware and rubble was placed in a pit under the floor, with a layer of dark sand, ashes, and lime spread over the top. The combination of materials absorbed heat from the sun and released it slowly after the sun had set.

The Greeks, Romans, and Chinese all knew that the power of the sun could be concentrated with mirrors. They proved through mathematical calculations that the parabolic curve was an even more effective device than the spherical mirror. Whether this knowledge was ever used for military purposes is left to legend, but we do know that other, more benign uses were made of the focused rays of the sun.

In the sixteenth century Leonardo da Vinci proposed that a mirror be built with a diameter of 4 miles (6.5 km) for the purpose of heating water to be used in a dye factory. The sculptor Verrocchio used a mirror to solder together sections of copper for a ball lantern holder for a cathedral in Florence. Alchemists made use of the technique by placing a water-filled vase at the focal point of a spherical mirror. The vase contained petals from flowers and the resulting heat caused the essence of the flowers to diffuse into the water.

A French optician of the seventeenth century built several mirrors up to 3 ft (1 m) in diameter which could melt tin in 3 seconds and cast iron in 16 seconds. The Baron of Tchirnhausen, Germany, hammered copper into thin sheets to make a mirror 5½ ft (1.7 m) in diameter. The thinner construction enabled it to be positioned by one person. The next development was to make the mirror out of sectional wedges of brass fastened to a wood frame of parabolic shape. This mirror, built by Peter Hoesen, could melt copper ore in 1 second and turned slate into a glassy material in just 12 seconds.

The expansion of developments known as the Industrial Revolution in eighteenth-century Europe brought out the desire to make practical applications of earlier developments. A French mathematics professor, Augustin Mouchot, refined the ideas of the Greeks and Romans with the development of a type of steam engine. He built a parabolic trough to concentrate solar radiation on a small tube filled with water which made steam. He also demonstrated how this principle could be used for cooking food, distilling water, and even pasteurizing wine.

Mouchot eventually made a conically shaped reflector, reducing the need to have the reflector track the sun. With this he demonstrated that a practical steam engine could be made for small applications such as powering a small printing press. He recognized that for large, industrial machinery uses, the reflector would need to be larger than the factory itself. He also experimented with separating water into its two components—oxygen and hydrogen—knowing that when recombined the process would release large amounts of heat. The principle of thermoelectric devices was known to him, and he planned to heat large numbers of metallic couplings with a solar concentrator to produce electricity to conduct the previous experiment. More efficient methods of electricity generation were being perfected and new techniques of mining and transportation made coal an inexpensive fuel source, so his efforts were largely ignored until more recent times.

THE NATURE OF SOLAR ENERGY

The sun represents the single physical source of fuel, food, and life on this planet. All physical energy sources on earth must trace their origin back to solar radiation. That radiation is the result of thermonuclear fusion taking place at the core of the sun, releasing high-frequency electromagnetic radiation. Whether the radiation is in the form of alternating electromagnetic fields or is in the form of waves is still under debate by scientists.

It is known that the energy traveling in all directions from the sun's surface is made up of different wavelengths, ranging from the longer wavelengths of radio frequency to very short x-rays and gamma radiation. The majority of this radiation is of shorter wavelengths, some of which can be detected by human beings as visible light. Wavelengths in the visible range make up 46% of the sun's radiation, 49% falls in the infrared band which is sensed as heat, and the remainder is in the ultraviolet band (Figure 7-1). All radiation travels at the speed of light, 186,280 miles per second (300,000 km/sec), reaching the earth in just over 8 minutes. The earth, at an average distance of 93 million miles (150 million km) from the sun, intercepts only a tiny fraction (2 billionths) of the total energy emitted by the sun. Even that small amount is equal to 35,000 times the total energy used on earth in one year.

The amount of energy that reaches the outside of the earth's atmosphere, called the solar constant, is 429.2 Btu per square foot per hour or 1.94 calories per square centimeter per hour. Of this amount, about 35% is reflected back into space and about 30% is absorbed by particles in the atmosphere. Some radiation is scattered in all directions by these particles but is not totally lost. This "diffuse" radiation provides light when the sky is completely overcast. The remaining portion, called "direct" radiation, is useful for heat and light. It has an effect on the movement of air currents in the atmosphere (see Chapter 8).

The total solar radiation received by the earth's surface (or any object in the path of the sun's rays) is the sum of both direct and diffuse radiation and is called insolation.

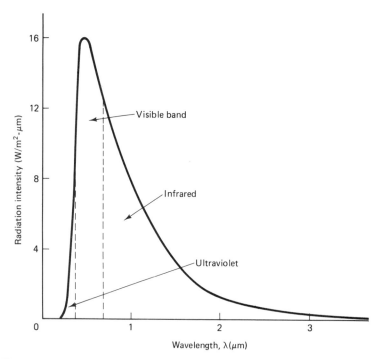

Figure 7-1 Solar radiation at the earth's surface. (From the U.S. Department of Commerce.)

Insolation, which affects the surface of any object on the earth, depends on several factors. The most important of these are altitude, azimuth, and the color of the object struck by the sun's rays. Energy density is greatest when the sun's rays strike the surface of an object at right angles and decreases as the angle becomes less than perpendicular.

Collector panels and walls designed as absorbers are most effective when they are oriented to true or solar south (Figure 7-2). Solar south will be several degrees west of magnetic south in the eastern part of the United States and several degrees east of magnetic south in the western United States. (Refer to a topographic map to determine the deviation between true and magnetic compass readings in your area.) A collector placed on an azimuth that is turned 15° away from solar south will decrease the total radiation received by less than 5%.

The collector tilt angle is very important in determining the "solar gain" (see Figure 7-3). The latitude at the location of a collector is the same angle measured from the horizontal at which a collector will be most effective at solar noon on March 21 and September 23. Only on those two days will the sun's rays be perpendicular to a fixed flat-plate collector surface oriented to solar south and tilted at the latitude angle toward the sun. At all other times the incident angle will be less than 90° and the effectiveness of the collector will be reduced.

The latitude angle is the best collector angle for year-round solar energy acquisition. Latitude +15° for winter and latitude −15° for summer are accepted as the best angles for solar gain.

Another factor to be considered is the amount of energy absorbed by the object compared with the amount reflected or transmitted to

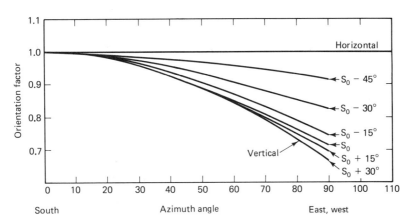

Figure 7-2 Effect of collector orientation. (From the U.S. Department of Commerce.)

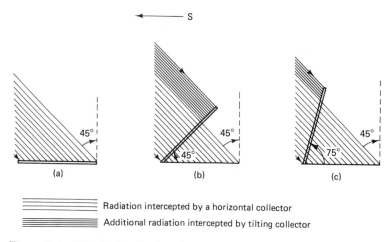

Figure 7-3 Effect of collector tilt on energy intercepted; collector tilt angle (a) $0°$, (b) $45°$, (c) $75°$. (From the U.S. Department of Commerce.)

other adjacent objects. As objects intercept solar radiation, some is reflected away as light and some is absorbed by the object and changed to heat. The darker and less reflective an object is, the more radiation will be absorbed and changed to heat. Enclosing the absorber surface helps to retain the heat that is developed.

CURRENT SOLAR APPLICATIONS

Solar heating techniques are well known and widely used. Both passive and active space heating systems are found on every inhabited continent and have been used in all climatic conditions except the most extreme cold of the polar regions. Although solar heating is comparatively well known, the ability of the sun's rays to produce cooling effects are less well known.

Water can be heated both by direct means and with the use of heat-exchange mechanisms and is an excellent, relatively low cost "first application" for both the individual and those engaged in commercial activities. Both small- and large-scale applications are finding acceptance.

Passive heating systems have two major components: glass or a transparent plastic cover over some type of absorber device, and thermal mass to store the heat for a period of time. Active systems add an electromechanical component which uses energy to move the heated air or water to storage and from there to where it is desired. The more components, the greater the cost and opportunity for system failure. In the northern half of the United States the most cost-effective method of using solar energy is to heat space passively and domestic water either actively or passively.

Somewhere between totally passive construction and active systems for gathering the sun's heat is the attached greenhouse. There is a "growing" interest in this system, which can provide up to half of the heat needed for a home and provide vegetables and flowers as well. There is ample evidence that this relatively low cost addition can provide warmth even in climates with severely cold winters.

Roof ponds can be used for both heating and cooling. When they are open to the sun during the day, they gather heat, and when covered, they radiate the heat into the structure. The process can be reversed to collect structural heat during the day and release it by radiation and evaporation during the night. Absorption chillers can use the heat of the sun to evaporate a volatile fluid which can absorb heat from other objects as it returns to a fluid state, cooling the objects in the process.

Photovoltaic (PV) cells produce electricity directly from sunlight, giving them the greatest potential for use of all systems devised (electricity is still the preferred energy form for ease of use). PV cells remain relatively high in cost for the amount of energy delivered; however, their cost is rapidly declining as the result of large investments in research and development. Many other applications and techniques are under investigation as we seek to exploit that most distant and safest of energy sources.

DOMESTIC AND COMMERCIAL WATER HEATING

Many processes in industry use water at temperatures of 100 to 180°F (38 to 82°C). Water used for cooking, cleaning, washing, bleaching, photographic work, electroplating, etching, and many other applications requires this temperature range. Water heating requirements make up about 2% of the total industrial process heat demand and 20% of the residential energy demand. Commercial as well as "homemade" solar collectors are capable of reaching the temperatures needed for nearly all these operations.

Much of the cost of domestic water heating can be met with solar equipment. Climate and the amount of available sunshine play an important role in determining savings. A study of more than 600 commercial domestic water heating systems installed on Long Island, New York, found savings over the use of natural gas during a 4-year period to be 33%; oil, 54%; and electricity, 57%.

Demand for hot water is relatively constant throughout the year in both industrial and residential situations. System size can be determined more easily than space heating and cooling needs, which are almost entirely seasonal in demand. The collector surface area needed for heating water is substantially less than that needed for space heating, so installations can be made with less regard for the space available for collector mounting.

Since solar radiation is available only during the daylight hours, and since it is variable with weather and length of day, it is often best to use solar heating as a supplement to existing or conventional means of heating water. Solar collectors are most efficient at lower temperatures. As they reach higher temperatures there is greater loss of heat from the collector as well as from associated plumbing. A domestic solar heating system is therefore installed as the initial source of heat and the conventional heating system is used as a backup or supplemental system to the solar unit. Due to climatic conditions, solar units are seldom used as the only source of hot water.

For new construction, the use of auxiliary heating with solar collectors has been mandated in many localities in the United States as well as other countries. The incentive for use in water heating is increased by these policies and should be encouraged in communities where such action has not been taken. There are further incentives in the form of tax credits for installation of solar equipment, and some states have passed legislation that prohibits an increase in assessed valuation for installed renewable energy equipment. See Chapter 13 for further information.

TYPES OF SOLAR COLLECTORS

Solar collectors have been constructed from costly materials with elaborate features and they have been made from the simplest of materials (often recycled materials). The range of heating efficiencies is great, but the fact is that almost any attempt at collecting the sun's radiation can be successful to some degree and will result in financial savings.

This should encourage anyone to investigate the potential for use of this renewable energy resource. Unfortunately, there have been some collectors built and sold with little regard for their ability to remain functional much longer than it takes the installer to leave town. Most of the manufacturers and installers of solar equipment have gone through a period of adjustment which seems inevitable in a new industry, and those who are currently in business generally have good reputations and stand behind their products and their work.

The simplest of solar collectors are batch heaters (Figure 7-4). When a faucet is opened, the heated water is drained out of the top of the tank and replaced by cold water entering the bottom. As the sun heats the water, the hotter, less dense water rises to the top ready for

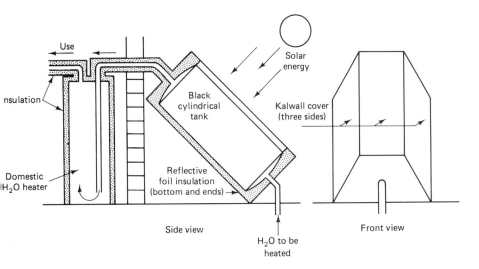

Figure 7-4 Batch heater.

use. Batch heaters are usually some type of tank enclosed in an insulated box with a glazed cover. The box is coated inside with reflective paint or foil to increase the solar gain. The container may be a metal tank, a plastic bag, or even a coil of garden hose. It is important that the surface of the container be coated with a dull black color to absorb rather than reflect the sun's rays.

FLAT-PLATE COLLECTORS

A more efficient but more complicated and expensive system uses a rectangular plate of metal housed in a frame and usually covered with one or more layers of transparent glazing to help contain the heat produced by the rays of the sun (Figure 7-5). The amount of heat is directly proportional to the surface area of the collector; so generally, the larger the area, the greater the gain. A blackened surface reflects the least amount of radiation, allowing the greatest proportion to be converted to heat.

Various techniques are used to increase the surface area without increasing the size of the frame. A corrugated or ribbed surface presents a larger "face" to the sun. One commercial collector uses a series of vertical fins crimped so that any radiation striking the surface is reflected downward rather than back toward the glazing, where it would be lost. This treatment increases the cost but increases the efficiency. The two factors must be considered when making cost-efficiency analyses.

The principal difference between air and liquid collectors is in the size of the passages, air passages being substantially larger than those for liquid. Air flows in contact with nearly the entire absorber plate surface, while liquid is circulated (mechanically or by thermosiphon action—through tubes soldered to the absorber plate. Airflow rates are typically 2 ft^3/min per square foot of collector surface area. Liquid flow rates are typically 0.02 gal/min per square foot.

The cover of the collector frame is usually a single pane of "low-iron" glass. Conventional glass contains iron compounds which prevent certain wavelengths of solar energy from entering the collector. Low-iron glass is transparent to the widest range of wavelengths and, although slightly more expensive, makes the collector more efficient.

Collectors must be insulated to be efficient. The greatest heat losses are through the glazing, but the only reasonable way to reduce losses is to use a second layer of glazing. The second layer of glazing will increase heat retention by about 10 to 15% but will decrease transmission of radiation by about 5%. The increased cost may not be worth the small increase in solar gain.

Both the sides and the back of the collector panel should be insulated to at least R-5. The insulation and materials from which the collector frame are made must be able to withstand a wide range of temperatures, from below freezing to highs of 250 to 400°F (120 to 200°C). The temperatures in collectors can reach this level in some instances, particularly if the collector components malfunction and the flow of fluid—either air or water—stops due to blockage or equipment failure. When this occurs, a condition called "stagnation" exists and may cause permanent damage to the collector.

Liquid collectors require temperature/pressure relief valves to maintain internal conditions below the boiling point of the liquid in case of stagnation. Air collectors are equipped with automatic venting devices to prevent damage to components in stagnation conditions. If the collector is mounted directly on an uninsulated wood surface, such as the sheathing of a roof, the wood could deteriorate as a result of the high temperature. If the collector is free-standing, insulation is just as important, since airflow around the outside of the collector could cause heat losses.

Fiberglass-reinforced plastics are often used as collector glazing because glass is heavy and has the unfortunate characteristic of being easy to break. The reduced potential for breakage is slightly offset by a loss in transmission of radiation, in the range of 5 to 10%. Various plastic films are also available.

Flat-Plate Collectors

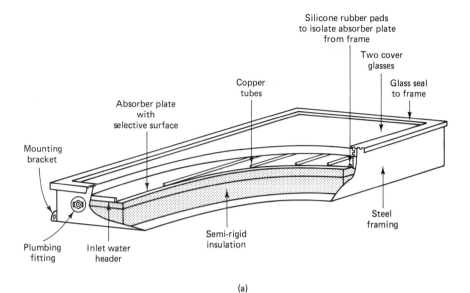

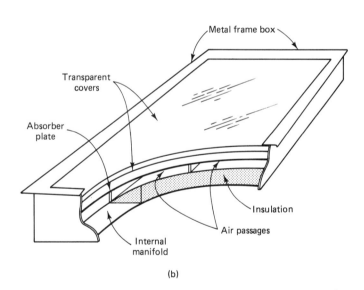

Figure 7-5 Flat-plate collectors: (a) liquid heating solar collector; (b) air heating solar collector. (From the U.S. Department of Housing and Urban Development.)

TYPES OF SOLAR WATER HEATERS

Solar water heaters are classified in several ways: by whether their operating principles are active or passive, whether they are open loop or closed loop, and whether they are direct heating or use a heat exchanger.

The simplest, least complex system (other than the batch heater) is the thermosiphoning (passive) system (Figure 7-6). It uses the principle that water and air both expand slightly when heated and will rise through the collector. No mechanical components are needed for this system. For liquid systems only a collector, a tank located 1.5 to 2 ft (45 to 60 cm) higher than the top of the collector, and the associated piping are needed. Normally a direct heating process is used, since this system is most effective in climates with little or no danger from freezing temperatures. Thermosiphoning air collectors require a heat exchanger near the top of the unit to transfer the heat to the water as the air cycles through the collector.

Direct active heating systems are more efficient than those using a heat exchanger, but water has a tendency to freeze if the temperature drops too low. In climates where freezing temperatures must be considered, a direct heating system may be used only during those months when such temperatures do not occur; otherwise freeze protection must be incorporated in the system. Since water expands when it

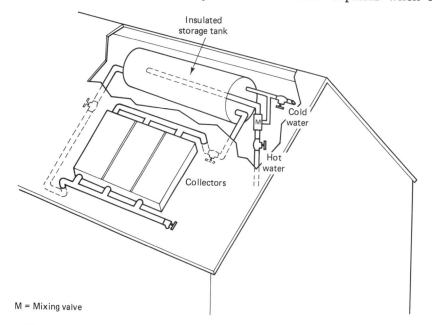

M = Mixing valve

Figure 7-6 Thermosiphoning system. (From New York State Energy Research Development Authority.)

freezes, damage to components and piping can be expected. Ambient temperatures of 32°F (0°C) are not needed for freezing to occur. If radiational cooling is great enough (as on a clear, cool night), radiation heat losses from the absorber plate may drop the water temperature below the freezing point even though the air temperature is higher.

One way to protect a direct heating active solar water system is to use one of the drain types. There are two different processes used, but both employ a more complex, and therefore more costly, set of components. These include a pump, a controller (which uses sensors to determine temperatures in various parts of the system), and a series of valves and vents.

A drain-back system (Figure 7-7) uses a heat exchanger to circulate several gallons of water through the collector. If freezing temperatures occur, a control mechanism stops the pump and opens a solenoid-controlled valve to allow the circulating water to drain back into a temporary storage container. When the danger from low temperatures is gone, the system begins to circulate water again.

A drain-down system (Figure 7-8) uses a similar process but eliminates two components, the temporary storage container and the heat exchanger. This system uses the direct heating method. When the sensor warns of freezing temperatures, the water circulating through the collector is dumped down a drain. When temperatures return to operating levels, the valves close and fresh, potable water is pumped into the system. Although water is lost from the system in the process, the reduction of two components will be less expensive and may make the overall operation more efficient.

When it is desirable to use a liquid system on a year-round basis, the most common usage is with indirect heating, that is, with the use of some liquid other than water and a heat exchanger. Heat-transfer liquids have the disadvantage that they all have heat-carrying capacities which are somewhat less than that of water.

Ethylene glycol, the common automotive radiator antifreeze, should never be used in a potable water system. It is toxic to human beings and will oxidize, creating corrosive problems in the system. The most commonly used liquid is polypropylene glycol, which is nontoxic. It has a specific heat of 0.85, compared to 1.0 for water. In the recommended concentration by weight with water, it has a freezing point of about −87°F (−31°C) and a boiling point of about 215°F (102°C).

Higher temperatures found in collectors tend to cause polypropylene glycol solutions to become acidic, so an inhibitor must be added to neutralize the organic acids that form. An annual check on the pH is recommended to make certain that the inhibitor is still active.

Other heat-transfer liquids include silicone oils, hydrocarbon oils, and fluids based on glycerine. All have specific heat values substantially

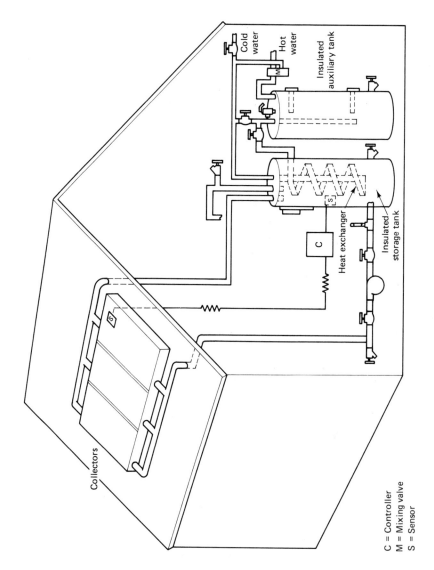

Figure 7-7 Drain-back system. (From New York State Energy Research Development Authority.)

C = Controller
M = Mixing valve
S = Sensor

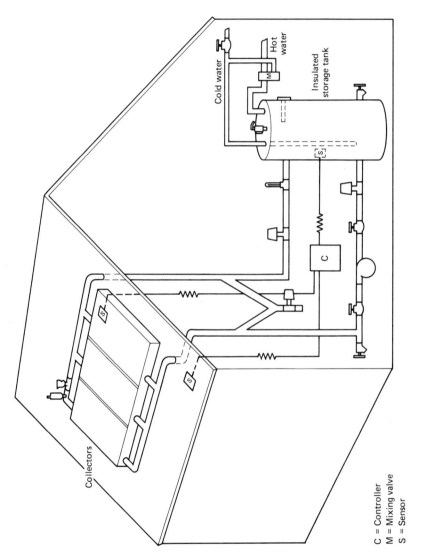

Figure 7-8 Drain-down system. (From New York State Energy Research Development Authority.)

C = Controller
M = Mixing valve
S = Sensor

lower than polypropylene glycol; all require flow rates and/or an absorber plate area 50 to 200% greater than does water in order to achieve the equivalent heat transfer. Other major considerations are price, viscosity, chemical stability, and protection against boiling. Silicone fluids have a low surface tension, which means that they can seep through the tiniest openings in a system. Their price is several times higher than those of competing fluids.

SIZING THE SYSTEM

Since solar energy is intermittent and variable, it is best to consider water heating as a supplement to a conventional heating system. Solar collectors can supply 75 to 90% of the residential hot water needs in a properly sized system on a regular basis, depending on geographic location and time of year.

The first step in determining the size of a solar hot water system is measuring the average daily hot water needs. One rule of thumb is that each person uses about 50 gal (190 liters) of hot water per day. For a family of four, then, the total daily requirement would be 200 gal (560 liters). For an apartment or condominium a slightly larger volume might be needed to compensate for the longer plumbing lines. For commercial needs, a careful measurement of hot water usage over a period of a week or month would serve to establish the base load.

The second factor relates to the times during the day when the need is greatest. An individual family can often adjust their use to match the time of day (usually from late morning through late afternoon) when the collector is producing the greatest amount of heated water. For a larger residential complex, or a commercial or industrial application, this may be difficult or impossible.

The temperature of incoming water varies nationally but is usually between 50 and 55°F (10 to 13°C). Conventional heaters raise that to usable temperatures. Most household applications need water no hotter than 120°F (50°C). The first "no-cost" energy conservation step is to reduce the temperature of the conventional water heater. Not only will this reduce energy consumption and fuel costs, but it also can provide a safety measure with reduced danger from hot water burns.

Although household water temperatures can be lowered, many commercial and industrial applications require that water be heated to higher temperatures. Restaurants are required to maintain temperatures of 180°F (82°C) for sanitary reasons, and industrial applications use water heated to temperatures either above or below this level. Solar-assisted heating can be of benefit in all instances.

Determining the size of absorber surface needed to deliver the total quantity of Btu needed to heat a volume of water requires that the insolation for the location of the collector be determined. Table 7-1 provides sample data for several locations, based on north latitude. A more accurate method is to refer to insolation maps for each specific area in the *Climatic Atlas of the United States*.

Several assumptions must be made: (1) that the collector surface is oriented to solar south and that the collector is fixed in that position, rather than tracking the sun, (2) that the collector tilt is equal to the latitude plus 15° to enhance the amount of radiation received during the winter months, and (3) that conditions are clear and no obstructions shadow the collector during the daylight hours.

The amount of radiation available is subject to climatic conditions and may vary greatly from one location to another. Phoenix, Arizona, receives as much as 77% of available sunshine during December and January (the lowest for that location) and a high of more than 90% during June and July, for an annual average of 86%. Portland, Oregon, by contrast, receives a low of 19% of available radiation during the same two winter months, and no more than 70% during the corresponding summer months, averaging 53% for the year. Minimal heating does occur as the result of diffuse radiation (if it didn't, we would be frozen and totally in the dark on cloudy days) but will contribute little to the heating effect of a collector.

The efficiency of the collector must be considered. Most commercial absorber plates are rated by the manufacturer, but with an eye toward sales, the figures used are usually on the optimistic rather than the conservative side. Experimental data rather than field-condition data are often used in advertising.

Collector efficiency is a function of the difference between the average temperature of the fluid in the collector (midway between the entering temperature and the desired temperature) and that of the ambient air temperature. As the temperature differential (ΔT) is reduced, the efficiency of the collector is lowered (i.e., a lower ΔT conducts less heat into the fluid). A generalized efficiency diagram is shown in Figure 7-9.

The point where the performance line intersects the vertical axis corresponds to the fluid inlet temperature at the same level as the ambient temperature, indicating the maximum efficiency of that collector. Where the performance line intersects the horizontal axis, the fluid entering the collector is at the same temperature as the collector plate (little or no additional radiation is being received), and the efficiency, any increase in fluid temperature, has dropped to zero.

Average daytime (ambient) temperatures can be obtained from the National Weather Service and often from local news outlets—radio, television and newspapers. Twenty-four-hour averages should not be

TABLE 7-1 Average Daily Solar Radiation: Horizontal Surface (Btu/ft^2/day)

Location[a]	Jan.	Feb.	Mar.	Apr.	May	June	July	Aug.	Sept.	Oct.	Nov.	Dec.
20	2349	2671	3019	3301	3421	3445	3423	3332	3106	2763	2421	2246
25	2103	2474	2891	3266	3463	3524	3485	3329	3013	2588	2192	1995
30	1851	2260	2740	3206	3482	3581	3526	3303	2877	2395	1950	1735
35	1590	2030	2570	3124	3479	3619	3546	3254	2759	2184	1698	1468
50	719	1275	1948	2746	3346	3621	3489	2979	2225	1470	910	669
55	535	1011	1769	2582	3209	3596	3441	2856	2012	1212	651	422
60	299	747	1459	2403	3185	3571	3389	2709	1784	950	405	200

[a] Location based on north latitude.

Source: *Solar Heating and Cooling of Residential Buildings*, U.S. Department of Commerce.

Sizing the System

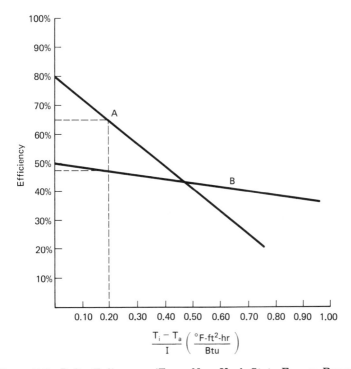

Figure 7-9 Delta T diagram. (From New York State Energy Research Development Authority.)

used, since only the daylight hours are useful for calculating collector performance.

If a preheat tank is used for storage in addition to the conventional heater tank, it should be 1.5 to 2.5 times the capacity of the latter. If a heat-transfer fluid other than water is used and therefore a heat exchanger is needed, a range of 0.5 to 0.9 should be calculated for the efficiency rating of that portion of the system.

Water heating needs can be calculated with relative ease in the following manner.

Example:
One hundred gallons of water at a temperature of 55°F is to be heated to 120°F.

heat required (Btu) = gallons × ΔT × 8.33
(weight of 1 gal of water)

Btu = 100 × 65 × 8.33 = 54,145 Btu

If a collector location of 40° North latitude is used (approximate latitudes for Philadelphia, Indianapolis, Springfield, and Denver) and solar radiation for September is taken from Table 7-1 (2600 Btu/ft²/day), a collector surface of 20.825 ft² would be needed to intercept 54,145 Btu per day (54,145/2600 = 20.825). Collectors are not 100% efficient, so a more realistic efficiency rating of 50% could be used. An efficiency rating of 50% would require 41.65 ft² of collector (double that of a 100% efficient collector) to deliver the heat needed to raise 100 gal of water from 55°F to 120°F.

Commercial collector modules are usually produced in sizes of 24 ft² (2.23 m²), so a decision would have to be made with regard to the installation of one module or two. One module would provide about 58% of the heating load under ideal conditions. Two modules would provide nearly 40% more heat than needed, assuming no further losses in the system. A more sophisticated analysis is needed to calculate the efficiencies for an entire system and the cost/return ratio. Most systems installed for domestic water heating use either two or three modules to compensate for additional losses in the system. Additional collector surface is needed for other-than-ideal conditions. Danger from overheating is minimized since most systems can operate at temperatures up to 180°F (82°C) for short periods without damage. In addition, adjustable tempering valves are usually installed to introduce cold water into the lines to prevent excessive temperatures at faucets.

FOCUSING COLLECTORS

One of the early techniques used to focus the sun's rays to generate greater heat was developed by the French mathematician Mouchot. His truncated conical reflector concentrated the sun's rays on a cylindrical boiler placed at the focal point of the collector. Water was turned into steam and used in a conventional steam engine. This was used by French colonists in Algiers to pump water and was demonstrated at the Universal Exposition in Paris in 1877. The next year Mouchot had a larger machine constructed which would pump 500 gal of water per hour. He also connected the collector system to a heat-powered refrigeration unit to make ice. The French in Algeria were particularly receptive to this invention, since there was little wood or fossil fuel to operate refrigerators, pumps, and other equipment.

Conical designs and parabolic configurations have both been used to concentrate the energy of the sun. Other designs which have been found to be usable include those of hemispherical or cylindrical shapes. Three components are necessary for any of these designs to function: (1) a highly reflective focusing surface, (2) a pipe to carry the heat-transfer fluid away for use or for storage and (3) insulation for the pipes.

Reflective surfaces can be made of any durable, easily shaped material. Lightweight, tough material such as aluminized Mylar can be used. The heat-transfer liquid must be able to withstand much higher temperatures than those found in flat-plate collectors. The focusing effect can increase operating temperatures to at least 300°F (150°C) and in some instances as high as 1000°F (540°C) at the focal point of the reflector. Insulation on the plumbing is especially important since the temperature differential with ambient air is so great. The insulation must also be able to withstand high temperatures.

The material used as a reflector should be durable, and although Mylar is easily shaped, it is not durable, lasting perhaps a year in an outdoor environment. Mylar W (a weather-resistant polyester) may last up to 3 years. Tedlar, a polyvinyl fluoride material, has a life expectancy of up to 5 years. Teflon, a fluorocarbon material, has a life of 20 years or more.

Metals have long been used as reflectors, but there are several inherent difficulties with most of them. First, they are generally heavier than plastics unless they are drawn to extremely thin sheets, and then they lose structural strength. Second, the best reflecting metals, such as silver, are expensive, which prohibits their use in all but the smallest application. Third, metals oxidize quickly, losing their reflective ability. Some metals can be anodized (electrically plated) on less expensive surfaces, but the process is costly. They can be coated with lacquers

Figure 7-10 Focusing collector. (Courtesy of Vulcan Industries.)

Figure 7-11 Performax™ HP-250, focusing collector tubes. (Courtesy of Energy Design Corporation, Memphis, Tenn.)

or other clear coatings, but this reduces their efficiency as reflective surfaces.

Locating the focal point for individual parabolic reflectors is difficult. Mass-produced units will have the assembly completed to this point. Maintaining the focus as the apparent "sun's path" changes throughout the day makes it vital to have an automatic (and expensive) tracking mechanism to follow the sun. If the site of the collector has substantial periods of cloud cover, this type of collector is severely limited in its application since direct sunlight is required to make it function. For trough or linear focusing collectors arrays, when the pipe carrying the heat-transfer liquid is large in diameter the focus point need not be exact, so spherical shapes are suitable and are more economical to fabricate.

Another technique for increasing the temperature at the focus of a collector array is to use a series of flat mirrors set up to track the sun and focus independently on a "target" area (Figure 7-12). Experiments at the solar thermal test facility near Albuquerque, New Mexico, have demonstrated that temperatures of at least 1950°F (1065°C) can be generated—high enough to melt plate steel in seconds. A similar solar furnace near Odeilla in Southern France uses 63 heliostats, each about 19.5 × 24.5 ft (6 × 7.5 m), to focus on an aperature in a tower to produce 6000°F (3300°C)—enough to melt any known substance.

A concept with perhaps wider applications and much less expense is the use of Fresnel lenses. These are made of transparent plastic with a set of concentric circles cut or pressed into one surface. Each ring refracts light rays at a slightly different angle, causing them to converge on a single point. Reflectors made of polished aluminum or aluminized

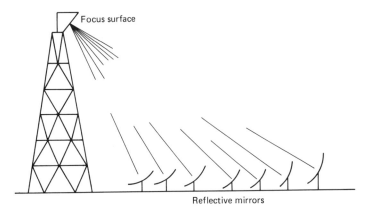

Figure 7-12 Tracking collector array.

plastic use the same principle and can be mass produced, providing a relatively inexpensive source for concentrating the sun's energy. This technique is being tested for possible uses to increase the efficiency of photovoltaic cells.

SPACE HEATING APPLICATIONS

When fossil fuel was in plentiful supply and prices were comparatively low, it was not considered necessary to build with the sun in mind. Designers of homes, offices, and factories could ignore solar effects, or if the rays of the sun were a "bother," more air conditioning was installed, shades were pulled, or tinted glass was installed. During winter, buildings with massive glass areas simply required more energy to heat.

Several circumstances have occurred which make space heating by solar means much more attractive from a financial as well as an aesthetic point of view. The cost of fossil fuels has gone sky (sun)-high, causing major impacts on individual incomes and business profits. It not only makes economic sense to use solar design, insulate, and conserve, but it is becoming socially acceptable as well. The Greeks were right all along.

A solar building must be designed with three principal features: It must collect solar radiation, it must store the heat collected, and it must be capable of releasing the heat when needed in the quantities desired.

To be a passive collector, the building should have its largest dimension oriented toward the sun. That surface needs to be designed and constructed in a manner which will most efficiently change the solar radiation received into heat. Heat could be transmitted directly into living/ working areas, but this is generally not the most desirable method, since overheating will occur during sunny, daylight hours, and rapid cooling will occur at night. Glass, which allows sunlight in, is also a very poor insulator and will transfer heat outward at a rate up to 15 times as great as that of a standard insulated wall.

In most climates it is important to incorporate architectural features such as overhangs or awnings which will shade south windows from the sun during the hottest months. The high (altitude) angle of the sun is kept from adding unwanted heat, yet during the colder months, the lower angle of the sun can project under the overhang and provide the heating desired. In warmer climates, it may be desirable to have east and west windows made of reflective glass to minimize the amount of radiation during the morning and evening hours.

Most materials absorb and store heat when they are in the path of radiation. When radiation stops or is lowered (as during cloudy periods) and the ambient air temperature is less than that of the material, the stored heat is released. The rate of absorption and release is dependent on many factors, but the most important is the property of the material called "specific heat."

The ability of materials to store heat is called the specific heat of the material. All materials are compared with water, which has a specific heat of 1.0, meaning that 1 Btu is required to increase the temperature of 1 pound of water $1°F$ (kJ/kg/°C). Most common building materials have a specific heat of 0.20 to 0.27, but they have a greater mass than water. The result is that they can store 30 to 40% as much heat per unit of volume as water can.

The structural components of buildings contain a great deal of mass, particularly when the construction is of concrete and masonry materials. Buildings have the ability to store and release large amounts of heat energy—the same type of energy we pay for if supplied by fossil fuels.

Adding thermal mass to a building during construction is relatively easy; retrofitting is sometimes difficult. New construction can take advantage of the use of concrete, brick, and stone in appropriate portions of the building, either for structural strength or for decorative purposes. One popular treatment uses black Vermont slate placed over several inches of concrete on the floor just below south-facing windows. No one said thermal mass could not have aesthetic value as well.

A well-insulated building with appropriate thermal mass will gain thermal energy slowly and lose it slowly, while a poorly insulated building will gain and lose heat quickly, so the judicious use of insulating materials can play an important role in using the sun as a long-distance heater.

Direct-gain applications rely on solar radiation falling directly on the storage medium, which is part of the structure, interior floors, and walls (see Figure 7-13). With this application, one-half to two-thirds of the total surface area in the space to be heated should be constructed of thick masonry materials. The thermal mass gains heat slowly and releases it slowly, maintaining a more uniform temperature within the building. The principle of delaying the twice-daily temperature shifts can be used to cool the same area during periods of hot weather, since the thermal mass is slow to heat up during the day.

Early direct-gain passive solar structures had problems with either too much glass or too much or too little mass. There are now formulas that can be used to calculate accurately the glass-to-mass ratio, depending on the structure volume, insolation available, degree-days, and structural Btu losses.

Storage of solar energy depends on the specific heat of the storage materials and on their mass per volume. For example, water has a specific heat of 1.0 and a mass of 62 lb/ft^3. Water will store 62 Btu/ft^3 per °F. Stone has a specific heat of 0.2 and a mass of 95 lb/ft^3. Stone has a heat storage capacity of 19 Btu/ft^3 per °F. To determine the amount of heat that can be stored, multiply the volume of the mass by the specific heat of the material by the amount of temperature increase, or

$$\text{stored energy (Btu)} = V_p C_p \Delta T$$

When architects have taken these factors into consideration and planned the heating and cooling loads for buildings, it has been found that the HVAC (heating, ventilation, and air-conditioning) systems can often be reduced in size by as much as 50%. In some instances, in schools and offices, the heating system is seldom needed; all the heat needed is supplied by solar energy, the occupants of the building, lights, and operating equipment.

When the thermal mass is located between the path of the sun's rays and the space to be heated, the approach is called "indirect gain." A masonry or water thermal storage wall (a Trombé wall) is built just inside a large area of south-facing glazing. Air is circulated between the glass and the thermal storage either by convection or with small fans,

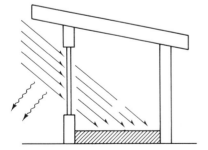
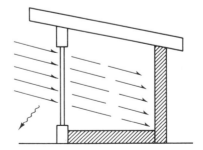

Figure 7-13 Direct gain.

with cooler air entering near the base of the air space and warmed air exiting near the top (Figure 7-14).

The mass of the storage wall also provides a time lag for heat transfer by conduction. As the sun shines on the wall, heat is conducted slowly through the wall toward the cooler interior of the structure. At night, when insulating drapes have been closed between the glazing and

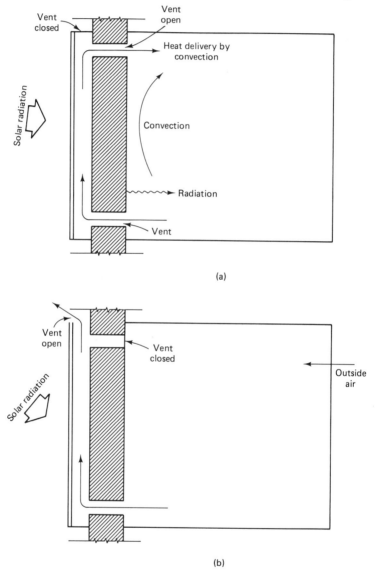

Figure 7-14 Trombé wall: (a) heating mode; (b) ventilating mode. (From the U.S. Department of Commerce.)

the Trombé wall, the heat that was collected during the day is radiated to the room. Trombé walls, therefore, provide space heating by using all three methods of heat transfer (conduction, convection, and radiation) and prevent overheating during sunny days because of the conduction time lag. Trombé walls have been found to be the best of the passive solar space heating system designs.

Since many people dislike the presence of a massive opaque wall obstructing their view, another approach may be used. "Water walls," transparent containers of water, can be used instead of the masonry Trombé structure (Figure 7-15). Light will still pass through but a clear view is not possible because of refraction. Some people have used the area behind the water wall to erect supports for planters, thus making a "plant wall" which they perceive as more pleasing in appearance. Direct-gain water walls are more commonly placed near the rear of rooms that have clerestory windows. The sunlight can then heat the water and the occupants will retain their "view."

Solar roof ponds can be used in geographic areas where both heating and cooling are necessary and where there are no extensive freezing periods or snow loads. The advantage of this approach is that it does not affect the appearance or physical arrangement of a building as does the masonry or water wall approach. The depth of water should be between 6 and 12 in. (15 to 30 cm) and can either be open to the elements or enclosed in plastic, metal, or fiberglass tanks.

Structural strength for the added weight of a large expanse of water must be carefully determined. In new structures this can be added to the building calculations, but special care must be observed if a roof

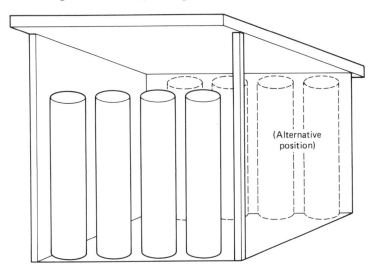

Figure 7-15 Water wall.

pond is to be added to an existing structure. Collapse of the building could occur if a large pond is placed on the roof, and leakage could cause damage to the contents and weaken the structure.

A uniform distribution of temperature is desirable throughout the roof pond. If stratification occurs, there is a greater heat differential between the surface and the ambient air, which will result in greater heat losses. Stratification is reduced if the bottom of the container is dark, preferably black, in color. Radiation will strike the dark surface, be converted to heat, and the heated water will rise to the surface.

Since the collector is located outside the building, some type of system is needed to control the heating and cooling modes of the solar roof pond. Movable insulation panels are placed in tracks or hinged frames above the water surface (Figure 7-16). During cold periods the insulation is moved to expose the pond to the incoming solar radiation.

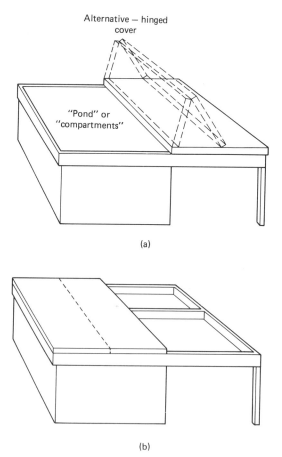

Figure 7-16 Roof pond: (a) open; (b) closed.

At night the panels are replaced and the heat is radiated to the building below. In summer the procedure is reversed. The insulating panels are removed during the dark hours, allowing evaporation to occur and heat to radiate to the cool night sky. During the day, with the panels replaced, the cool water provides a comfortable atmosphere for the building inhabitants. Evaporation from an uncovered water surface will dissipate about four times as much heat as will radiation alone.

GREENHOUSES/SOLAR ROOMS/SOLAR SPACES

There are many feasible designs for the construction of greenhouses, solar rooms, and solar spaces. They all have the same function—to serve as a collector for solar radiation and to maintain more uniform temperatures for personal comfort and/or for raising plants. The addition of a solar room is one of the most cost-effective methods of retrofitting an existing structure for solar space heating. A functional addition can be constructed for as little as $2.50 per square foot ($25.00 per square meter) (1980 prices).

Whether the unit is constructed as an addition to an existing structure or part of new construction, several conditions must be met. As with any solar collector, the orientation needs to be within 15° of solar

Figure 7-17 Sun room/solar greenhouse. (Courtesy of Garden Way, Charlotte, Vt.)

south for best performance. There should be some means of storing the heat with water walls, rock storage, sand, gravel, or masonry. If the space is to be used for growing plants, the inside walls should be of a light color to discourage plants from growing toward the glazing. If the space is used only for heating purposes, a darker color is needed to enhance absorption.

SPACE COOLING APPLICATIONS

Energy consumption for space cooling represents only about one-tenth that used for providing heat, although the demand is increasing more rapidly than the demand for heating. Cooling demand corresponds with the increase in sunlight, so efforts to increase the use of solar energy to produce cooling bear a direct relationship to one another. An increase in the use of solar energy to meet the cooling demand during the summer months would help reduce the peak electrical load in many communities caused by the increased demand for conventional air conditioning.

Although much has been written about the potential for solar energy to be used as a heat source, the ability of the sun to be used as a cooling agent is less well known. Before inexpensive portable coolers were readily available, canvas bags were used as "evaporative" coolers for drinking water. The temperature was never cold, but evaporation from the seepage out of the bag lowered the water temperature by a noticeable amount. Coolers placed in the windows of automobiles used the same principle when air rushed through a moist screen, giving up some of its heat to the evaporating water.

The principle of solar cooling has been known by desert dwellers for centuries. People living in hot, dry climates often construct their homes and public buildings with underground passages to replenish air for the building. As the building's temperature increases, the warmed interior air rises and is vented, creating a slight vacuum which draws in outside air through the cooler environment of the earth. The ducts are placed at least 3 to 4 ft (1 to 1.25 m) below ground, where the temperature remains relatively constant regardless of the heat on the surface.

The inflow of cooler air can be enhanced even more if a "solar chimney" similar to a flat-plate collector is used (Figure 7-18). The base is left open so that air from the building can enter, be heated further, and then exit more rapidly. The volume of air moving through the inlet ducts can be increased several times over a system that does not use this technique. No mechanical or electrical components are needed and the system is essentially self-regulating—the more the sun shines, the more rapidly the air moves; when the building cools, the air movement is slowed.

Space Cooling Applications

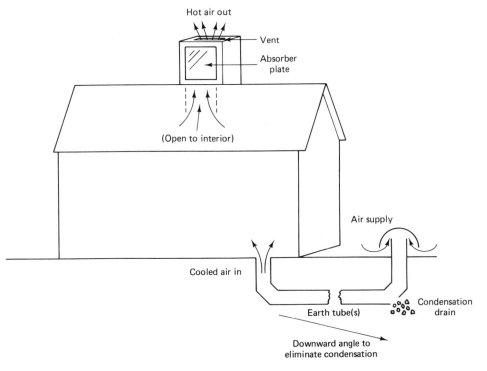

Figure 7-18 Solar chimney.

Solar-powered absorption coolers can save considerable energy by reducing electricity demand. An absorption unit contains a liquid mixture of refrigerant and absorbant (Figure 7-19). When heated, the refrigerant is vaporized and flows from the generator (heat source) through a condenser, expansion valve, and evaporator back to the absorber unit, where it recombines with the absorber liquid. One material used for this purpose is lithium bromide, which dissolves in water at low temperatures and is released with the addition of heat.

Heat to drive the system can be provided by a small gas flame or by a solar collector unit. The lithium bromide is driven off the water solution as a gas and circulates through pipes passing through a condenser unit. At this point it returns to liquid form, cooled by water from a cooling unit. The condensed fluid goes to the evaporator coil, which surrounds the container or material to be cooled. It is vaporized by absorbing heat from the chiller unit and then is brought back to the generator.

In this vapor–liquid–vapor procedure, the temperature limits are quite stringent. When lithium bromide is used, the range of operating temperatures must remain between 160 and 210°F (71 to 98°C). If

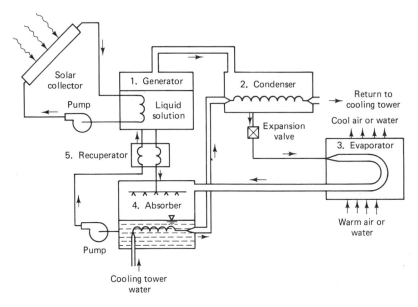

Figure 7-19 Absorption air conditioner/chiller. (From the U.S. Department of Commerce.)

the heat temperature drops below this level, the chemical may crystallize and block the tubing.

Other less complicated solar cooling systems rely on modifications of flat-plate collectors. Instead of using a collector during the day for solar heating, water is pumped through unglazed panels at night to provide radiational cooling. If this system is connected to a finned absorber in a warm area, a large amount of heat can be collected, redirected to the panel, and radiated outside, resulting in a cooling effect. In a system set up to generate electrical energy from wind power or photovoltaic cells, the cost of pumping the fluid could be substantially reduced.

SOLAR THERMAL STORAGE SYSTEMS

The design and construction of solar thermal storage systems must consider several factors. Cost will probably be the first consideration for the individual as well as the corporate planner. Cost factors must be rated against the performance of the system.

One of the first considerations is the amount of heat that must be collected and stored. Solar insolation for a given location is of primary importance in determining this factor. The size of the storage area that is available and the weight and volume of the storage medium are the next considerations. For liquid systems it is best to use liquid storage;

for air systems, rocks, water, or eutectic salts will each serve a particular need. Eutectic salts are "phase-change" materials, materials that change from a solid to liquid at a particular temperature, absorbing large quantities of heat in the process, then reverse the process to release heat when the temperature drops below the phase-change temperature. A smaller storage space can be used to store a large quantity of heat; however, these materials and their containers are considerably more expensive than water or rocks. Phase-change materials include many different chemical compounds. Those that are most often used are listed in Table 7-2.

TABLE 7-2 Eutectic Salts for Thermal Energy Storage

Hydrated Salts	Melting Point (°F)	Heat of Fusion	Density (lb/ft)
Calcium chloride ($CaCl_2$)	102	75	102
Calcium nitrate [$Ca(NO_3)_2$]	108	60	114
Disodium phosphate (Na_2HPO_4)	97	114	95
Sodium carbonate (Na_2CO_3)	97	106	90
Sodium chloride (NaCl)	1480	223	135
Sodium sulfate (Na_2SO_4)	90	108	97
Sodium thiosulfate ($Na_2S_2O_3$)	120	90	104

The cost of these salts ranges from just over 1 cent per pound to more than 5 cents, compared with rock, which can be installed for about 0.2 cent per pound, and with water, which costs essentially nothing. Eutectic salts must be mixed with other compounds to prevent clumping, increasing the cost further, and the salts must also be replaced or recharged after about 10 years. Unless there are severe restrictions on the size of the storage container, rock or water storage is still the least expensive storage medium for most applications.

Water is the most popular collector and storage medium. It has the highest specific heat of all commonly available materials, is cheap, easily available, and can be moved with ease with readily available techniques and equipment.

When using a water-base collector (Figure 7-20), care must be taken against freezing temperatures either by continuing to circulate warmed water (with some loss of heat) or by adding an antifreeze solution to the system. The latter approach makes it mandatory that a heat exchanger be used to prevent mixing chemical materials with the storage fluid. This component will reduce the efficiency of the system by several percentage points and increase the cost of the system.

Water systems require less space for both circulation piping and

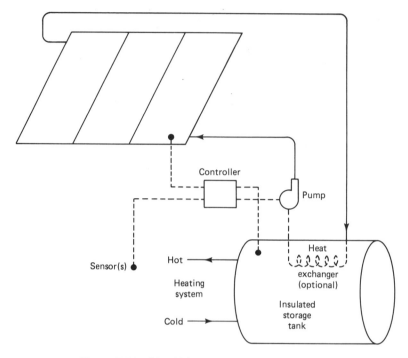

Figure 7-20 Liquid heat energy storage system.

for storage, since water has, volume for volume, twice the energy storage capacity of rock. There may also be corrosion concerns, since glycol solutions commonly used to lower the freezing point of water can become acidic and attack plumbing. Water systems are quieter during operation than air systems, which require fans and ducts.

Rock storage (Figure 7-21) is well suited for larger applications such as building heating (and cooling). Rock storage is heavy and difficult to move, but it is relatively inexpensive. The size of rock used for heat storage is still debatable, but the ratio of solid material to open space could be at least 2:1 with large rocks up to 6 in. (15 cm) in diameter and as much as 5:1 with closely packed small rocks, gravel, or sand. Many systems use stone 1 to 2 in. (2.5 to 5.0 cm) in size. Airflow through the storage bin will determine the best ratio. With 30% voids, the rock volume would have to be 2.5 times that of water in order to provide equivalent heat storage capacity.

When rock is used as the storage medium with air collector systems, it must be treated initially with a dilute solution of chlorinated material to prevent the growth of fungus, bacteria, and viruses. The dark, warm environment of a rock bin is ideal for the little "beasties" to flourish. It is this type of environment in a trap in ductwork which

Solar Thermal Storage Systems

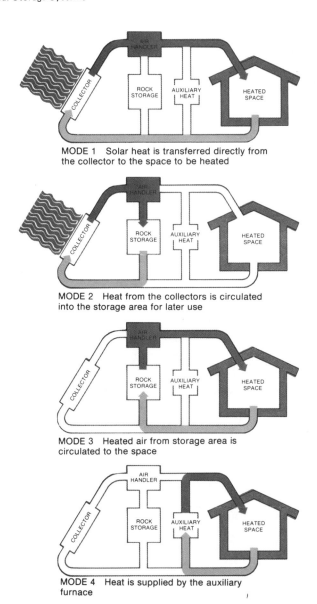

Figure 7-21 Rock heat energy storage system. (Courtesy of Research Products Corporation.)

is believed to have caused the outbreak of Legionnaire's disease and related respiratory infections.

Two heat circulation loops should be present in every solar heat collection system: the solar-to-storage loop and the storage-to-use loop. If an air collector is used, air could be allowed to circulate by natural convection, but this is usually adequate only for "solar walls" and then most often in new construction and with careful design. Forced-air circulation systems involve ductwork, fans, sensors, and controls to collect, store, and distribute the thermal energy properly, each adding to the cost and complexity of the system.

PHOTOVOLTAICS

Electricity directly from the sun—a simple idea, but difficult to put into reality. At noon on a clear day the energy density of the sun's radiation on the earth's surface in the temperate zone is 100 watts per square foot (or 1000 watts per square meter).

Conversion of light energy to electricity was observed by the French scientist Becquerel in 1839. In 1905, Einstein published three papers which changed conventional thinking in the scientific world. In addition to his work on the theory of relativity, he also described the photoelectric effect known to exist with certain materials. His third paper dealt with the behavior of light and described light as possessing discrete particles (photons) as well as having wave characteristics. It was the discovery of the two phenomena described in the latter two reports which provided the foundation for further investigation and development of the rapidly growing photovoltaic industry.

Semiconductors are materials that can capture and use the light of the sun to produce electricity. These materials are normally classified as insulators but can be made to act as conductors under certain conditions. When made very pure, semiconductors produce a crystalline lattice structure which can be represented as in Figure 7-22. When molten silicon is allowed to cool very slowly, the atoms align in a perfect cube pattern.

All elements contain electrons in one or more orbits around the nucleus. Each silicon atom has four outer-shell electrons. In the crystal form, it shares four with neighboring atoms to establish a stable material. The stability can be upset by injecting a minute amount of another material in a process called doping. Materials with either one more or one less valence electron than silicon are used. When materials such as phosphorus, arsenic, or antimony are used to dope the silicon, an extra electron is inserted into the lattice structure, making the material an n-type semiconductor. An extra electron is more easily moved when struck by a photon since it is a "free" electron without a place in the valence ring.

Boron, aluminum, indium, or gallium have only three valence electrons, so when they are used, a "hole" or vacancy is left in the valence ring, resulting in a p-type semiconductor. This creates a place

Photovoltaics

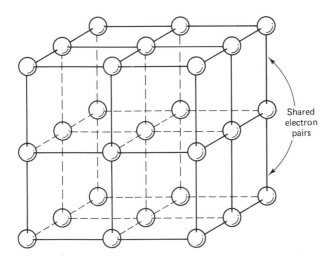

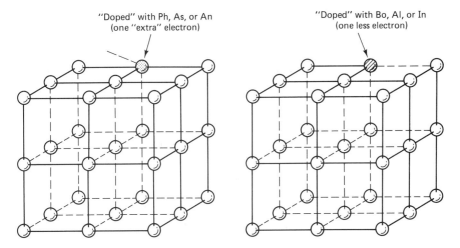

Figure 7-22 Crystal lattice structure. (Reprinted with permission of *Encyclopedia Americana*, copyright 1982, Grolier, Inc.)

for the "free" electron. The n-type has a surplus of electrons and is therefore negative. The p-type has a shortage of electrons (an excess of holes) and is therefore positive.

Placing the two types of materials in contact with one another creates a "conduction band." The energy needed to move an electron from the valence band to the conduction band is called the bandgap. Each semiconductor material has a characteristic bandgap, which determines the ultimate efficiency a solar cell of that material can achieve. The bandgap is a zone of static charge, developed when the electrons and holes immediately adjacent to each other in the unlike materials seek to neutralize their opposing charges. The result is a static barrier which electrons cannot cross.

When a photon of light of the proper wavelength strikes one of the orbiting valence (outer ring) electrons it moves to another atom, creating a hole or vacancy in the valence orbit of the first atom. Only light of a given wavelength can cause an electron to perform in this way. For silicon, a wavelength of 11,500 angstroms is needed (1 angstrom is 1 hundred-millionth of a centimeter). This corresponds to 1.12 electron volts (eV). Only about 45% of solar radiation falls within this range; the rest will only create heat within the silicon.

Electrons will not cross the bandgap, so some other pathway must be found for them to reach the holes in the p-layer. When electrodes are placed on each surface of the semiconductor sandwich and connected together, electrons flow readily through them to the p-layer. When an electrical load such as a light or other device is placed in the circuit, useful work is done.

Solar cells begin as an ultrapure molten mixture of semiconductor material with a minute amount of boron added to produce the p-type material. Single crystals are formed by inserting a "seed" crystal into the mixture and withdrawing it slowly. The resulting crystal may be as large as 6 in. (15 cm) in diameter and 3 to 6 ft (1 to 2 m) long. The crystal is sliced into wafers, commonly 0.010 to 0.014 in. (0.25 to 0.35 mm) thick. The wafers are then lapped and polished for smoothness. Up to 70% of the original crystal is lost up to this point. One face is then exposed to phosphorus vapor for doping. This develops the n-type layer to a depth of about 1/25,000 in. (1 micrometer; μm) and the n-p junction.

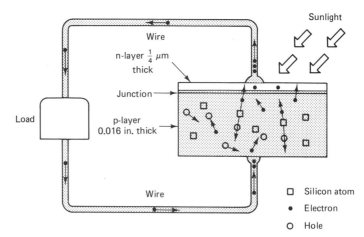

Figure 7-23 Silicon solar cell. (From the U.S. Department of Energy.)

Theoretically, solar cells should operate with an efficiency of 30 to 37%, but because of internal cell losses, reflection off the surface of the cells, areas covered by the contact electrodes, and other physical losses, the best efficiency that can be expected without reflectors or lenses is about 15%.

Other techniques have been developed to lower the cost of production. Single crystals can be cast as ingots, but a more promising approach is to produce flat sheets or ribbons of silicon. The process is shown in Figures 7-24 and 7-25. Cast materials have efficiencies as high as 15% and ribbon techniques are in the range of 11 to 12%. Another process, known as the Schottky junction, is formed by depositing a thin layer of a metallic conductor such as platinum onto a p- or n-type semiconductor. The same type of barrier is developed, and both free electrons and holes travel in opposite directions when the material is struck by light.

Photovoltaic modules are assembled by connecting individual

Figure 7-24 Silicon ribbon. (Courtesy of Westinghouse Electric Corporation.)

Figure 7-25 Ribbon cell and panel. (Courtesy of Westinghouse Electric Corporation.)

cells in a series–parallel array. This provides a unit that is large enough to generate usable amounts of electrical energy at the power levels necessary to do useful work. The usual size is a panel about 2 ft by 4 ft (0.6 m by 1.2 m), suitable for fastening on roofs or on isolated mountings. A unit of this size could be expected to produce from 35 to 65 watts.

The series–parallel connections provide protection—in case one cell fails, only that string will be affected; the other strings will continue to function. In addition, diodes, which allow current to flow in only one direction, are placed at the end of each string to prevent the flow of electrons back to the cells because of an imbalance in voltage.

Some units use concentrators similar to those used on solar heating systems: circular and parabolic reflectors and Fresnel lenses. When this technique is used, provision must be made to remove the heat that is generated, since higher temperatures can reduce efficiency. Tracking systems must be used with this application if it is to be economical. The heat that is generated can be used for other purposes, but this technique has not yet been well developed.

Energy storage is similar to that required with wind electric systems. Battery (chemical) storage is suitable for remote locations if the load capacity is less than the generating capacity. The system can be interconnected with the utility as with many other renewable electricity-producing units, according to the provisions of PURPA (see Chapter 8).

Amorphous (noncrystalline) structures have been found to be cost-effective solar cell materials. An "alloy" of silicon and hydrogen is deposited on a heated surface, then doped with either p- or n-type material. A thickness of only 1 μm is needed and the process is much cheaper than others. However, the efficiency is lower, on the order of 3 to 6%. As better knowledge is developed about the structure and physical characteristics of amorphous materials, better efficiencies are expected.

All of the above-mentioned processes provide solar cells that are more expensive than conventional electricity-generating facilities. A cost of $0.70 (1980 dollars) per installed peak watt for PV collector arrays is generally used as the point at which they will become competitive in general use. The cost per kilowatt-hour is another way to price PV arrays. When the cost of conventional electrical generation reaches $0.12 per kilowatt-hour, PV systems are competitive. Photovoltaic systems are, therefore, presently competitive in many areas of our nation.

Another factor in determining economic feasibility is for remote locations. When power lines must be installed from utility distribution systems over long distances, the cost of installation is often assessed against the customer. That cost can equal or exceed the price of a PV system installation. With tax credits, the cost of a PV system can be recovered within 5 to 10 years, and often less.

Most people equate the use of solar cells with the expansion of the space program. In fact, even with their initially high price tag, solar cells were the least expensive of the many options reviewed for generating electricity in space. These approaches not only proved the performance of the new technology but demonstrated the durability of the system under conditions of extreme temperatures and the vacuum of space. Current uses include "stand-alone" units such as those found in some pocket calculators, on mountaintops for powering meteorological stations, for navigation lights in remote areas, and even cathodic protection for bridges and pipelines.

Generation of electricity with PV cells ranges from supplemental power for homes (both in urban areas and remote sites) to larger applications such as pumping water. The latter function could make a major impact on many people in developing countries where conventional fuels are in short supply but where there is abundant sunlight for most of the year. Fossil fuels will continue to increase in price as they become less available, so PV can only become more attractive from a cost standpoint.

Public utilities consider the generation of electricity with solar cells of enough importance that they are installing large arrays. Several units have been built in California. One in San Bernadino County has a

Figure 7-26 Solar photovoltaic. (Courtesy of Consolidated Edison.)

capacity of 1 MW, and another in San Luis Obispo County is rated at 6 MW. The first uses a flat array composed of 256 modules mounted with computer-controlled trackers. The larger system has 756 tracking units, each with 128 modules for a total of 96,768 modules, with laminated glass reflectors to increase the energy density striking the cells.

The largest central receiver PV electric facility built to date is operated by Southern California Edison, the DOE, the Los Angeles Department of Water and Power, and the California Energy Commission.

Economists have estimated that the future looks promising for solar electric systems. They predict that by 2010 more than 131,000 MW of solar-produced electricity will be available worldwide, more than 1000 times that produced in the mid-1980s. Nearly 90% of that is expected to come from PV cells.

A report from Brookhaven National Laboratory judged the PV industry to be one of the safest and cleanest of all the energy options. Photovoltaics can also provide an additional option for decentralization and diversification of energy resources.

There are, however, several safety constraints as PV production increases. The major component in the manufacture of semiconductors is silicon made from silica, one of the most common materials found on earth. The breathing of silica dust by miners is a recognized occupational hazard about equal to the danger of coal dust for coal miners. The extraction and refining of ores for the dopant materials could result in large amounts of mercury and alumina sludge residues. Any pollutants produced are projected to be less than those currently found as a result of conventional electrical production.

Toxic gases are used during the production of semiconductors. In

Figure 7-27 10-MW photovoltaic facility. (Courtesy of Southern California Edison.)

a 1980 study, the California Department of Industrial Relations found that the PV industry had 1.3 illnesses per 100 workers, more than three times higher than for manufacturing in general. Workers in PV manufacturing are exposed to several acids and gases that have been found to cause burns, respiratory illness, and are in addition potential carcinogens. Great care is required in handling, transporting, and storage of gas cylinders used in the industry.

Solvents such as sulfuric, hydrofluoric, and acetic acids are used in several stages of PV cell manufacture. Gases used in the etching and doping processes include arsine, phosphine, and diboran. A process called "dry plasma etching" is being used in a closed-chamber system, considered to be a less hazardous procedure. High-frequency alternating current is used in the process, creating another potential danger.

An additional concern is that the technical advances are so rapid in this field that health and safety agencies have difficulty keeping up to date on developments. The proprietary nature of many of the developments adds another barrier to maintaining a close watch on the health of workers.

Waste disposal must be monitored carefully, since landfill techniques may not be adequate for the proper disposal of the wastes from the toxic materials used. Air pollution must be carefully monitored, since reactive organic gases are a large part of the processing sequence. One of the gases, arsine, will destroy red blood cells if inhaled, requiring a complete blood transfusion.

THE FUTURE OF SOLAR ENERGY

Solar energy is expected to become an even more important factor in the total energy picture in the near future. Home buyers are becoming more sophisticated in demanding energy-saving features in newly constructed homes and are retrofitting existing homes for better energy efficiency. Both passive and active systems will increase in popularity as energy prices continue their upward climb. Architects will increase their use of building features that make use of principles of solar gain, and contractors will need to take those principles into consideration in their building practices.

Both private construction and commercial buildings will make greater use of solar technologies. Some applications will be used to provide power during peak-load times, others will be used to supplement existing heating/cooling/lighting systems. Sunspace attachments to existing structures and "green space" incorporated into new buildings have already made some impact on building design and energy consumption. Combination systems (wind/PV, solar/hydro, solar/wind) will be used to a greater extent than before, with one technology supplementing the other.

The installation costs for photovoltaic systems have already reached levels which are competetive with nuclear generation and costs are expected to continue to decline with additional developments in research and development. The use of PV systems in isolated locations and at reasonable costs will develop new markets in all parts of the world. The demand for electricity will increase in developing countries.

Continued tax credits by both the states and the federal government should continue, creating an incentive for private individuals as well as businesses and corporations. The dollars spent in energy-saving equipment and design have been shown to create additional jobs and increased economy.

Commercial utilities have seen the advantages of solar energy and have launched massive building programs in cooperation with the DOE. This effort will continue to grow.

SUMMARY

Countless experiments over thousands of years have shown the variety of ways in which the sun's energy can be concentrated and made to do useful work. Mathematicians, physicists, philosophers, and national leaders have all tried to make the sun do their bidding—some with a great deal of success. Through all the efforts, the nature of energy has become better understood.

It is known that the intensity of insolation is dependent on the latitude and climatic conditions of a particular site as well as the angle at which the rays strike a surface. We know that the sun's rays can be concentrated using the principles of reflection and/or refraction. The rate at which objects absorb and release heat energy has become the basis for entire industries and countless applications.

Heating of water and living/working space is the most common application for solar energy. Although highly sophisticated systems have been developed to perform this task, simple devices and approaches have been found to perform well in most situations. The principles that govern the satisfactory performance of systems are easily understood by nearly anyone.

Light and heat have been found to be suitable energy producers when used in conjunction with solar radiation. Photovoltaic cell systems and other solar applications are becoming competetive with conventional means of producing heat, light, and electrical energy. The impetus of high costs for fossil fuels makes the future of solar energy very "bright" indeed.

ACTIVITIES

1. From the *Climatic Atlas of the United States* or other references, determine the average daily solar radiation for your geographic location for each of the months of the year. Take a set of readings with an insolation meter and compare the difference (if any). What are the possible causes of differences compared with the *Atlas*?
2. Calculate the energy needed to provide the heating and cooling needs of a home, office, or work site.
3. Determine if the sun at your geographic location is adequate to provide a major portion of the heating and cooling needs by means of flat-plate collectors (select either an air or a liquid system).
4. Construct a solar cooker, oven, or other heating device for a single heating function, using low-cost, easily available materials.
5. Determine the needs of a home, office, or industry that could be met with a photovoltaic system. What factors would need to be considered for PV to be a cost-effective approach?
6. Construct several solar air collectors as test models. Use various glazing materials, absorber plates, and depths of air space. Experiment with different angles of placement. Record the results of the tests and determine the best combination for your location.
7. Build a passive solar collector from a Styrofoam cooler. Experiment with various materials as energy storage media. Describe your results.
8. Hold a design competition among members of your group to see what design will boil a given amount of water most quickly (aluminum foil, Fresnel lenses, glazings, etc.).

BIBLIOGRAPHY

Anderson, Bruce, *Solar Energy: Fundamentals in Building Design*. Harrisville, N.H.: Total Environmental Action, Inc., 1977.

———, *The Solar Home Book: Heating, Cooling and Designing with the Sun*, Harrisville, N.H.: Cheshire Books, 1976.

Butti, Ken, and John Perlin, *A Golden Thread, 2500 Years of Solar Architecture and Technology*. Palo Alto, Calif.: Cheshire Books, 1980.

D'Allessandro, Bill, "Solar Heaters Prove Out in Utility Test," *Solar Age*, March 1984, pp. 29-31.

Daniels, Farrington, *Direct Use of the Sun's Energy*. New York: Ballantine Books, 1964.

Flavin, Christopher, "Photovoltaics, International Competition for the Sun," *Environment*, Vol. 25, April 1983, pp. 7-11, 39-43.

Glasstone, Samuel, *Energy Deskbook*. New York: Van Nostrand Reinhold, 1983.

Iverson, Wesley R., "Amorphous Silicon Wins Place in the Sun," *Electronics*, April 5, 1984, pp. 95-96.

LaDou, Joseph, "The Not-So-Clean Business of Making Chips," *Technology Review*, May-June 1984, pp. 22-33.

Leckie, Jim, G. Masters, H. Whitehouse, and L. Young, *More Other Homes and Garbage*. San Francisco: Sierra Club Books, 1975.

Marbach, William D., and others, "Solar Energy Lights Up," *Newsweek*, May 16, 1983, pp. 88-89.

Maugh, Thomas H., III, "Catalysis in Solar Energy," *Science*, Vol. 221, September 30, 1983, pp. 1358-1361.

Maycock, Paul, "The Expanding PV Industry," *Alternate Sources of Energy*, May-June 1984, pp. 18-19.

Mazria, Edward, *The Passive Solar Energy Book*. Emmaus, Pa.: Rodale Press, 1979.

McCullagh, James C., ed., *The Solar Greenhouse Book*. Emmaus, Pa.: Rodale Press, 1978.

Meyers, Robert A., ed., *Handbook of Energy Technology and Economics*. New York: Wiley, 1983.

Montgomery, Richard H., *The Solar Decision Book: A Guide for Heating Your Home with Solar Energy*. New York: Wiley, 1978.

Reif, Daniel L., *Solar Retrofit, Adding Solar to Your Home*. Andover, Mass.: Brick House, 1981.

Rosch, Winn L., "Polymer Film Could Replace Silicon Chip," *The Electron*, Vol. 11, No. 3, May-June 1984, p. 1+.

U.S. Department of Commerce, *Design of Systems*. Washington, D.C.: U.S. Government Printing Office, 1980.

———, *Solar Heating and Cooling of Residential Buildings, Sizing, Installation and Operation of Systems*. Washington, D.C.: U.S. Government Printing Office, 1980.

Chapter 8
Wind Energy
Energy from Thin Air

CONCEPTS

1. The wind is an inexhaustible source of solar energy, ensuring unlimited amounts of energy.
2. Generally, wind-driven systems produce no pollutants and are environmentally safe.
3. Wind systems are decentralized and capable of being tailored to individual requirements.
4. Wind systems can be independent of large-scale utilities or merged with them at the owner's option.

GLOSSARY

AIRFOIL—any surface that uses the pressure of the air through which it moves to create lift.
ALTERNATING CURRENT (ac)—electric current that reverses polarity (direction) at regular intervals.
BETZ COEFFICIENT—the maximum fraction of the power in a wind stream that can be extracted by a turbine (the Betz limit).
CUT-IN SPEED—the wind speed at which the WECS begins to generate electricity.

CUTOUT SPEED—the wind speed at which the WECS shuts down, either automatically because of blade design, or by means of electronic sensors.
DIRECT CURRENT (dc)—an electric current of constant direction which does not vary, or varies without changing polarity.
HERTZ—a unit of electrical frequency; cycles per second.
NACELLE—the housing used to protect the gearbox, pitch controls, and generator mounted at the top of the WECS tower.
POWER COEFFICIENT—the efficiency of the WECS as a ratio of power extracted from the wind stream, limited to a theoretical maximum of 59.3% (the Betz limit).
PURPA (Public Utilities Regulatory Policies Act)—provides for payment by commercial utilities for electricity generated by private individuals or small independent companies which is fed into the utility grid.
STALL—a condition created when an airfoil goes beyond the angle of maximum lift, blocking or spoiling the airflow pattern so that no lift is created.
TIP-SPEED RATIO—the ratio of blade tip speed to wind speed.
TURBULENCE—rapid changes in velocity and/or direction of the wind.
WECS (wind-energy conversion system)—the total system of components that change wind energy to a more usable form—including the rotor, generator, or other conversion device; controls; tower; power transmission device(s); and energy storage device(s).

INTRODUCTION

The invisible, moving molecules of the atmosphere have energy that can be tapped for useful purposes. The use of wind energy is as old as the first uses of sails for boats and as "wind catchers" to raise water and grind grain. Babylonian records show the use of wind machines for water pumping as early as 1700 B.C. During the Middle Ages, wind machines were used to grind grain as well as do a variety of other tasks. In the early twentieth century wind machines were used to pump water and power electricity generators in many rural locations (Figure 8-1).

The operating principles of these early-twentieth-century wind-powered generators remain the basis for the design of modern wind turbines. Among the best known of the earlier units are the Baker and Aeromotor windmills used for pumping water, and the Windcharger and Jacobs used for electrical generation. The latter is usually referred to as the "Cadillac" of the line, and the former as the "Chevrolet," based primarily on the relative cost of each machine. All these companies are in business today.

There is renewed interest in generating electricity in decentralized locations. In addition, factors such as conservation, financial savings, self-sufficiency, and the need to supplement and eventually replace petroleum-based electricity-generating equipment have created a desire

Figure 8-1 Small WECS, 1920s-1950s. (Courtesy of Winco, Division of Dyna Technology, Inc.)

to increase our understanding of the systems and operating principles of wind machines.

PRINCIPLES

All wind turbines use some form of airfoil to convert the linear motion of moving air to rotary motion, then make that motion produce electrical or mechanical energy. The primary components of the wind-energy conversion system (WECS) used for electrical generation are the tower, rotor, generator/alternator, the speed control system, and the wiring to transmit electricity generated to the load or to the storage system.

The rotor/generator is the actual conversion device, consisting of a blade assembly and an electrical generator. When purchasing a WECS, the power output curve of each potential system must be evaluated to determine which is most suited to the electrical needs of the user. Identify a supplier who has a reputation for high-quality equipment and service, just as you would with any other major purchase. The cost of a generating system will range from a few hundred dollars for a very small unit capable of generating a few hundred watts, to $50,000 or more for units capable of generating 15 kW or more. The average single-family residence will require a unit of 10 to 15 kW. Reliability is a key factor, since repair of a tower-mounted machine weighing from several hundred pounds to several tons for some commercial units is a major endeavor. Wind machines should be capable of operating continuously for 24 hours a day for up to 20 years with minimal maintenance or repair.

The speed-control mechanism is a vital part of the system. Most come with the rotor/generator and are designed to shut down or otherwise modify the operational characteristics of the blade assembly when wind velocities exceed the design limits of the turbine. This must be done to prevent structural damage to the blades, the tower, or the turbine itself.

The "storage" system may take one of three forms. Batteries are commonly used in dc systems; the electricity in excess of immediate on-site needs may be sold to public utilities (under provisions of PURPA); or the electricity generated may be converted to another form of energy which may be stored, such as thermal or mechanical energy forms. The excess may also be used in other ways, such as heating or pumping water, compressing air, for electrolysis, and for other purposes to be discussed later.

Small WECS usually use a dc generator to send electricity to a battery storage system and from there to dc loads. Typically, this is of lower voltage (24 V) than conventional house wiring using ac (120 V). Dc systems require an inverter to change the dc to ac for small motors and other ac appliances. The induction generator is commonly used with commercial, industrial, and utility systems. Either 120 or 240 V ac can be produced.

SITE CHARACTERISTICS

Critical to the effectiveness of the system is its siting. Sites must be evaluated by means of a monitoring process to determine the most effective location and height for the tower, which must not be so high as to make the cost prohibitive.

Wind velocity is generally higher as the altitude increases above the surface of the earth. This is especially critical, since the power of the wind is proportional to the cube of the wind speed. Another important factor in site consideration is the effect of adjacent structures or objects near the top of the tower. The tower should be at least 20 ft (6 m) higher than nearby obstructions; if it is closer than this, turbulence from air passing over or around the objects will reduce the wind's velocity and may cause gusts or shear forces that could cause damage or destruction to the blade or tower. The greater the distance between the top of the tower and any turbulence-causing objects, the less likely such problems are to occur.

Before a WECS is purchased or erected, wind velocity is usually monitored by means of an anemometer, which should be connected to a recording device such as a strip chart or magnetic tape to provide data about round-the-clock wind velocities. Some areas have already been tested for wind velocity, or an estimate of wind velocity can be made from existing data, usually from nearby airports or the National Weather Service. When hills, valleys, and structures are near the site, off-site estimates may be seriously in error. To be sure that wind velocity data are accurate, determine the high, low, and average wind velocities for best siting of the tower.

Types of towers are as varied as the needs and availability of materials. Simple towers such as utility poles will serve well in areas of relatively flat land for a small WECS (under 1 kW). More costly steel or concrete towers are more often used. Some are self-supporting, some are guyed, and some are attached to other structures. There is a precaution to be taken when mounting a WECS to an existing structure. Vibrations and/or harmonic oscillations frequently develop which may cause structural damage to the original structure, or noise may be generated which is irritating or harmful to people or animals. This type of mounting is generally not recommended. Support structures may be as inexpensive as $500 to $800 for wood or as expensive as $5000 to $10,000 for a self-supported steel tower.

BLADE DESIGN

Blades of every WECS are airfoil surfaces designed to create lift and to use that lift to drive a rotor. Horizontal shaft wind machines use several variations of blad configuration. Most often used are the two-, three-, and four-blade designs. There are also models that use a single blade with a counterweight to maintain balance and momentum.

Both up-wind (blades on the windward side of the tower) and down-wind versions are usable. Advocates of the down-wind type feel that in the event of blade breakage, less damage will occur to the tower and generator. Proponents of the up-wind type feel that there is less loss of power because the "tower shadow" is not a factor as in downwind designs. Down-wind designs must use blade and hub designs which are stronger to withstand the increased turbulence of the shadow. When damage has occurred, it has most often been due to mechanical problems, linkage failure, or electronic control malfunction rather than to any design flaw in the blade system itself.

Blades for WECS produced in the 1930s were made of solid hardwood. Because it is easily shaped, wood is still the favorite material for small systems, but if the blades are not sealed against moisture, they can become unbalanced and vibration will occur. Many different materials are now used, including laminated Sitka spruce, fiberglass, and

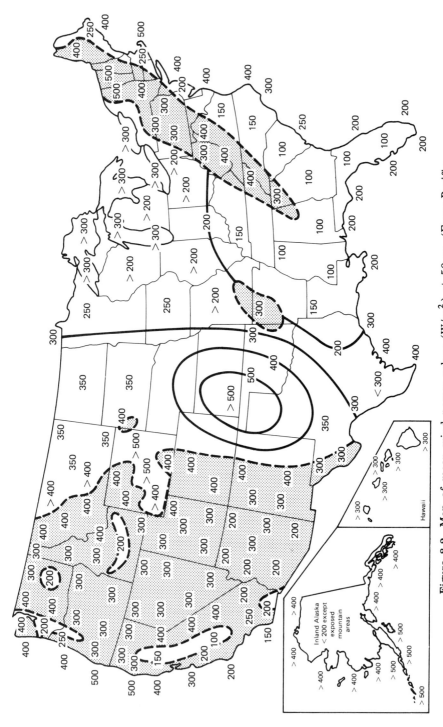

Figure 8-2 Map of mean wind power values (W/m^2) at 50 m. (From Pacific Northwest Laboratory, which is operated for the U.S. Department of Energy by Battelle Memorial Institute.)

aluminum. Metal blades tend to cause radio and television interference. Just as buildings and trees can cause reflections (ghosts) on television screens and static interference with radio waves, the rotation of the blades sometimes coincides with a harmonic of the electronic signal and causes interference in radio and television receivers. Combinations of materials are being used in many current designs both to reduce the interference and to make the best use of weight-to-strength ratios.

CONTROL MECHANISMS

There are several types of speed-control mechanisms used to prevent horizontal-axis wind turbines from rotating too fast in high winds. The most common applications use variable-axis control, spoiler flaps on the blades, blade-pitch control, coning, or various braking mechanisms.

Variable-axis control uses the pressure of the wind either to swing the rotor assembly to one side or tilt it upward to move it out of an effective attitude with the wind. With side-tilt control, the rotor and generator are hinged so they may turn out of the wind stream, with the tail vane still parallel to the wind direction. The tip-up mechanism uses a spring-loaded design to limit the tip-up until the rotor can again turn at a safe speed.

Flaps at the ends of the blades actually "spoil" the smooth flow of air over the blades, reducing their speed and maintaining a safe speed. There are also spring-loaded controls which are designed to begin to operate as the cutout speed is reached.

Some units use fixed-pitch blades, some use a variable-pitch system. Lift is a function of the angle of the blade (pitch). If the velocity of the wind changes, there is an automatic change in the pitch, thereby changing the lift. When the pitch is too high, the performance of the blade system deteriorates, creating a condition called "stall." In aircraft, this can be a serious problem, but with wind machines, it can be useful in regulating the speed of the blade assembly when the system has had this factor incorporated into its design.

Blade-pitch control is used on small rotors of under 20 ft (6 m). Weights attached near the hub of the rotor are set so that centrifugal force causes them to vary the pitch of the blade. Only a small change in the attack angle of the blade is necessary for it to "spill" the air and thus become a less effective airfoil. One version uses the weight of the blade itself as the control mechanism. The blades move up and out from the hub in a helical motion to change the pitch.

Coning is the action used in some down-wind machines to cause the blades to furl, as the ribs in an umbrella pull together when they are partially closed. This technique is obviously not feasible with upwind designs if you wish the tower and blade tips to remain intact.

Several types of braking systems are being tested, but the inherent complexities and lack of fail-safe designs with mechanical and/or electronic controls would appear to make them less reliable than those mentioned above.

ELECTRICITY-PRODUCING UNITS

Electricity-producing units require a higher operating speed than that supplied by the rotor, so some means must be used to acquire an added speed [revolutions per minute (rpm)] differential. Either a gear system or a chain or belt drive is used to interconnect the rotor with the generator. A typical top windmill speed would be in the range of 125 to 225 rpm in wind speeds of 7 to 20 mph (11 to 32 km/h). Gearing ratios would convert that to common generator speeds of 1500 to 3600 rpm. Belts and chains are less expensive than gearboxes, but they are less reliable and generally have a shorter operating life than gears. They are also subject to jumping off pulleys and sprockets when there are sudden changes in wind velocity or wind direction.

GENERATOR TYPES

Generators with outputs as low as 200 W have been used for decades. More recently developed designs have produced units rated as high as 4.8 MW. There is also a giant 7.5-MW unit on the drawing boards. The more common sizes are in the range from 10 or 12 kW to over 100 kW.

Regardless of size, these units are usually of two basic types: generators which produce dc and alternators which produce ac. Direct-current generators are used primarily on small-scale systems.

Alternators similar in design to automobile alternators are also used on some small-scale systems. They have the advantage over direct-current generators in that they are lighter in weight per unit output and have lower maintenance costs since there are no brushes to replace. The ac output is converted to dc with rectifiers containing electronic diodes (one-way controls).

The induction generator is the type most often used in commercial systems and by WECS owners who interconnect with the local utility. The induction generator has a rotor and stator just like a standard ac generator, but there is no physical connection between them and thus no need for brushes, slip rings, or diodes. Instead, the unit makes use of the principle of magnetic induction. The rotor's magnetic field is "induced" through the magnetic interaction with the stator. Induction generators produce ac directly and use minute amounts of grid electricity to synchronize the rotational speed of the rotor to produce a frequency of 60 hertz, just as in the grid.

Automatic protection must be provided in case of grid outages. If the utility shuts down, as during a power failure, the load on your system is removed and serious problems may develop. Since the field current for the induction generator is supplied by the utility, if it shuts down, the controlling frequency for your system and the load is also removed. Without a load, the rotor may continue to increase in speed unless automatic controls have been built into the system. If the velocity of the wind is great enough, the rotor can destroy itself through centrifugal force. With an automatic disconnect, the WECS is not feeding into a "down" utility and creating danger to linemen working on the system. All of these factors must be designed into any system that is tied to the utility grid.

INVERTERS

An inverter changes the direct current available from the storage batteries to the 115 V ac required for most appliances that plug into standard household outlets. The advantage of such a system is independence from utilities. The major disadvantage is the cost. Inverters cost about $1 per watt of rated power. Furthermore, they are most efficient at their rated power, so their use must be planned carefully. Since the cost may be as much as $3000 to $5000 for the battery/ inverter system, a cost that must be repeated after about 10 years due to battery life, the type of system purchased and the sizing needs must be calculated with care.

Most inverters in current use are of the electronic or solid-state type. The lower-cost inverters produce a square-wave ac signal. Although this is suitable for most appliances, it may produce an audible hum or distortion in electronic gear. A more expensive type can be used which develops a true sine wave similar to that of the utility grid.

"STORAGE" SYSTEMS

Electricity cannot be stored as electricity. It must be converted to another form, such as chemical energy stored in batteries, used to pump water for future use in hydroelectric systems, or used to do other work which can be reconverted to produce electrical energy at another time.

Storage batteries require large amounts of capital, a safe place for locating the system, and a control mechanism to maintain proper voltage and amperage levels. Deep cycle batteries are the only suitable type to use since the charging/discharging cycle is apt to be much more severe than that of an ordinary auto battery system. Deep cycle batteries are those found on forklift trucks and golf carts. At least 20 batteries are necessary to supply the needs of the average home. Batteries are normally rated at 12 V dc, so a series-parallel array must be used either to accept or to provide the power necessary for appliance use. An inverter must be used because appliance motors require that electricity be supplied at designated voltages.

Storage systems other than deep cycle batteries can be used. Water pumped into a reservoir or storage tank constitutes one method. The principal advantage is that the electricity generated can easily be sent via conventional wiring to the site of the well or other supply to drive an electric motor and turbine pump, rather than using a water-pumping windmill at the well site, where wind power might not be as favorable. The water stored in the reservoir can then be released and allowed to run through the turbine which turns the motor (now used as a generator) to produce electricity.

Water can be treated in a quite different manner through electrolysis—separating it into its two components, hydrogen and oxygen. This process is very expensive at the present time using conventional electric generation. It can be considered economical only when there is direct use of the hydrogen as a fuel or in a fuel cell.

Compressed air is another storage medium that is easily used to meet short-term high-energy demands. With this system, WECS-produced electricity powers a motor to drive a compressor. The compressed air may then be used during low-wind periods to turn an air turbine, which turns the motor (now acting as an alternator) to produce electricity. Compressed air may also be used directly to power pneumatic motors.

Flywheel storage devices are being investigated, but as yet do not provide an effective alternative. They must be large and capable of high rotational speeds to be effective storage devices, yet these factors make them subject to centrifugal forces that tend to destroy them. Studies are under way to improve the strength of materials used in flywheels and to redesign the wheels so that if catastrophic failure does occur, only minimal damage will be done.

To eliminate the battery system entirely, it is possible to use a synchronous inverter which converts variable dc electricity produced by the WECS to ac of the same frequency and voltage as the utility grid. Any electricity that is not required by the producer can be fed directly into the grid. When the wind is not blowing, the utility can return electricity to meet power demands. The system must be metered to show

how much wind-generated electricity has been provided to the utility, as well as how much electricity has been used from the utility.

Under provisions of PURPA, the utility must pay back to the independent producer the "avoided costs" it would have incurred had the WECS not been contributing to the grid. The avoided costs reflect only the costs of producing electricity and do not include transformer or distribution costs of the utility.

Another "storage" system converts the electricity produced to thermal energy. The simplest way is to connect directly with some type of resistance-heating device to heat either space or water. No costly inverters are needed, since resistance heaters use either direct or alternating current without modification. The water may be stored in well-insulated tanks for use as a supplement to space heating or as preheated water for domestic or commercial uses. This is currently the most economical storage system.

SMALL/INTERMEDIATE VERSUS MEGAWATT SYSTEMS

Wind systems described above refer most often to the small residential and small commercial sizes. The range of sizes of the various wind machines in use or planned for construction has been divided into five categories:

Small (0 to 4 kW)
Small commercial (8 to 15 kW)
Large commercial (25 to 40 kW)
Industrial (100 to 300 kW)
Utility (>1 MW)

One of the most famous of the megawatt machines is the Smith-Putnam machine, which was built on a 110-ft (34-m) tower on Grandpa's Knob near Rutland, Vermont, in 1940 (Figure 8-3). It had a rotor blade measuring 175 ft (53.5 m) from tip to tip, and generated 1.25 MW of electricity in winds of 30 mph (48 km/h) or more. It successfully supplied energy to the Vermont Public Service Corporation between October 1941 and March 26, 1945, when fatigue failure in one of the two blades forced a shutdown. During operation its turbine speed was 28.7 rpm, with a blade-tip speed of 185 mph (300 km/h). Gearing increased the generator speed to 600 rpm, with output at 2400 V. After shutdown it could have been repaired, but it could not compete with the cheap petroleum and hydropowered electrical plants of that time.

Figure 8-3 Smith-Putnam 1.25 MW WECS. (Courtesy of the Vermont Public Service System.)

As petroleum prices have gone up, the economic advantages of electricity from large-scale wind systems have become more evident. Several countries have built or are constructing megawatt-size wind machines using materials and designs that will withstand the enormous forces generated by the momentum of blade lengths of 100 ft (30 m) and longer. Among these are Denmark, Sweden, West Germany, Great Britain, and the United States. They have each embarked on a developmental strategy which is designed to produce functional MW wind generators at suitable sites as part of the energy supply system. At the present time, units of this capacity are primarily experimental, but several show promise of being functional in essentially their present design.

Since 1973, the U.S. government has sponsored a program of research and development in wind energy to test the feasibility of the large WECS. The first of these projects was named Mod-OA (200 kW) and was used as a test model at various sites where the larger units were later installed. The Mod-1 was a 2000-kW unit; the Mod-2 is 2500 kW

(Figure 8-4). A 4.8-MW unit (Figure 8-5) has been installed at Medicine Bow, Wyoming, as a feasibility test unit for the megawatt wind-farm concept. Although preliminary design data indicate an impressive cost/ size improvement with the giant machines, smaller WECS (under 100 kW) may be more feasible in terms of financing.

A more feasible system, "wind farming," is under development in several parts of this country. Several sites are under development in northern and southern California operated by private wind developers in conjunction with public utility systems. Sites for future installations are being reviewed in Montana, Wyoming, New Hampshire, and elsewhere. The island of Oahu, in Hawaii, was to have been the site for several 2.5-MW units, but high interest rates and the "oil glut" put that proposal on hold.

Several operational wind-farm sites in California, recently com-

Figure 8-4 DOE/NASA 2500 kW experimental wind turbine. (From the U.S. Department of Energy and the National Aeronautics and Space Administration.)

Figure 8-5 4.8-MW wind turbine. (Courtesy of Hamilton Standard, Division of United Technologies Corporation.)

pleted, have produced nearly 2 million kilowatt-hours of electricity in a 3-month period. Although this amount is not a large percentage of the total electrical energy needed, it has proved the value of the windfarm concept as a useful supplement to the existing power-generating capacity of each utility. It has also demonstrated the value of separate, smaller units which are capable of being operated independently yet can function as a partial utility system. This is in contrast to the use of a single MW unit which if shut down is totally "down."

Each of the wind-farm systems uses WECS with two- or three-blade rotors to drive generators that produce between 44 and 56 kW each at maximum wind speeds of 18 to 22 mph (30 to 35 km/h). Blade diameters vary from 50 to 69 ft (15 to 21 m). Control for one system is achieved by means of wind sensors which activate an electric motor that turns the nacelle and rotor assembly away from the wind. Pop-out flaps at the blade tips activate at 60-mph (97-km/h) wind speeds, stalling the blade to prevent damage.

Figure 8-6 Wind farm, Tehachapi Mountains. (Courtesy of Southern California Edison.)

A second system uses the same type of controls plus an electrically activated brake to protect the machine further. A third type is a free-yaw design that needs no motor drive. It shifts with the wind, but the blades are pitch controlled to stall at 55 mph (90 km/h).

Some damage has occurred in wind-farm systems, due in one case to an internal blade spar that failed. Fatigue cracks developed in another unit, necessitating redesign of a casting and a more rigorous quality control program. Each of these problems demonstrates that further testing is necessary before high reliability of the entire system can be achieved. That the units can be repaired with relatively short downtime indicates that the design factors are fundamentally correct.

OTHER ROTOR DESIGNS

Several other types of wind machines are in use. One is the drag collector type, most commonly referred to as the Savonius (Figure 8-7). Of the vertical-axis class of wind turbines, the basic design is that of an S-curve with a vertical axis at the midpoint, creating, in effect, two semicircles. A more efficient treatment is to offset the two halves by a distance equal to the radius of the original circle, allowing the air to flow through the unit. One or more units may be stacked, with three providing the most efficient arrangement. The S-rotor is limited to a maximum efficiency of 33%, based on wind-tunnel tests, so that while it may be mounted without concern over orientation to the wind, its effectiveness is just over half that of a blade-type WECS.

Another type of rotor, the Darrieus, is currently undergoing tests

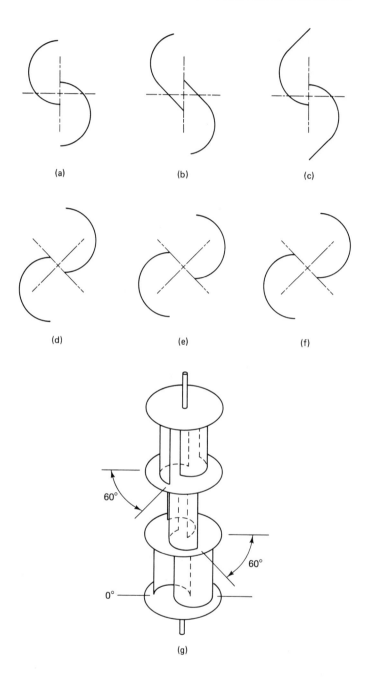

Figure 8-7 Savonius designs: (a–f) possible configurations; (g) three-tier array. [Adapted from Michael Hackleman, *Wind and Windspinners* (Mariposa, Calif.: Earthmind, 1974), pp. 66, 122.]

to determine its suitability as a commercial design. This airfoil-drag collector possesses the advantages of the airfoil (blade) type and the ease of orientation of the S-rotor. Since it also has a vertical axis, the tower requirements are much less than those of horizontal-axis units. The Darrieus does not require a variable-pitch mechanism to protect itself from damage in high winds. It uses aerodynamic characteristics to go into stall when high winds occur. The Darrieus has the disadvantage of not being able to begin to rotate without some type of "starter" unit. Sometimes a Savonius unit is added to the axis to provide the necessary startup torque or a small electric motor may be used for this purpose. The cross-sectional shape of the Darrieus blade is that of a fixed-wing aircraft, with the leading edge thicker than the trailing edge, thereby causing lift. This causes the blade to turn into the wind regardless of the wind's direction. The curve of the blade is the same shape that a rope would take when spun around a vertical axis.

The airfoil shape of the Darrieus causes the blade to turn into the wind at a speed greater than the wind. For each rotation of the blade it has a one-half turn of lift followed by one-half turn without lift. The rapidly changing torque has caused fatigue problems to develop in the

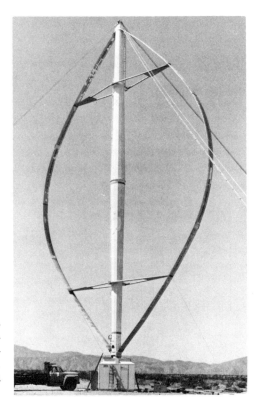

Figure 8-8 DAF series 6400, 500-kW vertical-axis wind turbine, located at Southern California Edison's site in Palm Springs, Calif. (Courtesy of Southern California Edison.)

axis shaft, with the larger Darrieus units experiencing failure because of the problem. Smaller units do not appear to have the problem.

Another undesirable feature of both the Savonius and Darrieus types is that their rotational speeds are lower than that of the blade turbine. More costly gearing mechanisms must be used to bring the rotational speed up to that required by a generator. Since the generator system is at or near ground level, this expense may be somewhat offset. The Savonius is very adaptable to water pumping where complex gearing is not needed.

The cyclogyro is a variation of the vertical-axis machine (Figure 8-9). It uses straight, vertical blades of uniform airfoil design, allowing for easier manufacture. Wind-tunnel tests indicate a better power coefficient for this design than for either the Savonius or the Darrieus. The blades are designed to change pitch automatically to obtain the most effective angle as the wind speed changes. In addition this approach makes the rotor a self-starter, eliminating one additional cost in its construction.

Concentrating collectors, sometimes referred to as augmented designs, operate on the principle of the venturi, or funnel effect. By

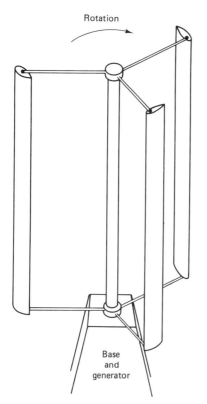

Figure 8-9 Cyclogyro.

constricting the flow of a fluid, in this case air, the velocity is increased. This effect may be dramatically realized when walking between two tall buildings on a windy day! The effect is established by building a shroudlike cylinder around the blade, causing the airflow to be channeled, increasing the velocity of the airflow by as much as two times.

A modification of the augmented design technique, called the dynamic inducer, uses blade tips set at right angles to the blade, creating in effect a moving partial shroud. This has the effect of joining the shroud to the blade, using about 1% of the area a shroud would take but creating about the same lift. Some drag is created by this method.

Since wind power is proportional to the cube of wind velocity, doubling the velocity increases the wind power by eight times. The fundamental equation by which to determine wind power is as follows:

$$P = 0.005 A V^3$$

where A is the area in square feet swept by the blade, V is the velocity of the wind in miles per hour, and P is the power generated (in watts). For example, if wind velocity is 8 mph (the lower limit for useful operation of most WECS), a doubling of the velocity by means of the concentrator would increase the power output from a rotor with a 10-ft diameter from 200 to 1608 W maximum power.

The theoretical maximum power conversion of a wind system has

Figure 8-10 DAWT installation concept with 18-ft (5.5-m)-diameter rotor. (Courtesy of K. M. Foreman, Research and Development Center, Grumman Aerospace Corporation.)

been calculated at 59.3% (the Betz limit—some air movement must remain downstream of the blade to maintain airflow). A more reasonable maximum efficiency is closer to 50%, providing an actual efficiency for the system of 20 or 30% when all losses in the conversion system are taken into consideration. It is easy to see from this formula how even a small increase in the velocity of the air can make a large change in the power output of a WECS.

The tornado wind turbine is based on the concept of air flowing into a chimney through adjustable vertical vents on all sides (Figure 8-11). The vanes channel air into a tightening spiral, increasing its velocity. Air turbulence over the top of the chimney augments the vertical movement of the spiraling air column. Upward movement of air draws additional air into the base of the tower, where it passes through blades or vanes attached to generators. A tower of this design, constructed using conventional techniques, would be strong, stationary, and omnidirectional. Since the generator and its associated blade and shaft system would be close to the ground, it would be easier to mount, to access in case of repair, and could be designed with fixed pitch

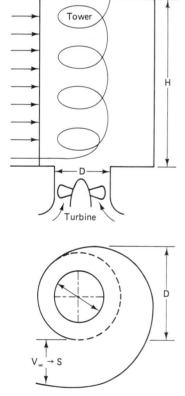

Figure 8-11 Tornado wind turbine, experimental. (From E. W. Jacobs, "Research Results for the Tornado Wind Energy System: Analysis and Conclusions," *Solar Engineering—1983*, Proceedings of the ASME Solar Energy Division Conference, Orlando, Fla., April 18-21, 1983; ASME Publication H00253. Courtesy of the Solar Research Institute, Golden, Colo.)

blades, lowering the cost still further compared with tower-mounted units of comparable size.

The German Ministry for Research and Technology has developed a unique chimneylike wind generator, consisting of a tower 33 ft (10 m) in diameter and 650 ft (200 m) in height, being tested on the La Mancha Plain south of Madrid, Spain. Surrounding the base of the tower is a "greenhouse" 800 ft (245 m) in diameter constructed of plastic sheet material supported on a steel grid. The effect of this arrangement is to heat the air underneath, allowing it to escape up the tower, thus creating a strong updraft. Housed in the base of the tower is the four-bladed turbine. Blade rotation begins when airflow reaches

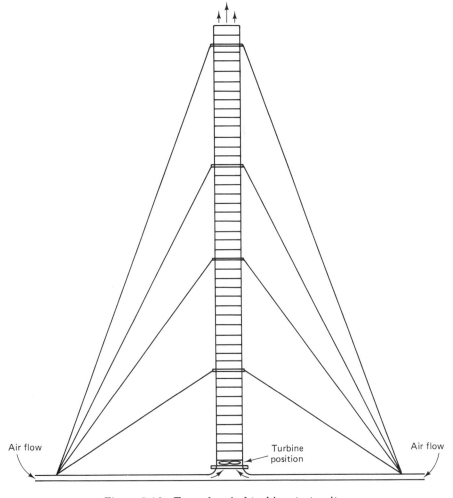

Figure 8-12   Tornado wind turbine, test unit.

13 ft/sec (4 m/sec) and is held at a constant 150 rpm to maintain the 50-hertz frequency used in the European grid.

The arid climate and flat terrain of the plain in Spain make an ideal site for testing this concept. The unit was put to the test prior to its completion when storms with 100-mph (162-km/h) gusts lashed the area. Power lines were knocked down and roofs torn off, but the tower was undamaged and the plastic glazing sustained only minor damage. Testing for at least one year is planned with the Spanish utility, Union Electrica, the recipient of electricity generated by the system.

Another novel wind generator design uses four pairs of blades, arranged in a modified fan array (Figure 8-13). The blades are of triangular cross section, with knife edges and a $12\frac{1}{2}°$ fixed pitch, but no twist. The blades are arranged in depth along the same horizontal axis, with each pair slightly offset from the pair just in front of it. Wind-tunnel tests have shown that this design can extract more than 56% of the wind's energy at low velocities of 8 to 12 mph (13 to 20 km/h). The Betz limit is theoretically 59.3%, giving this unit an efficiency of 94.4%.

The array has generated electricity in winds of 4 to 50 mph (6.5 to 80 km/h). A two-blade array creates a "disk" of air with a cone of pressure in front of the swept area of the blade. This has the effect of spilling up to 40% of the total energy of the wind. The eight-blade configuration uses some of that spillage to turn the blades behind the front pair, in effect using the airflow more efficiently. It runs at about one-half the speed of a conventional two-blade unit, extracting more energy from the airflow, and works effectively over a wider range of wind speeds, especially those of lower velocities. It has the further advantage of being constructed using existing technology, and it could be mounted in arrays on towers of simple design.

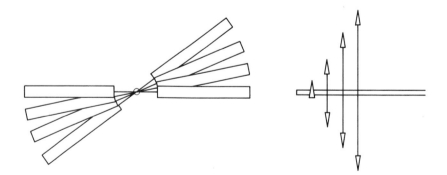

Figure 8-13 Allison eight-blade fan array. (Courtesy of William Allison, Environmental Energies, Inc., Copemish, Mich.)

The effect that makes a baseball curve or a golf ball slice or hook can make one type of wind turbine turn and produce energy. An aerodynamic force, the Magnus effect, discovered by Gustav Magnus, is created when a cylindrical object spins around an axis perpendicular to its flight path (see Figure 8-14). Research efforts by Thomas Hanson led him to apply the principle to a low-cost windmill design. This experimental design is also being used on ships and boats to reduce fuel consumption. The well-known marine biologist Jacques Cousteau wants to incorporate the principle on his experimental catamaran, Moulin à Vent.

Wind turbines usually use a complex airfoil blade shape which is expensive to manufacture. Instead of blades, Hanson uses three aluminum cylinders, each 24 ft (7.3 m) long and 45½ in. (115 cm) in diameter. Each barrel spins on its own axis as well as rotating as part of the three-barrel rotor. Each barrel is attached to a trolley-track system in the nacelle with rollers to transmit the rotary motion, then through gears to the main output shaft. The generator is designed to produce 225 hp in a 40-mph (65-km/h) wind. It is estimated that 10% of the output will be needed to keep the barrels rotating. The primary advantage is the reduced cost of the rotor. The machine is also a free-yaw design, eliminating any need for directional control. At wind speeds over 40 mph (65 km/h) the rotor speed remains constant, thus reducing the need for speed-control mechanisms.

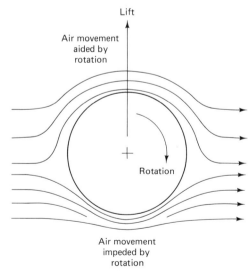

Figure 8-14 Magnus effect turbine.

INSTITUTIONAL BARRIERS TO WECS

Despite the advantages of the various types of WECS, there are serious considerations which can be classified as barriers to the use and expansion of such systems. The first of these are current building codes and related building safety regulations, which may prevent or at least restrict the installation of WECS in locations where they would be technically usable. Closely related to this are zoning regulations which prevent the installation of towers.

Federal and many state energy tax laws encourage installation of WECS. This encouragement may take the form of reduced property assessment, reduced income tax rates, or a tax credit on annual income taxes.

Electrical use in the modern residential sector has increased greatly on a per capita basis over the small demand required in the 1920s through the 1950s. The many electrically operated appliances and other equipment currently in use create a high electrical demand, requiring a much larger, more expensive WECS than before. Energy conservation efforts, high utility electric rates, and improved designs are making WECS attractive again.

"Wind rights" have not been established for WECS owners. Fast-growing trees on adjacent property may seriously reduce the effectiveness of a small WECS. Recent court decisions have prevented the construction of tall buildings which shade solar collector installations. This has opened the door to further energy-related rights.

Construction of WECS is more labor intensive than light-water nuclear reactors per Btu output: 470 person-years compared to 261 person-years for nuclear. With proper planning, WECS are compatible with agricultural and grazing uses of land and could conceivably make isolated land, which is now of little value, into "paying" real estate.

ENVIRONMENTAL IMPACTS

Improvements in the designs of earlier WECS and the development of composite materials have helped to enhance the progress of WECS as a feasible energy source. Once WECS are in place, there is no additional cost for fuel except for an insignificant amount used by synchronous generators for frequency control and field excitation. Maintenance costs are budgeted for no more than 10% of construction costs in current wind farms. By reducing the nation's dependence on imported oil, greater national security is established.

Overall pollution levels should be reduced to the extent that

WECS are added to the electricity-generating grid, since no pollutants are emitted by WECS. Wind machines are relatively small producers compared to conventional electrical-generating systems at the present time, so it is not likely that large-scale electricity outages will result if equipment fails. The construction of wind farms has demonstrated that many commercial-size WECS units can be backups to each other. Reliability of WECS equipment is increasing rapidly with the added experience of manufacturers and installers, and has been known to reach 95% reliability with some units.

Improvements in blade design have reduced noise so that the sound of a rotating WECS is no greater than the background levels associated with moderate to high winds. The noise may consist of no more than a muffled "whoosh," although earlier designs of large-scale WECS did create high decibel levels and some radio and television interference.

Single towers or large arrays of WECS create a visual impact that is offensive to some people. The potential sites for location of WECS units are often in areas that are natural flyways for migratory birds. The impact of this will not be known until more experience has been acquired. Any concerns regarding changes in wind-flow patterns or in microclimate modifications have been estimated to be very small and probably trivial compared to normal shifts in wind patterns over time.

The problem of ice buildup on blades and tower structures was once considered to be a potential problem. When icing occurs, the airfoil shape is changed and will cause the blades to lose lift and go into stall. A stationary rotor will turn until the heaviest blade is at its lowest point. Any ice shedding will be at or near the base of the tower, as will the ice dropping from the tower itself. The shedding of ice is no more a problem than that from tall buildings found in any urban area. The DOE has compiled research regarding the nature and extent of potential WECS impacts in its Environmental Factors Report. The findings are based on the experience of the DOE with its experimental large-scale WECS program.

SUMMARY

Moving air as an inexhaustible source of energy is being harnessed by improved versions of wind energy conversion systems. The development of the new and often larger systems is based on principles of design that have been used for centuries but have been improved as new materials are found to be better suited to the use of larger systems. Both small- and large-scale WECS units have become familiar as use of the systems has become more widespread. Better information regarding wind velocities, wind flow patterns, and seasonal variations has provided the basis for selection of sites for successful installations as additional data has been collected by improved monitoring equipment.

Blade designs include those made of wood, metal, synthetic materials, and combinations of materials to meet different operating conditions. A greater variety of control mechanisms for each of the proven wind systems has improved the reliability of systems and thereby increased the expected output of electricity.

In addition to the conventional two- and three-bladed propeller-like airfoils, several alternative rotor designs have been tested and found to be suitable for certain applications. Other designs are currently undergoing testing and may be found to be usable in the future.

Storage of electricity is not possible, so several storage mechanisms have been designed. These include the conventional battery units which store chemical energy and release it in the form of direct-current electricity, water pumped into reservoirs for release at peak-load times to operate hydro systems, conversion of water into hydrogen and oxygen, its two gaseous components, as well as flywheel storage, thermal storage, and the use of electricity to compress air. The most feasible means to "store" electricity from any but the smallest systems is to connect the system into the utility grid which makes each system a backup to the other.

Federal legislation provides for payment for the electricity generated on an "avoided cost" basis. Other legislation in the form of investment incentives and tax credits has aided in the development of this industry. As each WECS design is proven, the installation cost tends to be reduced, making wind an ever more popular consideration for one of the forms of energy production for individual households and farms, small businesses, and for investors in systems that supplement conventional utility companies.

ACTIVITIES

1. Obtain the average annual wind-speed readings for your community, based on the data provided by the National Weather Service.
2. Using an anemometer mounted at the appropriate height, 33 ft (10 m) and located away from buildings, trees, and other surface obstructions, record the wind speed over a period of one week or one month. Compare your findings with the data provided in Activity 1.
3. Determine the feasibility of installing a wind-powered generating system in one or more locations in your community based on the availability of wind energy.
4. Are there currently any wind-powered generating systems in use in your

community? Are they commercial or privately owned? Do they supply power to the utility grid? If so, at what rate of financial return compared with the rates the utility charges?

5. Determine the electrical needs for one household and estimate the size of a WECS needed to provide the electricity for that household. What would the cost be, including installation, grid connection, and/or storage system? (Remember to base your calculations on realistic efficiencies for the system.)
6. What restrictions are there to the construction of a wind-powered generator system in your community?
7. Invite a representative of the local utility to discuss WECS and the regulations and barriers to grid connection.

BIBLIOGRAPHY

Alternate Sources of Energy, bimonthly (various issues), Milaca, Minn.

Dennis, Landt, *A Book of Windmills and Windpower*. New York: Four Winds Press, 1976.

Eldridge, Frank R., *Wind Machines*, 2nd ed. New York: Van Nostrand Reinhold, 1980.

Foreman, K.M., "Economics of DAWT Wind Energy Systems, Grumman Aerospace Corporation." Paper presented at the DOE/SERI Fifth Biennial Wind Energy Workshop, Washington, D.C. October 5-7, 1981.

Hackleman, Michael, *The Homebuilt, Wind-Generated Electricity Handbook*. Mariposa, Calif.: Earthmind, 1975

———, *Wind and Windspinners*. Mariposa, Calif.: Earthmind, 1974.

Hunt, V. Daniel, *Windpower: A Handbook on Wind Energy Conversion Systems*. New York: Van Nostrand Reinhold, 1981.

Leckie, Jim, G. Masters, H. Whitehouse, and L. Young, *More Other Homes and Garbage*. San Francisco: Sierra Club Books, 1981.

Marier, Donald, *Wind Power for the Homeowner*. Emmaus, Pa.: Rodale Press, 1981.

Noll, Edward, *Wind/Solar Energy*. Indianapolis, Ind.: Howard W. Sams, 1975.

Park, Jack, *The Wind Power Book*. Palo Alto, Calif.: Cheshire Books, 1981.

Popular Science, monthly (various issues), Times Mirror Magazines, Inc., New York.

Renewable Energy News, monthly (various issues), CREN Publishing, Ottawa, Canada.

Simmons, Daniel M., *Wind Power*. Park Ridge N.J.: Noyes Data Corporation, 1975.

Solar Age, monthly (various issues), SolarVision, Inc., Harrisville, N.H.

Wegley, Harry L., et al., *A Siting Handbook for Small Wind Energy Conversion Systems*. Richland, Wash.: Pacific Northwest Laboratory, March 1980.

Chapter 9
Geothermal Energy
Heat from Down Under

CONCEPTS

1. Geothermal energy is inexhaustible.
2. Heat is caused by the slow decay of radioactive elements within the earth.
3. Being site specific, geothermal energy is not a national resource.
4. Geothermal energy has been used for centuries around the world for domestic heating.
5. Geothermal fluids can have negative environmental effects.
6. Geothermal energy will play an important role in the future energy picture.

GLOSSARY

HEAT OF VAPORIZATION—the quantity of heat necessary to change a liquid to a gas without increasing its temperature.
HYDROFRACTURING—injecting cold water into superheated bedrock to create cracks and pockets through which water can be circulated.
IMPERMEABLE—not permitting the passage of fluid.
MAGMA—a layer of semimolten rock below the earth's crust.
THERMAL GRADIENT—the change in temperature in relation to change in depth.

INTRODUCTION

The position of some leaders in the energy field is that the feasibility of any alternative energy source depends on the ability to capitalize on its development and use. Some energy sources being used to reduce our dependency on conventional fossil fuels are diffuse in energy intensity and/or geographical location of the source. Although geothermal energy sources are accessible only in specific geographic locations, the concentration of energy at these sites is often such that large-scale commercialization holds considerable promise. Several energy leaders are calling geothermal energy the safest, cleanest, and most economical of the energy alternatives being considered for commercialization.

BACKGROUND: GEOTHERMAL HEAT AS AN ENERGY SOURCE

Hot springs emerging from openings in the earth's surface have been sources of awe, recreation, and enjoyment for centuries. Bathing in and drinking mineral waters from some of these springs have been said to provide improved health, long life, and even magical powers.

Our planet is a vast natural reservoir of heat energy. Deep within the earth is a molten core believed to be made of iron and nickel. Estimated to be approximately 1800 miles (2900 km) below the surface and possessing temperatures of several thousands of degrees, this core, coupled with the natural radioactive decay of such elements as uranium, thorium, and potassium, provides a constant source of thermal energy which slowly migrates to the surface. Surrounding the core is the mantle, which is a layer of semimolten rock called magma. This magma layer is generally found approximately 20 miles (32 km) below the surface, although in geothermal resource areas it may be only 5 miles (8 km) underground. Above the mantle is the cool outer layer of the earth's surface termed the crust.

The earth's crust is not a uniform, evenly distributed layer of rock but actually many separate plates floating on the mantle. Where the edges of these plates meet or drift apart (folds or faults) or at isolated thin spots, hot magma may be forced through the crust to the surface in the form of volcanic eruptions. The term "lava," which we associate with volcanoes, is actually magma which has reached the earth's surface. Temperatures of lava have been recorded as high as 2000° F (1193°C).

Although references vary depending on the data base used, the average thermal gradient or increase in temperature with drilling depth is approximately 90°F per mile (30°C per kilometer) in the United States. Gradients of 74 to 196°F per mile (8 to 50°C per kilometer) are common around the world. Underground temperatures of nearly 932°F (500°C) are generally found worldwide at a depth of approximately 22 miles (35.4 km). Needless to say, drilling wells to reach this thermal energy is beyond practicality and current technology. Maximum practical drilling depths are around 13,000 ft (4 km) with currently operating geothermal electrical power plants using wells 3280 to 6560 ft (1 to 2 km) deep. As of 1983, the deepest well known projected 6.2 miles (10 km) below the earth's surface.

Some geothermal sources are easy to identify by hot surface springs, geysers, or steam rising through fissures, cracks, or rifts in the ground. Without these indicators, geological, hydrological, geochemical, and geophysical surveys must be conducted. These surveys include such activities as measuring temperature gradients, electrical resistance, and chemical analysis of soils as well as gravitational, electromagnetic, and active and passive seismic surveys of a potential area. These activities will often lead to drilling of exploratory wells in the most promising locations, since these tests are indicators, not guarantees, of geothermal energy presence.

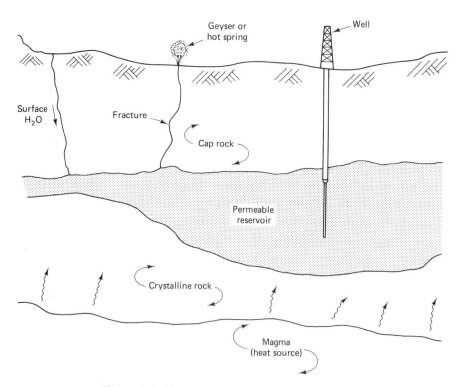

Figure 9-1 Features of a geothermal resource.

Although there are approximately 500 active volcanoes in the world today, and drilling geothermal wells near them may appear logical, the low thermal conductivity of rock makes these efforts currently impractical. When exploration is done along faults and near known or newly discovered "hot spots," thermal gradients are often found to be large enough to warrent exploitation. An example of one of the best hot spots in the United States is near Marysville, Montana, where the gradient is 114°F/mile (75°C/km). Other super-hot areas in the world may contain gradients as high as 341°F/mile (110°C/km).

GEOTHERMAL SITE LOCATIONS

Where vents and fissures exist in the earth's crust, internal pressure forces heat effects close to the surface where they can be utilized. Various hot springs have been used by people for over 2000 years. For example, the hot mud sites of Balaruc-les-Bains, France, have been popular since Roman times. In the United States, hot springs of the intermountain west were used by native Americans long before they were "discovered" by pioneers.

Although use of geothermal energy for hot baths, domestic hot water, space heating, mineral recovery, and agricultural growth stimulation has a long history, it was not until 1904 that the first electricity was generated using this energy source. At Larderello in Tuscany, Italy, where boric acid was being extracted from the boron-laden geothermal fluids, Prince Piero Ginori Conti devised a small steam-powered reciprocating engine which, coupled with a generator, produced approximately 15 kW for illuminating the boric acid factory. This unit was replaced in 1913 with a 250-kW turbo alternator establishing the first commercial-scale electricity-generating plant from geothermal energy sources.

In the United States, William Bell Elliot discovered a geothermal resource area near Calistoga, California, in 1847. Soon becoming The Big Geysers hot-springs resort, the area was a flourishing tourist attraction until the mid-1920s, when interest in the site declined. A 250-kW electricity-generating plant using the geothermally produced dry steam was erected, but it was not financially competitive with the more economical hydroelectric plants that were also being built at that time.

The electricity-generating potential at The Geysers remained unexploited until the early 1950s, when B. C. McCabe leased the area from The Geysers Development Company and established the Magma Power Company. Magma began drilling and soon joined with Thermal Power Company in developing six geothermal wells. In 1958 Magma-Thermal contracted with Pacific Gas and Electric Company (PG&E) to supply steam and the first plant began producing electricity in September 1960. As of 1983 there were 22 plants on line at The Geysers. Seventeen of these plants have been constructed by PG&E, and their projections are to complete two more in each of the years 1985, 1987, and 1988. The Geysers area of northern California is now the largest producer of electricity from geothermal energy sources in the world.

Farther south in California's Imperial Valley are two additional plants, one at Brawley (1980) and the other at the Salton Sea (1982). Southern California Edison (SCE) plans to continue development of this area, which is estimated to have sufficient potential to supply the electrical needs of the entire southwestern United States for at least two centuries. With 24 plants currently producing electricity and several others soon to be built by PG&E, SCE, and others, California appears to be a real "hot spot."

The United States Geological Survey (USGS) has identified 118 sites over 3.3 million acres in the western United States as Known Geothermal Resource Areas (KGRAs) and another 99 million acres which have potential (see Table 9-1). Of the states encompassing these areas, California and New Mexico are believed to outrank all others as having high-temperature resources, while Texas and Louisiana, along the Gulf of Mexico coast, lead as sources of hot water at high pressure. All or part of 13 western states have hot dry rocks with 550°F (288°C) average temperatures within 3.5 miles (5.6 km) of the surface.

Space and water heating for residential, commercial, and industrial buildings using geothermal fluids have been utilized for decades in individual as well as group applications. Although it is difficult to determine when and who the first settler was to use this natural re-

TABLE 9-1 Designated KGRAs in Western United States

State	Number of KGRAs	Acres	Percent of Total Acres
California	23	1,475,641	43.6
Nevada	30	635,462	18.8
Oregon	13	431,936	12.8
New Mexico	8	327,853	9.7
Idaho	9	178,018	5.3
Utah	9	124,594	3.7
Alaska	3	88,160	2.6
Montana	4	58,655	1.7
Washington	3	35,612	1.1
Colorado	4	20,825	0.6
Arizona	2	3,700	0.1
Wyoming	0	0	0.0
Hawaii	0	0	0.0
Total	108	3,380,356	100.0

Source: U.S. Geological Survey.

source for heating, the first U.S. group of users in an officially designated "heating district" was recorded in 1894 at Boise, Idaho. In the 1930s more than 400 homes were being heated geothermally, although only about 240 are using the source today. Huge reserves are available in the Bruneau–Grand View area just south of Boise.

Geothermal district heating is also in current use at Monroe, Utah, and Klamath Falls, Oregon. In Klamath Falls, there are approximately 400 wells from 88 to 1900 ft deep (27 to 580 m), providing heat for businesses, residences, swimming pools, and schools. A down-hole heat-exchange system is used to heat municipal water, which then provides space heating for most of the eastern portion of the city, including the Oregon Institute of Technology. In a down-hole system a heat exchanger is located in the geothermal well and water or other fluid is circulated through a closed loop and heated by the geothermal formation.

In addition to the aforementioned states, nearly all of the western states are currently involved in exploratory efforts toward geothermal energy development. Hawaii currently has a well 6500 ft (1.98 km) deep which has the estimated capacity to provide electricity for the entire state. Although most sources are not expected to produce the bonanza found at The Geysers, many are projected to reduce demand on fossil fuel utilization significantly, for both electrical generation and thermal energy applications. Figure 9-2 illustrates the location of geothermal energy sources in the United States.

Figure 9-2 U.S. Geothermal sources. (Courtesy of National Science Teachers Association.)

Worldwide development of geothermal energy resources is increasing as fossil fuels become more expensive. Even though the United States is the leader in electrical energy output, Italy boasts the greatest number of geothermally activated electrical plants. Table 9-2 shows worldwide geothermal energy use for generating electrical power.

New Zealand, Iceland, and the USSR have also been world leaders in the use of geothermal energy. Since 1958, New Zealand has placed 14 electrical generating plants on line and is building others. Considerable use is also being made of the thermal resources for heating schools, hospitals, and residential structures. Iceland's resources have been used for heating since 1928. The capital city of Reykjavik, with over 11,000 commercial and industrial buildings and 88,000 homes, is 99% heated by geothermal energy sources. Residential users pay less than $200 per year for all their space heating and hot water needs supplied by the insulated pipe distribution system. In the USSR, primary use is for space heating and agricultural applications. Geothermal sources provide 60% of the heat for Makhachkala, a city with a population of 250,000.

TABLE 9-2 Worldwide Geothermal Capacity (MW) in 1981 and Plans for the Year 2000

Country	1981	2000
United States	932	5,824
Philippines	446	1,225+
Italy	440	800
New Zealand	203	382+
Mexico	180	4,000
Japan	168	3,668+
El Salvador	95	535
Iceland	41	68+
Kenya	15	30+
USSR	11	310+
Azores	3	3
Indonesia	2	92
China	2	2
Turkey	0.5	150
Costa Rica	0	380+
Nicaragua	0	100
Ethiopia	0	50
Chile	0	15+
France	0	15+
Total	2,538.5	17,649+

Sources: R. DiPippo, "Geothermal Power Plants: Worldwide Survey," U.S. Department of Energy (Washington, D.C.: Government Printing Office, 1981), and United Nations Conference on New and Renewable Sources of Energy, "Report of the Technical Panel on Geothermal Energy.

Active worldwide use of geothermal energy as a low-level heat source and as a high-level energy source for conversion to electricity will undoubtedly continue. Table 9-3 lists countries that are currently developing this inexhaustible energy source.

GEOTHERMAL FEATURES

There are three types of geothermal resources: hydrothermal, hot dry rock, and geopressurized. The majority of these resources lie far below the water table, but surface or near-surface thermal areas are common where folds or faults allow the internal heat to reach the water table. Subsurface water systems are also frequently involved.

The most prevalent type is hydrothermal, and as the name implies, it is water–temperature related. At lower temperatures, hydrothermal sources are termed liquid dominated because they contain mostly hot fluids with very little steam. Where temperatures are higher, proportionally more of the liquid in the system becomes steam and the source is known as vapor dominated. There are estimated to be 20 times as many liquid-dominated as vapor-dominated sites. At wells where temperatures have been recorded as high as 570°F (299°C), superheated water vapor flashes to highly pressurized steam when brought to the earth's surface. These are termed dry steam sites and are the least common and the most desirable for commercial development. Other than in National Parks such as Yellowstone, the only one known to exist in the United States is at The Geysers in northern California. Hydrothermal sources are generally found less than 11,000 ft (3048 m) below the surface under impermeable rock layers.

Hot dry rock formations are not associated with subsurface water. To extract the available heat, cool water is circulated through the hot rock formation, where it becomes hot water or steam. The thermal energy is then brought to the surface to do work. Where no cracks or fissures exist in the hot dry rock, hydrofracturing is necessary to allow the injected water to contact the rock and become heated.

TABLE 9-3 Countries Actively Developing Geothermal Energy Resources

Chile	Indonesia	Panama
China	Italy	Philippines
Costa Rica	Japan	Portugal
El Salvador	Kenya	Turkey
Guatemala	Mexico	USSR
Honduras	New Zealand	United States
Iceland	Nicaragua	

The process of hydrofracturing involves pumping and pulsating cold water down a well at high pressure to create cracks and pockets in the hot rock formation. A second well is required so that water injected into one well will circulate through the hot fractured zone and return, heated, to the surface through the other. Once the heat energy is utilized the water can be recycled, thereby reducing environmental concerns.

The first experimental plant using hot dry rock is currently under development in New Mexico and, if cost-effective, could provide initiative for commercialization. This plant will also provide opportunity for additional investigation into the technology necessary to acquire energy from hot dry rock so that thermal sources associated with volcanoes can be tapped.

Geopressurized formations hold great potential for development along the Louisiana and Texas Gulf coast. Test wells were drilled in 1979 and commercialization is expected in the late 1980s or early 1990s at sites both on- and offshore. The intense interest in this geopressurized area is due to its extensive size and its offer of three energy forms instead of one. In addition to the 350°F (177°C) water, geological pressures of approximately 15,000 lb/in^2 (1055 kg/cm^2) and dissolved methane are available. The current concern is to design facilities that can utilize the thermal, mechanical (pressure), and chemical energy forms safely and efficiently.

Regardless of which type of geothermal resource is being considered, there are several characteristics which are required to make energy extraction feasible:

1. The source temperature must be adequate for the use intended.
2. A reservoir of sufficient size must be available.
3. Water must be present or able to be injected, removed, and recycled as a heat-exchange medium.
4. A layer of cap rock must be present which can seal or contain geo or thermal pressurized fluids in the permeable reservoir.

GEOTHERMAL ENERGY CONVERSION TECHNOLOGY

As with any product, successful commercialization of an energy form requires not only sales but appropriate delivery to the consumer. The problem with thermal energy is that it transfers its heat rapidly, via conduction, convection, and radiation, to a cooler environment. In some instances where liquid-dominated hydrothermal sources are located near urban areas, fluids directly from the well can be pumped through large insulated pipes to the users for industrial process heat, crop drying, greenhouses, aquaculture, air conditioning, water heating, or space heating. Most geothermal fluids, however, contain high levels of salt and and/or other corrosive minerals that prohibit this simple technique. Heat exchangers, located directly in or near the well, are most often used to heat potable water or other fluids which are transported through insulated pipes to customers who remove the heat via their own heat exchangers. The liquid is then returned for reheating at the well exchanger.

Unfortunately, many geothermal sites are not located in close proximity to large urban centers. If the direct heating method were attempted in such situations, the heat energy in the fluid would be dissipated or reduced to unusable levels. One method of solving this problem is to build plants for energy-intensive industries, such as aluminum and chemical manufacturing, near the geothermal sources. Where this is impractical the only other alternative is to convert the thermal energy to another form which can easily be transported while not significantly losing its value. Examples of this would include the electrolysis of water to form hydrogen (chemical energy) or the conversion of heat energy to electrical energy. The hydrogen would be piped or transported in tank vehicles, while by using transformers, the electrical energy can be transported to users over a large geographic area. Even though many consumers will convert the electrical energy back to heat energy with a resulting loss of efficiency, the "free" fuel used to turn the turbine and the vastly greater number of consumers makes the conversion of geothermal energy to electrical energy commercially attractive. In the mid-1980s there are not enough users of hydrogen to warrant massive efforts of its production at the wellhead, but there is great potential for the future.

All three types of geothermal resources can be used to supply the minimum 300°F (149°C) heat energy necessary for conversion to mechanical energy via a steam turbine which is connected to an electrical generator. All three of the hydrothermal forms are being used commercially. Liquid-dominated sources are the most numerous, but three-fourths of all geothermally produced electricity in the world is produced from dry steam fields. Electricity produced through use of the hot dry rock and geopressurized types is expected to be available by 1990.

Well-drilling technology is not unlike that which has been used for decades in efforts to exploit petroleum and natural gas resources. Bits with water or fluid injection systems designed especially for high-temperature drilling and electrical discharge heads are being used for fast, efficient establishment of exploratory and production wells. Not all drilling efforts are successful, and well costs varied between $900,000 and $2,600,000, in 1983, depending on depth and/or site characteristics.

The components for geothermally driven electricity-generating plants are readily available commercially and the technology is well established. Production wells providing dry steam at high temperature and pressure are the easiest to exploit, since the steam, stripped of any liquids, can be used to turn turbine blades directly (direct-steam system) (see Figure 9-3). Used, partially condensed, steam can then be fully condensed and returned to the well, or vented into the atmosphere. Where condensing turbines are employed instead of standard, low-pressure units, nearly twice as much electrical energy can be converted from a given volume of steam—greatly increasing production efficiency. Dry steam plants have a resource utilization efficiency between 50 and 60%.

Figure 9-4 shows PG&E's hot dry steam unit, which began operation in February 1983.

Electrical production from liquid-dominated sources is technically more complex and consequently more expensive. If the source is hot enough, flash steam or separated steam can be used, as in Cerro Prieto,

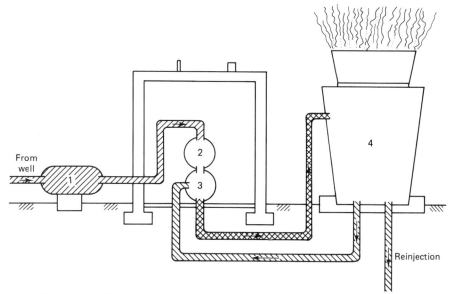

Figure 9-3 Schematic of a dry steam geothermal electricity generating plant. (1) Geothermal fluid from the well enters centrifugal separators to remove rock particles, thereby preventing damage to turbine. (2) Steam rotates turbine connected to electrical generator. (3) Steam is condensed in a direct-contact condenser. (4) Condensed steam is further cooled in the cooling tower and may be recycled to the condenser for cooling incoming steam or reinjected into wells to replenish groundwater levels.

Figure 9-4 Pacific Gas and Electric Company's Unit 17. (Courtesy of Pacific Gas and Electric Company.)

Mexico. Pressurized hot water rising from the well undergoes pressure reduction and "flashes" into steam, which is used to operate a turbine.

A variation of this process is to keep the hot water pressurized. Instead of being allowed to flash into steam, it is circulated through a heat exchanger, where it vaporizes a secondary fluid such as isobutane or Freon, which then operates the turbine. This is termed a binary system and has an advantage over the flash-steam system in that lower geothermal fluid temperatures can successfully be utilized because the binary fluid has a lower boiling point and lower heat of vaporization than the well fluid. Additional advantages are that the binary system is smaller, less expensive, and has higher turbine efficiency. Figure 9-5 schematically illustrates typical geothermal plants.

The total flow system uses the full flow of liquid, vapor, and dissolved solids and gases from the well. Energy is extracted when the material rapidly expands to the surrounding atmosphere through a single-stage axial-flow impulse-type condensing turbine.

A number of hybrid systems using both geothermal energy and fossil fuels have been advanced. The preheat system, proposed for Burbank, California, uses geothermal energy to preheat water which must be turned into steam before being directed through the turbine in a conventional electricity-generating plant. Fossil fuel is used to "top off" the preheated water. It is estimated that a 42% fuel conversion efficiency can be reached—a 7% increase over a regular oil-fired plant. Another hybrid system, called fossil-superheat, uses fossil fuel to superheat geothermal steam for turning the turbine, and a third,

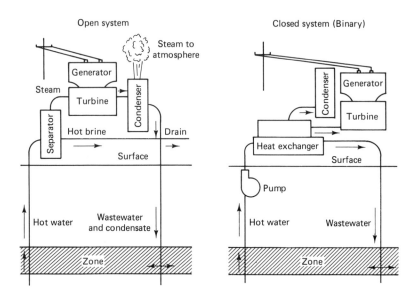

Figure 9-5 Schematic illustrations of typical geothermal electricity-generating plants. (From the U.S. Council on Environmental Quality.)

compound hybrid is a "topping" system for providing both hot water and steam.

Development and construction of plants to convert geothermal energy to electrical energy are progressing slowly, with the growth rate of installed capacity over the last five years being nearly 12%. The first 1000-MW plant constructed at The Geysers cost PG&E approximately $200,000,000, which is $4,000,000,000 less than the estimated cost of a new nuclear plant of the same capacity.

TECHNICAL PROBLEMS

A major problem with using geothermal energy, from the technical standpoint, is the large quantity of dissolved gases, minerals, and sediments contained in the fluids. These contaminants make the fluids highly corrosive to piping, turbine blades, heat exchangers, and other parts of the system, as well as being potentially explosive where gases such as methane are present. Table 9-4 lists several of the contaminates found in geothermal fluids.

Geothermal plants can be noisy and smelly, especially when the design of the energy conversion system requires continuous venting to the atmosphere of low-pressure steam containing H_2S. This creates problems of protecting all personnel on the site from hearing damage

TABLE 9-4 Materials Found in Geothermal Fluids

Materials/Compounds	Gases
Arsenic	Ammonia (NH_3)
Boron	Argon (Ar)
Calcium compounds	Carbon dioxide (CO_2)
Heavy metals	Carbon monoxide (CO)
Lithium salts	Ethane (C_2H_4)
Salt	Hydrogen (H_2)
Silica	Hydrogen sulfide (H_2S)
Sulfur	Methane (CH_4)
	Nitrogen (N_2)
	Oxygen (O_2)
	Radon (Rn)
	Others

caused by the continuous roar of the steam release as well as respiratory damage from noxious gases such as hydrogen sulfide.

Geothermal resources are frequently located in arid climates where the availability of water is a concern. Some plants process 400,000 gal (1,514,000 liters) of mineral-laden water per day and the place of discharge and opportunity for replenishment is vital. Flash-steam plants require additional water to make up for the evaporation from the cooling towers, while the use of condensing turbines at dry steam sites results in an excess of water at the site. There is also a need to determine maximum rates of extraction so that groundwater levels can be preserved while the hottest possible temperatures are maintained.

Sites of high-level geothermal energy are few in number and accessibility is often difficult. Where conversion to electrical energy is necessary, the plant must be located as close as possible to the production wells so that the thermal energy is not dissipated. One mile (1.6 km) is considered maximum between wells and the generating facility.

Additional problems related to geothermal energy use are primarily financially oriented. Drilling, plant, and equipment costs must be such as to make the output competitive with conventional fossil-fueled energy systems.

ENVIRONMENTAL ISSUES

A number of issues have been raised in both theory and practice which are concerns for the environment adjacent to geothermal energy facilities. Unlike fossil-fueled electricity-generating plants, which emit combustion by-products, most of the potential environmental impacts of geothermal energy are either relatively minor, are short lived, or affect only the immediate site of the facility. In comparison geothermal energy is clean, but each of the systems, whether hydrothermal, hot dry rock, or geopressurized, has its own associated problems. The hot dry rock system has fewer associated environmental problems because it is a closed system.

As might be assumed, the development of a site for the generation of electricity is going to change the land use, alter the view, disrupt the normal residential routines, and/or affect adjacent agricultural efforts. Noise will be inevitable while drilling, testing, and venting the wells. Constructing roads, piping systems, and the plant itself will also be disruptive. Mufflers or silencers will be required for units that vent pressurized steam to the atmosphere. Probably the most unfavorable impacts during operation of the plant will be the associated odors and potential for water pollution and water table alteration.

In a majority of cases, use of the hydrothermal vapor-dominated type of system results in water, air, hydrogen sulfide (H_2S), and a multitude of other gases. The H_2S is dissolved in the water and can be treated with soluble iron compounds which change it to water and free sulfur. There is a new system available (Stretford process) which is 99% effective in removing the gaseous H_2S.

The liquid-dominated system liberates carbon dioxide (CO_2), H_2S, and water vapor when the geothermal fluid is flashed. The H_2S, which has an objectionable odor, noxious properties, and is corrosive, can be treated as in the vapor-dominated system. Additionally, minerals such as arsenic, boron, and mercury potentially contribute to pollution. When valuable mineral content is at an adequate level, reclamation processes are used and the mineral by-products attained.

Removal of large quantities of mineral-laden fluids from geothermal wells potentially have several repercussions. Continuous removal could cause the fluid level to be reduced, ultimately affecting the area groundwater level and perhaps the output of the well. Fluids and solids in the earth provide support for weight above. A plant at Wairakei, New Zealand, has settled 25 ft (7.6 m) and is descending at a rate of 1.3 ft (0.4 m) per year [Meyers]. Additionally, mineral-laden fluids deposited on the surface have very high potential for groundwater and air pollution. A plant in El Salvador, after cooling and settling the fluids, discharges the effluent into the ocean. Hydrogen sulfide and other noncondensable gases are vented into the atmosphere.

Returning the geothermal fluids to the well after the heat energy has been extracted is the best way to solve or eliminate most of the associated problems. Reinjection of the effluent has to be done, however, under pressure into a separate well hole at a depth differing from that of the supply well. This procedure eliminates cooling of the in-coming fluids, helps to maintain well fluid levels, reduces the potential for groundwater contamination, reduces the possibility of "sinking areas," and reduces air pollution, but related costs are high.

Seismic activity (earthquake) is a concern especially in hot dry rock regions. The potential for inducing such activity is present in all geothermal areas due to drilling operations which are conducted along fault lines. Since the hydrofracturing method of acquiring the thermal energy from superheated rock could create movements along the fault as the cold water hits the rock and flashes to pressurized steam, it further increases the environmental concern in these formations.

Table 9-5 lists these and additional environmental issues associated with geothermal exploitation. The primary concerns exist at all sites. Whether a concern is primary or secondary is dependent on the size, location, and type of source, the structure of the geologic area as well as the technology used, and the operator's regard for the environment. A positive attribute of geothermal energy is that there appear to be no environmental hazards that cannot be handled technologically.

TABLE 9-5 Environmental Issues

Primary concerns
  Noise
  Odors
  Land-use alteration: recreational, residential, agricultural, aesthetic
  Groundwater changes: level, quality
  Hydrogen sulfide (H_2S) emissions: noxious, corrosive
  Economic impact: public, private
  Radiation exposure of employees

Secondary concerns
  Landslides
  Soil erosion
  Accidental spills
  Pipeline ruptures
  Well blowout
  Population impact
  Settling (sinking) of the site area
  Induced seismic activity: earthquake
  Particulate emissions
  Weather changes due to cooling tower drift
  Endangered species
  Alteration of area sites: hot springs, geysers

THE GEOTHERMAL FUTURE

Use of geothermal energy will continue to grow at sites throughout the world where access to the resource is relatively easy and large cities or other energy consumers are in close proximity. By 1990 the state of

California alone is projected to multiply fivefold its 1984 geothermally supplied electricity-generating capacity. In the nation, 3 to 5% of our total energy needs may be supplied by geothermal energy.

Small-scale use of liquid-dominated systems will continue to increase as the price of fossil fuels continues to escalate. Residential applications for space and domestic water heating, industrial and commercial use for process and space heating, and municipality use in government buildings and schools will continue to increase.

Without federal incentives, however, large-scale commercialization will be slow unless there is a dramatic long-term up-swing in petroleum and coal prices. Utility companies are in business to make a profit, but without financial support and encouragement, plants to convert the "free" thermal energy to electrical energy will not be constructed in great numbers. Emphasis will continue to be placed on the use of energy forms which are receiving tax benefits, price supports, other economic subsidies, and political support. Developers are finding that it has become more economical to build a geothermal facility than to erect a conventionally fueled steam-generating plant or prepare a new hydroelectric system for operation.

Recent developments in the design of low-temperature turbogenerators are providing the opportunity to use geothermal resources which were previously not applicable for producing electricity. Temperatures below 300°F (150°C) were considered impractical prior to 1983. There are three of these newly designed turbogenerators in Lakeview, Oregon, using 219°F (104°C) sources to produce 800 kW of electricity. It is projected that even lower temperatures could be used with this new system if reduced cooling-water temperatures can be acquired. This installation could provide encouragement for many other developers in the very near future.

The nation's first commercial geothermal hot water plant is expected to be completed in 1985. Located at Heber in California's Imperial Valley, the $125 million, 47-MW plant will supply enough electricity to serve 45,000 area residents.

A new binary process was developed in 1983 which dramatically improves heat extraction. Composed of an efficient double-loop closed heat exchanger using brine and isobutane, this process could make available for exploitation approximately half of the nation's geothermal resource now uneconomical to develop because of its moderate temperature.

Logic, if applied to a national energy policy, would encourage the development of geothermal energy resources where they are available. Utilization of this energy form for local needs would therefore reduce demand for conventional energy forms (e.g., natural gas, petroleum, and coal). These forms of energy, which are transportable, would then be available in larger quantities for use in areas of the nation where alternative, renewable energy forms are inadequate or not available. An estimated 20% of U.S. electrical energy needs could be acquired from geothermal sources with competitive prices and expected technological advances.

Developing countries that have geothermal assets will be exploiting them at a faster rate than will the industrialized nations, especially if they have exportable minerals and energy forms such as petroleum and natural gas. Industrialized nations will assist in the exploitation by sharing technological developments and knowledge in exchange for trade. It makes little sense for an emerging nation to place its energy development efforts on fossil fuels and subject itself to the same energy predicament found in leading industrial nations. By using the renewable and inexhaustable energy forms for their own domestic needs, these countries can then export their fossil assets to benefit themselves economically.

SUMMARY

The inexhaustible thermal energy form found beneath the earth's surface has been used for thousands of years and holds significant potential for future utilization. Since ease of acquisition is essential and transportation for long distances is not feasible, geothermal energy use is restricted to the immediate area or the general region if converted to electricity.

Geothermal formations are classified as hydrothermal, hot dry rock, and geopressurized. Hydrothermal systems are prevalent, but hot dry rock is considered the most valuable form. All formations and technological methods of exploitation hold potential environmental degradation hazards, but they are not as significant as the hazards associated with most fossil-based commercial energy forms.

Significant development of geothermal energy is expected in developing countries and some areas of industrialized nations as the availability of conventional fuels decreases and costs increase. The Geysers in Northern California is the world's largest site for the generation of electricity from geothermal energy.

ACTIVITIES

1. Review the literature on the famous geysers of Yellowstone National Park. Explain the theories of why they erupt, eruption intervals, and temperatures contained.

2. Design and construct a model that will demonstrate how and why a geyser works.
3. Explain how a pressure cooker functions. Describe the interactions of time, temperature, and pressure to affect energy utilization.
4. Study the geologic potential of your area to provide geothermal energy. Identify possible applications in terms of their impact on the economy, environment, and employment.

BIBLIOGRAPHY

Boslough, John, "Hot Rocks," *Technology Illustrated*, December–January 1983.

Di Pippo, R., *Geothermal Energy as a Source of Electricity*. U.S. Department of Energy. Washington, D.C.: U.S. Government Printing Office, 1980.

———, *Geothermal Power Plants: Worldwide Survey*. U.S. Department of Energy. Washington, D.C.: U.S. Government Printing Office, 1981.

Greenfield, Aryeh, "Breakthrough in Low Grade Geothermal," *Renewable Energy News*, May 1983.

Lengquist, R., and F. Hirschfeld, "Geothermal Power: The Sleeper in the Energy Race," *Mechanical Energy*, Vol. 98, January 1976.

Meador, Roy, *Future Energy Alternatives*. Ann Arbor, Mich.: Ann Arbor Science, 1978.

Meyers, Robert A., ed., *Handbook of Energy Technology and Economics*. New York: Wiley, 1983.

Pryde, Philip R., *Nonconventional Energy Resources*. New York: Wiley, 1983.

Schwaller, Anthony E., *Energy Technology: Sources of Power*. Worcester, Mass.: Davis, 1980.

Sheahan, Richard T., *Alternative Energy Sources*. Rockville, Md.: Aspen Systems Corp., 1981.

Turner, Wayne C., ed., *Energy Handbook*. New York: Wiley, 1982.

Chapter 10
Energy from Waste and Biomass
Trash to Energy

CONCEPTS

1. The accumulation of "waste" materials is often offensive to human beings and is frequently harmful to the environment.
2. Most materials contain components that have the potential to be converted into usable energy forms or used to reduce the energy required in converting virgin materials into finished products.
3. The process of chemically converting or extracting energy from waste materials or from biomass may be more costly than burning or burying the materials.
4. The process of converting materials to more easily used energy forms may result in a net energy loss.

GLOSSARY

ANAEROBIC DIGESTION—a bioconversion process that controls the decay of organic material in the absence of oxygen for the purpose of producing methane.

BIOMASS—organic materials that capture and store solar energy through the process of photosynthesis.

CATALYST—a chemical or other material that aids in a chemical reaction without becoming a part of it.

ETHANOL (ETHYL ALCOHOL; CH_3CH_2OH)—an alcohol derived from organic synthesis.

DIRECT COMBUSTION—the simplest and most used biomass conversion process, burning to produce heat.

GASIFICATION—the conversion of solid biomass materials to gaseous form in limited amounts of air or oxygen to produce synthetic natural gas (SNG), methanol, ammonia, hydrogen, carbon monoxide, or synthetic gasoline.

METHANOL (METHYL ALCOHOL; CH_3OH)—an alcohol used chiefly as a solvent and in organic synthesis.

pH—a symbol for the hydrogen ion concentration in a solution, thus an indication of its acidity or alkalinity. For a neutral solution, the pH is 7.0; an acid solution has a pH lower than 7, and an alkaline solution has a pH higher than 7.

PYROLYSIS—the process used to break down biomass with heat in the absence of oxygen and at a lower temperature than that required for gasification. This process yields oil, charcoal, and synthetic gas.

SLURRY—a mixture of animal wastes and water.

INTRODUCTION

When we use materials to produce goods, some waste is inevitable. Some of the waste can properly be called by-products, but much of the material "left over" after processing can only be called "waste material," insofar as no profitable use for it has yet been found. What can be done with the mounting piles of waste materials? We too often flush, bury, or burn what we no longer can tolerate, but disposal practices are often wasteful of resources; they create pollution and use personnel, transportation, or water resources that could be better used in other ways.

Waste sources are variable in both type and availability. The largest amounts of waste materials are generated from farming and crop production totaling about 390 million tons (358,800 million kg) annually of which only a small portion, about 12.3 million tons (11,160 million kg), is readily collectible. Next in potential for waste utilization is animal manure, with about 200 million tons (181,440 million kg) produced each year and about 26 million tons (23,600 million kg) readily available for processing into methane and compost. Urban refuse is produced at the rate of at least 220 million tons (200,000 million kg) per year—about 1 ton per person per year (900 kg per person per year), with at least 55 million tons (50,000 million kg) available for treatment by various technically suitable means. Logging, industrial wastes, municipal sewage solids, and other sources provide a total of about 200 million tons (181,440 million kg) annually, of which less than 20 million tons (18,144 million kg) is available for treatment. Logging waste can be used economically only within a 20- to 25-mile (32- to 40-km) radius of the source, since transportation is often difficult and costly.

Some industrial wastes, including paper, aluminum, glass, and scrap metals, already have a market for conversion. The enormous quantities of some industrial and commercial wastes have created the need for industrial waste exchange networks. Such networks perform a valuable clearinghouse function in identifying sources of waste materials and in locating potential customers for them. The networks can also serve to make others aware of the need for using rather than disposing of certain waste materials. They then serve a dual purpose in helping to resolve the problem of waste disposal and in locating potential sources of materials to be reclaimed. A third purpose is also served, that of reducing the volume of materials that must be placed in landfill sites or lagoons, or otherwise treated at great expense in order to prevent contamination of the environment. After all other recycling efforts have been exhausted, many industrial wastes can be converted to thermal energy. These steps allow us to realize the greatest amount of potential use from materials before they are finally discarded.

Other sources of biomass include waste and residues left after food-processing operations. This avenue would seem to offer the most promising source of underdeveloped potential biomass energy. The material is already concentrated at the processing site, and it creates a disposal problem compounded by the potential for pollution of water and land. One major problem that prevents immediate conversion of this type of material to usable energy is the lack of suitable small- and medium-scale conversion systems. Large-scale systems, which function well when the input is consistent in type, moisture content, and particle size, have been found to be cost-effective and result in a net energy gain. Units for small-scale operations have not resulted in positive returns.

THERMAL CONVERSION OF WOOD

Increasing prices for fossil fuels have created a growing market for wood as a substitute fuel in many parts of our nation. Wood lost its popularity in the United States first to coal, then to petroleum fuels, then to natural gas and electrical-resistance heating. Each of these sources presented advantages in addition to cost for either the homeowner or the commercial user. The first advantage was one of convenience and ease of handling. The increase in particulate emissions with coal and high-sulfur content oil was not considered a serious environmental hazard until recently. In addition, fossil fuels were handled with automatic equipment to a greater degree than was wood, a real advantage when labor costs began to increase.

Wood is still the primary thermal energy source for the nonindustrialized nations. The recent "rediscovery" of wood as a fuel in industrial countries has generated research and development efforts for more efficient stoves for homes and burner units for commercial plants. Much progress has been made to increase performance, to the point where efficiency levels of 50% on a sustained basis are common. This percentage refers to that portion of the energy in the wood that is released during combustion and is actually delivered to the room or material to be heated. This compares to fuel oil at 65% and natural gas at 75% conversion efficiencies.

Wood use, whether in homes or as a commercial fuel, seems to add a sense of independence to the user. As a renewable resource, wood burning has led to an increased emphasis on forest management practices, resulting in improved woodland quality. It has also increased the price of wood and woodlots. Wood is bulky in relation to its heat (Btu) content, and a dry place for storage is frequently a problem and an increased expense. Proper seasoning of wood (to 20% moisture content) is a necessity for maximum return in heat value. Proper drying requires from 6 months to 2 years, depending on the species of wood and the drying conditions. Wood debris, insects, and fungi are common problems for both home and commercial users.

Wood is measured in units called cords. A standard or full cord is defined as a stack that is 4 ft high, 8 ft wide, and 4 ft deep (approximately 1.2 m × 2.4 m × 1.2 m) [Figure 10-1(a)]. A more common home wood measure is the face or short cord; it is 4 ft by 8 ft on its face, with split wood or standard lengths from 12 to 24 in. (30 to 60 cm), the most common length being 16 to 18 in. (40 to 46 cm) [Figure 10-1(b)]. Care must be taken by persons purchasing wood because of the wide range of sizes called a face cord.

Wood heating units for home use are of several types. The most common type in use in the past was the fireplace, and it is still popular in some modern homes. The open fireplace may be aesthetically pleasing, but it is the most inefficient way to heat with wood. The heat produced is transferred to the room primarily by radiation, but most is lost up the chimney because there is no way to control the supply of air for combustion. The efficiency of fireplaces has been shown to be in the range of 10% on down to as low as −10%, indicating that a fireplace chimney may draw more warmed air out of a room than it supplies. A fireplace insert with glass doors which control combustion airflow increases the unit's efficiency, but the primary heat transfer is still by radiation. The addition of convective air circulation loops within the insert will increase performance up to 35%.

Thermal Conversion Of Wood

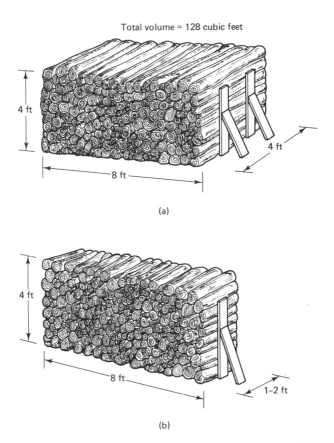

Figure 10-1 (a) Full (standard) cord; (b) Face (or short) cord. (Courtesy of Northeast Regional Agricultural Engineering Service.)

The air-circulating stove, which has a double wall containing controllable air vents, allows air from the room to circulate between the walls of the stove and return to the room. The advantages are better transfer of heat to the air in the room and lower surface temperature of the stove, reducing the danger of burns. As with the fireplace inserts, these units may operate with natural convection or forced (fan-driven) airflow.

Another recent development is the airtight stove, which is tightly sealed at all joints and has a small draft opening to control the amount of air entering the firebox (Figure 10-2). Some stoves need tending only once or twice a day, a major convenience to the homeowner. A few manufacturers add a commercial ceramic or natural soapstone outer layer to such stoves, increasing the mass, which provides an even release of heat.

Some manufacturers of airtight wood stoves have been able to increase efficiency by adding a secondary air supply. These units supply fresh air (oxygen) after primary combustion has occurred. The introduction of air permits the volatile gases driven off during initial combustion to be burned rather than exhausted up the chimney.

Another technique to increase the efficiency of wood-burning stoves is the addition of a catalytic combuster. This is a ceramic honeycomb unit that creates secondary combustion by reducing the combustion temperature of the volatiles in the smoke, allowing them to burn. In addition to increasing the efficiency of the stove to as much as 80%, such units reduce smoke and therefore air pollution and creosote buildup.

Safe installation of wood-burning devices is a prime requisite. The manufacturer's instructions should be carefully followed. Chimneys should be cleaned at least once a year and flue pipes checked for creosote deposits and thin spots several times during the heating season. Flammable materials (wall coverings, furniture, rugs, and drapes) should be a minimum of 3 ft (1 m) away from the stove unless proper heat shields are installed, and then the distance should not be less than 18 in. (45 cm).

A study of forest reserves in New England has been made by the U.S. Department of Agriculture's Natural Resource Economics Division. The results indicated that many homeowners in New England rely on wood as a primary heating fuel, with as much as 30% of the total

Figure 10-2 Defiant parlor stove. (Courtesy of Vermont Castings, Inc.)

residential heat being provided in this manner. Approximately 70% of Vermont residents heat their homes with wood. The study also concluded that woodlots are poorly managed in terms of the long-term utilization of forestland. Lumber-grade trees are often burned as fuel because market prices favor cord wood rather than lumber. At the current level of use, there does not seem to be a major concern about the ability of the forest to renew itself. The use of wood as a fuel displaces nearly $300 million worth of petroleum and almost $125 million of electrical heating per year.

Wood varies greatly in its Btu content, the major factor being the type of wood, hardwood or softwood. Species of hardwoods such as hickory will yield as much as 40 million Btu per cord, whereas aspen and balsam fir yield less than 13 million Btu per cord. The Btu value of wood is of importance to anyone who purchases wood for fuel as well as for those who have access to low- or no-cost wood from their own woodlots or from "waste" materials.

The moisture content of wood is an important factor since unseasoned wood requires heat to drive off the water that remains in the "green" wood. Newly cut wood should be split and stacked to allow air circulation around each piece. Exposing the interior portions of the wood to the atmosphere allows the moisture content to be reduced to a level of about 20%. (See Appendix C for the Btu values of various wood species.)

Both the increasing costs of petroleum-based fuels and the impact of environmental legislation have made wood-fired units attractive to industry. When wood is brought to processing plants, it is cut or chopped into smaller pieces, either for lumber or as components of plywood or pulp for paper. Some waste is inevitable. Much of the waste is in the form of chips, bark, and sawdust, formerly piled and forgotten or discarded in a manner that was not environmentally sound.

Although several large lumber producers have been using their wood waste for many years, smaller producers could not afford to do this with previously existing equipment. The recent development of dependable small-scale wood-fired boilers, wood storage, and automatic conveyer systems makes the conversion of wood waste much more desirable and more economically feasible. Equipment similar to that used to pelletize animal feed has been adapted to convert wood and plant waste to fuel. The waste is either extruded or compressed to form uniform pellets, which are easy to store, transport, and use in existing boiler equipment. The development of pelletized wood has aided in the design of feed mechanisms for automated combustion units (see Figure 10-3). Bagasse, the residue from the processing of sugarcane, is especially adaptable to the pelletizing process. Its conversion to a fuel material reduces a major waste problem and provides

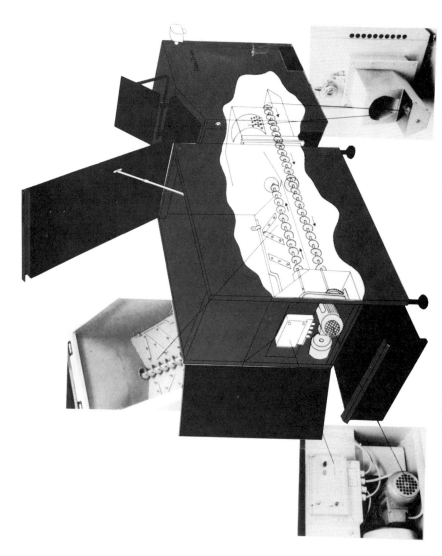

Figure 10-3 Stoker for pelletized fuel. (Courtesy of Passat, Inc., Lennoxville, Quebec.)

energy for on-site use. Three tons of pellets can provide as much energy as 1 ton of coal.

Chip wood furnaces seem to hold promise as an alternative to fossil fuels for some industries. Furnaces designed to use this fuel form must have a closely controlled rate of fuel injection and a means of storing the excess heat. Chip wood is "processed" from sticks, branches, and other relatively unusable wood but has the advantage of having a large surface-to-volume ratio which can be acted on by the fire in the combustion zone, making the unit more efficient and more pollution free.

FORESTRY MANAGEMENT AND ENVIRONMENTAL IMPACT

In spite of some serious mismanagement and overcutting of woodlots and forests, wood is the fuel of choice (or necessity) for many people. While fossil fuels are "renewable" on a cycle of about 1 million years, wood is renewed in less than a generation. Trees are the most efficient solar storage batteries known. Trees grow with great variety due to natural selection, covering vast areas on all the habitable continents. Over half of the wood produced as a result of photosynthesis is burned within a few miles of where it grew. As a rule of thumb, about one full cord of wood can be collected from a properly managed acre of land each year without depleting the woodlot.

Land-based energy farms have been developed only on a pilot-scale size. The varying species, uncontrolled growing conditions, and variations in energy needed for drying biomass make expanded use of this approach somewhat uncertain. The costs of development of a wood-burning electricity-generating plant, for example, are higher than those of comparable coal-fired plants but about one-third those of nuclear power plants.

Plants that are large enough to be economical require vast land areas to provide the wood necessary to sustain them. A plant built in Vermont will require at least 10,000 acres of forest with a minimum of a half million tons of wood per year to remain operational. Owners of woodlots and residents of rural areas will need to be convinced of the value of the plant to allow such heavy, continued logging to occur (not an enviable job for the public relations person in an area where one state's motto is "Live Free or Die").

Large-scale forestry operations have been shown to alter local ecosystems even over the short term. Removal of trees may reduce the level of soil nutrients, and where the nutrient supply is near the critical minimum, may cause drastic reductions in further tree growth. Soil compaction from logging may reduce the ability of the soil to absorb water. This will also tend to reduce growth, but more important, it increases erosion and loss of thin topsoil, and may cause subsequent flooding. Clear-cutting operations, even with replanting, change the habitats of many flora and fauna, which often depend on a complex microenvironment of mixed species to survive.

Forest and woodlot management and lumbering operations can be made more cost-effective in some cases with the use of bark, chips, and small branches used to generate on-site or near-site heat and steam. Particular caution needs to be taken in the removal of logs and other materials from woodlands in order to maintain the amount of ground cover necessary to prevent erosion and to assure continued retention of rainwater. Removal of the entire tree prevents renewal of the soil, since bark and tops no longer are available to nourish future growth.

The renewability of wood as a fuel source depends primarily on analysis of the forest to replenish itself. Chlorophyll, the green coloring matter in plants, is the original solar collector and as such does the job well. Nature's work has been improved with the development of hybrids, in particular hybrid poplars and other tree species. Grown intensively, hybrids are capable of producing and storing 250 million Btu per acre per year based on an alternate-year harvest. This is equivalent to 40 barrels of oil or 2500 therms of natural gas. Shipping costs for 100 miles (160 km) would raise the cost to $2.50 per Btu, in contrast to No. 2 fuel oil at $6.30 per million Btu, coal at $2.17 per million Btu, and regulated natural gas at $4.76 per million Btu (1984 industrial user average prices).

The environmental impact of wood combustion is somewhat different than that of fossil fuels. Burning wood in carefully controlled burners results in insignificant sulfur and nitrogen oxide emissions compared with those of fossil fuels. Carbon dioxide levels are essentially the same when wood is burned as when it is allowed to decay on the forest floor, although the amounts are produced in a much shorter span of time. Increased use of wood-burning stoves in homes, where controls are apt to be less stringent, may cause significant increases in particulate matter released into the atmosphere. An additional concern is that polycyclic organic matter (POM) may make up to 4 or 5% of organic emissions, some of which have been shown to be carcinogenic. Airtight stoves achieve a higher overall heating but not necessarily higher combustion efficiency. As a result, POM and other organic emissions (the products of incomplete combustion) may increase the relatively low background levels found throughout the atmosphere. This is especially a problem in valley areas, where wind patterns do not disperse the particulate matter. Wood is not an appropriate source of heat when firewood must be transported long distances. Fuel for transit may exceed the thermal content of the wood. In such cases it would be less costly and less harmful to the environment to use petroleum products directly.

URBAN AND INDUSTRIAL WASTE

Urban waste consists of solid waste (most often labeled "garbage") and sewage sludge, much of which is now being treated in municipal sewage treatment plants. Estimates of the total amount of urban waste range from 200 million to over 800 million tons per year. Estimates are difficult to verify, as a nationwide reporting system has not been established.

The majority of this mountain of refuse consists of paper and paper products. Solid waste, of which at least 70% by weight is combustible, may be treated in several ways. It may be burned in facilities that generate process steam for nearby manufacturing plants; used to produce electricity or other energy forms; buried in landfills; or subjected to a combination of these treatments. Thermal processing for steam production ranges in efficiency from 65 to 80% depending on the type of materials burned and the design of the conversion system. A more effective treatment is to use municipal solid wastes (MSW) in conjunction with conventional fuels to reduce dependence on fossil fuels such as coal or oil. Either process reduces the volume of ash or residue from such wastes by as much as 80 to 90%, thereby greatly increasing the useful life span of landfill space devoted to the disposal of waste material. This is no small saving, since land suitable for this purpose is increasingly difficult to acquire and more expensive as well.

The economic conditions of the 1970s and 1980s have forced many industries to review their manufacturing and disposal practices. Extensive efforts are being made by most companies to reduce energy use and waste in order to stay in business. Recycling, selling wastes to other industries, or conversion to thermal or other energy forms have become for many companies a new form of expense reduction, and for some an additional income source.

Waste: Thermal Treatment—Pyrolysis

Biomass materials may be burned at temperatures starting at 390 to 575°F (200 to 300°C) in the absence of oxygen in a process called pyrolysis. These materials are converted to usable energy products such as combustible gases, oils, and char. They are of complex composition, and processing must be designed carefully and controlled according to the input and desired end products.

The gases formed in the combustion process can be used to aid in the combustion or may be used as a chemical feedstock in other processes. Generally, the higher the temperature of the process, the greater the amount of gas formed. Gaseous components include CO_2, CO, CH_4, H_2, and others in smaller amounts.

The oil that is produced has a high viscosity and must be heated to flow easily. It compares with No. 6 heating oil in this quality, but has less than 60 percent of the Btu value of No. 6 oil (10,500 Btu/lb versus 18,200 Btu/lb).

Techniques to convert char to a useful gaseous component continue along the lines of coal gasification research; however, since biomass may contain as much as 50% excess water, the equations are not directly applicable, and further study of the actual reactions is needed. Some toxic wastes can be safely disposed of in this manner, but others require temperatures above 1830°F (1000°C) to be treated effectively.

Pyrolysis reactor designs vary widely. Both fixed- and moving-bed designs are used, as well as suspended, fluidized, and entrained systems. They may be of vertical arrangement, inclined rotating or horizontal shaft kilns, and other configurations.

One experimental application is a mobile pyrolysis unit that can be transported on two trailers to fields and forest areas for use in processing agricultural and woodlot wastes. It may be used on cotton and rice waste as well as wood waste. After they have been shredded and dried, wastes are subjected to 1100°F (650°C) to produce the three components (combustible gases, oils, and char) indicated above. The unit will process 100 tons (90.7 kg) of waste per day and operates on a portion of the material produced. About 30 tons (27.2 kg) of oil and 20 tons (18.1 kg) of char can be produced. It is expected the cost of the oil will be 10 to 15% less than crude oil prices.

Waste: Thermal Treatment—Gasification

Several gasifier plants are in operation in the United States, ranging from research and pilot plants to commercial systems. They are classified under the types of gasifier inputs used: air, oxygen, pyrolysis, pyrolysis gasifier, steam, and char combustion. Several methods are used to manipulate the waste materials, including up-draft, down-draft, sloping bed, moving or oscillating grate, and fluidized bed, as well as other experimental methods.

The products are further classified according to the levels of energy contained. The gases produced are either low-energy gas (LEG), with a rating of 150 to 200 Btu per standard cubic foot (SCF), or medium-energy gas (MEG) produced in oxygen and pyrolysis gasification and rated at 350 to 500 Btu/SCF. This compares to pyrolysis oil and char at 1200 Btu/SCF. Low-energy gas is used in industry for heat-

Urban and Industrial Waste 239

ing, and medium-energy gas may be used in several ways: as a chemical feedstock or to synthesize methanol, ammonia, and gasoline.

One example of this process is the Gengas unit, originally developed during World War II in western Europe when scarce petroleum resources were relegated to war activities. Wood cubes or charcoal were the base fuels for Gengas, with hardwood materials being preferred over softwoods because of their higher Btu content.

The components of the wood gasifier unit (Figure 10-4) are the

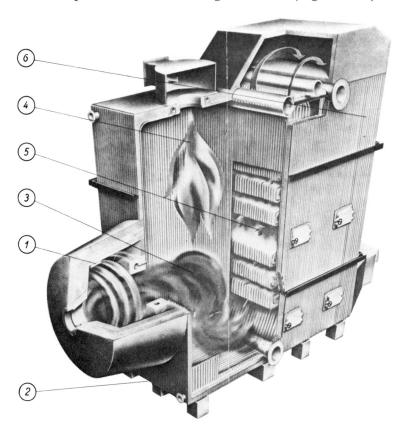

Peat cyclone installed in HOGFORS-55 boiler

1. Fuel—air compound and secondary air are blown into the cyclone in the direction of the tangent
2. Molten slag runs out through the hole at the bottom of the cyclone
3. Combustion gases flow to the water cooled furnace which operates as a space for after-burn
4. The cyclone boiler can even be equipped with an oil-burner
5. Convection-zone
6. Preheater for air

Figure 10-4 Wood gasifier unit. (Courtesy of Mountainstream Energy Systems, Windsor, Ontario, Canada.)

generator, a gas cleaner, a radiator for cooling the gas, and a mixer to introduce the gas into the carburetor. The generator is a modified down-draft type, loaded from the top. Air is introduced just above the grate through jets, where the combustion zone is created. The heat produced dries and charcoals the wood chips or wood cubes located just above this area. Excess water content evaporates and is bled off the top of the unit.

In the combustion zone, carbon reacting with air produces both CO_2 and CO. The down-draft pulls this mixture along with some water vapor into the area just below, where a reduction atmosphere exists. The CO_2 is reduced to CO and the water vapor is reduced to CO and hydrogen. The wood gas has the following contents:

Wood Gas	Volume (%)
Combustible element	
Carbon monoxide (CO)	17-22
Hydrogen (H_2)	16-20
Methane (CH_4)	2-3
Heavy carbon hydrogens	0.2-0.4
Noncombustible element	
Carbon dioxide (CO_2)	10-15
Nitrogen (N_2)	45-50

Some problems associated with chip compaction may occur, but since these units are often mounted on moving vehicles, the normal vibration of motion will keep such problems to a minimum.

The gas produced, containing small quantities of chips and soot, must be passed through a filter, usually fiberglass, to remove any particles that would result in carburetor blockage or engine damage. It is necessary to cool the gas (1) because gas at high temperatures has low weight by volume and therefore low energy by volume, and (2) because it also contains water vapor, phenols, and wood acids. Cooling removes these three components before they can create problems in the engine. Provision must be made to drain these by-products from the cooling unit. Manual-control mixers are employed to regulate the amount of air going into the ignition chamber since an automatic system would be very costly and complicated.

Gengas has a lower combustion speed, and thus ignition must be advanced compared to that with conventional fuels. Spark plugs must have a higher heat rating than with liquid fuels. Gengas has good antiknock characteristics, with an octane rating of approximately 100. Engine compression ratios can be as high as 10:1.

Fuel consumption varies greatly depending on the species of wood used, size and loading characteristics of the wood pieces, engine size and rpm, and driving characteristics such as speed and acceleration. When used in conjunction with diesel fuel, mileage has been shown to double in some cases, although trip time was increased by 40 to 50%. One producer calculated that if wood cost $100 per cord, the fuel would still cost only 60 cents per gallon. When these units use only scrap wood, the operating cost per mile is reduced substantially.

Since the four generator components comprise a considerable amount of weight, only medium-size to large vehicles would tend to be suitable for this application. Farm tractors may have only a limited amount of space for mounting if other types of implements are also to be mounted. Pickup trucks and flatbed trucks are ideally suited since the unit would take up a limited amount of space and would be easily accessible for loading and servicing. Buses could also make use of this system as a fuel source.

Several safety precautions must be taken. Gengas contains CO, which has no taste or smell and could be breathed with serious or fatal consequences. Opening the unit to charge it with wood could expose the operator to high heat and flare-up. Units cannot be operated in buildings or enclosed spaces without adequate mechanical ventilation. Fire extinguishers should be standard equipment during operation.

Mechanical Systems

Mechanical systems for the separation of urban wastes use a combination of several techniques to remove and isolate the different materials. Both dry and wet processing are used. The dry process is used primarily for the preparation of urban waste to use as refuse-derived fuel (RDF). Several steps are required to make the most efficient use of this source of energy (see Figure 10-6).

Unsorted waste is delivered by trucks to a temporary storage area, then sent to shredders which reduce the material to more manageable size for the sorting process. Magnets remove ferrous materials in the initial stages, then air classifiers help to separate the lighter and heavier materials.

Heavier particles are sent through a sorter, which removes aluminum and other nonferrous metals by means of an eddy-current separator, leaving principally glass to be separated before the remainder is sent to the landfill. This system has the advantage of recovering materials that may easily be recycled to nearby plants and reduces landfill needs substantially.

The ligher materials in urban waste are separated from the heavier materials at the primary air classifier and used as refuse-derived fuel. The materials may also be pelletized by means of compression to make handling and storage easier. This RDF is most often mixed with con-

Figure 10-5 Refuse-derived fuel plant, Monroe County, New York. (Photo by M. H. Kleinbach.)

ventional fuel and burned to produce industrial process steam and/or electricity, although it can serve as the primary fuel for an industrial steam plant (see Chapter 14 for more information).

The heavier materials are first sent through a magnetic separator which removes additional ferrous materials. The remainder is sent through a screen which separates glass and other ceramic materials from nonferrous metals.

Wet-processing systems have met with limited success, as have some of the dry-RDF-fueled plants. Emission regulations have prevented full-scale operation of some plants, and uncontrolled and sometimes dangerous materials entering the plants have reduced or stopped operations at others.

Biological Systems

The two major products of biological systems in refuse treatment are fuels (primarily methane and alcohols) and soil conditioners. The use of microorganisms to produce various types of materials has long been known and the practice of using living cells to produce alcoholic beverages is nearly as old as humankind itself.

Biological transformation occurs naturally, without the need for human intervention. Two processes predominate in the bioconversion of various organic materials. In the first stage, no methane is formed. Complex organic materials are ingested by a group of bacteria called acid formers. Fats, proteins, and carbohydrates are converted to soluble

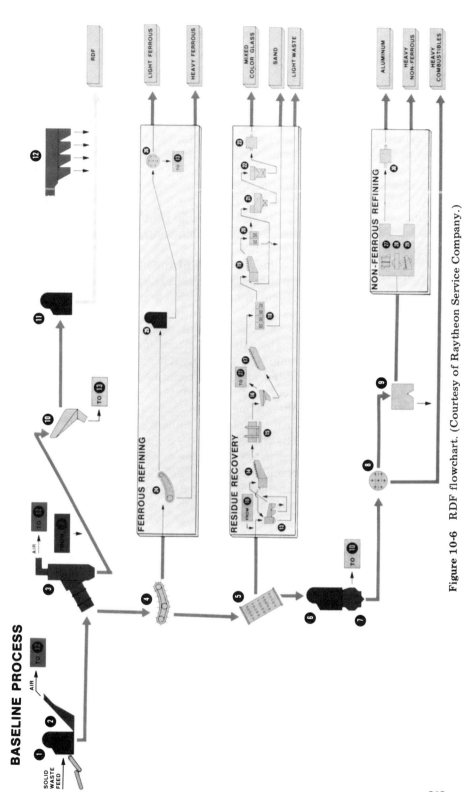

Figure 10-6 RDF flowchart. (Courtesy of Raytheon Service Company.)

volatile acids such as acetic acid. In the second stage, methane formers convert the organic acids into CH_4 and CO_2 through their digestive processes. The latter process is primarily responsible for the presence of the vast amounts of natural gas found in various parts of the world. Methane formers which perform the anaerobic function are known to have existed for 4 to 4½ billion years. During that time they have been busy producing the methane we find so useful as a fuel.

During the time of cheap energy, natural gas was considered a by-product of oil wells and was flared (burned as "waste") to rid the drilling site of a potential fire and explosion problem. In the past 10 to 15 years we have begun to develop techniques to retain and utilize the gas, primarily right at the well site.

Another "problem," similar to that of the natural gas found in oil wells, is that of gases produced from landfill areas. These gases have been found to be an excellent source of LEG and MEG. The gas, often flared as is the natural gas from oil wells, is composed of varying parts of CH_4 and CO_2 as well as other gaseous components. Although the energy content is not as high as that of natural gas (50 to 60% methane compared with 90% methane for natural gas), it has several vital positive characteristics. It is generated in close proximity to urban areas, where the need for fuel is higher; thus transportation costs are substantially reduced. It can be used in existing natural gas burners with little or no modification, and access is relatively easy since it can be recovered from comparatively shallow wells with existing technology.

The Poenti Hills plant in the Los Angeles Sanitation District is the first in the nation to be economically successful. Two turbine-generator units are fueled by methane gas from an adjoining landfill. Los Angeles is realizing nearly $1,000,000 annually in income from the sale of electricity from the plant.

Some communities have experienced problems with methane gas migrating from landfill sites to nearby homes or other structures, creating an explosion hazard. Removal of the gas for productive purposes substantially reduces that hazard as well as removing or reducing the odor caused by escaping hydrogen sulfide, a common component of gas produced at landfill sites. Where the landfill site is used for vegetation, or for a park, golf course, or other public use, the removal of this gas not only reduces odor but allows better growth of plants, trees, and shrubs.

There are several important criteria necessary for CH_4 recovery to be successful from an urban landfill: a minimum in-place volume of at least 1 million tons (907 million kg) of refuse, a minimum surface depth of 40 ft (12 m), a minimum surface area of 40 acres (16 hectares), and a minimum annual waste disposal rate of 150,000 tons (136 million kg). There are at least 1000 and perhaps as many as 10,000 landfills in the United States with a potential for gas recovery operations. Recovery rates of experimental and pilot plants range from 80 to 280 cubic feet of gas per ton of municipal solid waste (MSW) per year. About 200 billion cubic feet of CH_4 could be produced per year, with a value of $400 to $500 million. It is projected that many larger cities will be tapping their former landfill areas as a source of this relatively inexpensive renewable chemical energy.

BIOMASS CONVERSION: METHANE

Anaerobic digestion is used to produce a usable product—methane (CH_4)—from organic wastes. Organic wastes, a major cause of pollution in the environment, are the result of the disposal of waste products from farms, homes, and other sources of human and animal habitats. The scientist Louis Pasteur recognized the presence of a flammable gas which resulted from decaying organic matter. He suggested that the gas be used to light the street lamps of Paris.

Current waste disposal practices often result in the loss of elements essential to plant growth, an increase in insect and rodent infestation from improperly managed systems, and in hygienic and aesthetic concerns as well.

"Methane (or biogas) is a renewable fuel that requires no expensive discovery, production, refinement and transportation processes, each with its own concomitant pollutants and expenditures of energy, labor, and other resources" [Auerbach, p. 1]. Methane production occurs when microorganisms digest sewage or other effluent containing carbon, nitrogen, and other elements. The organic material is decomposed to form two major components: methane gas and compost, rich in plant nutrients. The fertilizer value of manure is essentially unchanged by the anaerobic digestion process and odor, insect, and weed seed problems are substantially reduced.

Just as methane formers work in natural settings, systems have been developed to control and use these biological processes to produce renewable energy. The basic components of the system consist of an airtight chamber (digester) into which the slurry can be dumped or pumped, an agitation system (either hand cranked or power-driven), a collection system for the gas produced, and a discharge system for the compost or used slurry (Figure 10-7). Auxiliary accessories would include a pressure gauge, a meter, and piping to the fuel-burning equipment.

Several conditions are necessary to complete the process.

1. Digestion normally takes place in a temperature range of 85 to 95°F (29 to 35°C), although temperatures as low as 60°F (16°C) and as high as 140°F (60°C) can be used with lowered efficiency. Beyond these two extremes the anaerobic organisms cease to function and no gas is produced. In temperate and cold climates, it is necessary to provide insulation around any portion of the digester that comes in contact with the air or ground around it. In some cases it may be necessary to burn some of the methane produced to heat the digester and its contents in order to maintain the proper operating temperature.
2. Different types of organic matter determine the carbon/nitrogen (C/N) ratio as well as the levels of plant nutrients to be found in the compost at the end of the cycle. The C/N ratio can have a varying range depending on the type of organic material initially used, ranging from more than 500:1 to less than 1:1. Ideally, the closer the ratio approaches but does not exceed 30:1, the more efficient will be the operation. The bacteria require both elements in order to live and function, but they convert carbon about 30 to 35 times faster than they do nitrogen.
3. Slurry conditions are important for optimum gas production. The slurry should contain 7 to 9% solids, with water being added to materials that have a moisture content lower than this, and sawdust or straw added when the water content is too high.

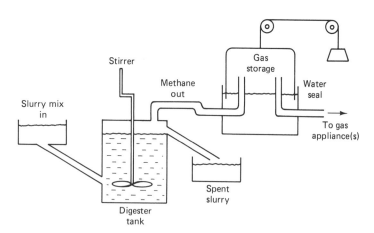

Figure 10-7 Methane generator unit.

4. The pH level of the system must remain in the range of 6.6 to 7.6, with optimum results at 7.6, an alkaline environment. Gas production drops off rapidly after this point.
5. Mixing of the slurry during operation is necessary to prevent the formation of a crust which would prevent the gas from rising to the top of the unit for withdrawal and subsequent use. Since this process must exclude oxygen, the mixing or movement must occur without opening the unit. This is also desirable from an olfactory standpoint!

A balance in the growth of the two types of bacteria is important. When either the acid formers or the methane formers grow more rapidly than the other, an imbalance can occur which slows the generation of biogas and may stop the process entirely. In the latter case, the unit must be emptied and the process started again.

Methane formers are more sensitive to environmental conditions and will cease to function if one or more of a variety of changes occur. Temperatures outside the normal operating range, pH changes if acid-former levels are not controlled with alkaline buffers, and levels of ammonia, nitrogen, and solids concentrations above the range of 7 to 9% are all contributors to a potential imbalance between the two types of bacteria.

Various styles of digesters have been found to operate effectively. The most common type consists of an open-top tank buried at least three-fourths of its depth in the ground. (This will be the approximate depth of the slurry contained inside.) A second, slightly smaller open-end tank is placed in an inverted position inside the first tank so that its opening is just below the surface of the slurry. This provides a seal for the process. As gas is produced, the upper tank rises with the pressure, but not above the top of the slurry, thus providing a sealed container for the methane. Counterweights allow the top tank to remain suspended and to provide a slight pressure for direct use in nearby burners.

The slurry contents may be added as a single input (batch feed) or fed in daily (continuous feed) via a submerged tube in proportion to the amounts used to replenish the mixture as the biogas is produced. The gas collected can be used directly or stored as compressed gas in an auxiliary storage tank.

A second type of digester is a relatively low-cost, earth-supported digester which uses a "plug-flow" process feeding into a flexible rubber or plastic bag (*Cornell University Agricultural Bulletin No. 397*). This unit can be fed batches of material on an intermittent basis, with each batch forcing the preceding one further into the system until the compost or expended slurry moves out the opposite end.

Biogas produced by either of these digesters consists of a mixture of various gases: CH_4 ranging from 55 to 80%; CO_2 (having no heat value) ranging from 20 to 45%; and trace amounts of hydrogen sulfide (H_2S), ammonia (NH_3), and other gases. The higher the CH_4 content, the higher the heat value, ranging from 600 to 800 Btu per cubic foot. By comparison, natural gas will provide from 1100 to 1200 Btu per cubic foot.

Digester sizing will depend primarily on potential and use, although the anticipated volume of waste to be used will set some practical limits as well. Farm units, dairies, feedlot operations, and broiler or egg-laying operations are the most likely users of biogas systems. A rule of thumb for beginning the sizing calculations is that 1 ton of fresh manure produces in excess of 2000 ft^3 of gas, or that 1 lb = 1 ft^3. Batch digesters operate on a 1- to 3-month cycle depending on temperature variations and the type of waste input. The size of the slurry container is determined by the loading capacities of manures, vegetable matter, and water volume to achieve a slurry value of 7 to 9% (solids-to-liquids ratio).

A simplified process for determining digester size is based on the following: initial volume of manure or other wastes, plus water to develop a slurry consisting of from 7 to 9% plus 15% additional volume for preliminary gas collection. Where a single type of waste is used, the C/N ratio can be read from existing tables such as Table 10-1.

The space above the slurry (for nonflexible containers) is generally held at a maximum of 15% of the slurry volume. Storage facilities are sized to about 50% of the slurry volume. Gas usage is calculated on the basis of the appliance or use that has the highest gas consumption. Pressure is essential to provide movement from storage to point of end use and is most commonly provided by means of the weight of the

TABLE 10-1 Carbon Content, Common Organic Materials

Materials	N (%)	C/N Ratio
Cornstalks	0.84	53
Grass clippings	2.2	19
Leaves	0.2	203
Manure		
Dairy cow	2.6	16
Horse	2.3	25
Pig	3.8	14.4
Sheep/goat	3.8	20.1
Turkey	4.2	8.3
Sawdust (uncomposted)	0.11	511
Straw, oats	0.53	83
Straw, wheat	0.33	124
Sugarcane (bagasse)	0.35	113

upper movable container counterweighted so that incoming gas is compressed on a water seal.

Some helpful relationships:

$$1 \text{ gal water} = 8.3 \text{ lb}$$
$$1 \text{ gal water} = 0.134 \text{ ft}^3$$
$$1 \text{ ft}^3 = 7.481 \text{ (7.5) gal}$$
$$\text{specific gravity of slurry} = 1.19$$

The relatively high CO_2 percentage is not a contributor in the combustion process. Bubbling biogas through limewater can decrease the CO_2 content and consequently raise the Btu level of the gas. This may not be necessary with open burners or other direct heat-producing devices. Although very small, the H_2S component has a tendency to corrode metal parts. If the gas is to be used in an internal combustion engine, it should be "scrubbed" by passing it through iron filings as it leaves the digester.

The initial costs of setting up a digester can be quite high. A farm-size unit (100-cow dairy, for example) could involve an expenditure of more than $30,000. The value of biogas produced could amount to as much as $4000 per year compared to the fuel it would replace. Payback periods could be as long as 7 years or shorter depending on the size and efficiency of the facility, as well as the effects of tax credits and inflation.

Safety considerations must not be ignored. Methane gas is colorless, tasteless, and odorless when there is no H_2S component. Small amounts of the latter can serve as a check on possible leaks of the system. Biogas, with its low CO_2 content, is nontoxic, but like many other gases it can cause asphyxiation if allowed to accumulate and replace oxygen in an enclosed space. A 5 to 14% mixture of CH_4 and air is explosive when ignited. Since it is lighter than air, it will collect at high points in buildings and hence is less noticable until a potentially serious situation has developed. Inexpensive gas detectors can be used to warn of dangerous gas–air mixtures. Digester and storage tank construction should follow the same precautions as for gasoline use.

COMPOSTING

A closely related biological process called composting is being used at an ever-increasing rate. Composting is a controlled microbial decomposition of residues from livestock, crop processing, and other materials, including human wastes. The process is site-specific in that it is not cost-effective to transport large amounts of waste for processing into useful nutrients and soil conditioners.

Nitrogen fertilizers (i.e., commercially produced artificial fertilizers) require large amounts of natural gas during their production. To conserve this increasingly valuable nonrenewable resource implies better utilization of waste materials. Past practices have been to spread livestock manures on open land without any type of processing. Such practices result in the loss of 50 to 75% of the nitrogen before it can be absorbed and used by crops and grasses.

The process of composting farm animal wastes has been shown to reduce the volume of the manure by about 25% and the weight by as much as 60%, thereby reducing fuel costs and time in spreading the resulting compost on fields. Livestock waste is usually piled in rows up to 4 or 5 ft (1.25 to 1.5 m) high by 10 to 12 ft (3 to 3.5 m) wide. The material must be turned to allow for aeration, since the action is caused by aerobic bacteria. Temperature conditions should not exceed 150°F (65°C). If temperatures are reduced below 60°F (15°C), the bacterial action is reduced. Moisture conditions are usually controlled by the addition of bedding material or other carbon-content materials. This also increases the carbon proportion, bringing it closer to the 20 or 30:1 C/N ratio necessary for the best bacterial action. One sign of low carbon content is a strong ammonia odor, indicating that nitrogen is escaping as a gas rather than being absorbed and "stored" by the microbes. Compost in its completed state has no ammonia odor.

Although this process is usually considered suitable for agricultural situations, many urban communities are considering its use as a sewage treatment step. The resulting compost has been shown to reduce the pressures on landfill operations, ocean dumping, and incineration. Approximately 10% of the municipal sludge in the United States is now being processed for compost. In addition, other materials, such as leaves collected by municipal cleanup crews, pharmaceutical by-products, and some industrial wastes, are candidates for processing. The U.S. Department of Agriculture reports there are currently 30 to 50 composting projects using the windrow method. This method is suitable where land is relatively easily acquired and where large operations produce 100 to 600 tons of waste per day.

Several related developments indicate the growth and increased interest in this waste control technique. There has been an increase in the number of consulting firms that provide technical expertise to communities. Several companies have begun the development of equipment for the handling and processing of materials used in the composting process. Materials with high carbon content are needed to maintain the C/N balance, so "wastes" such as sawdust and bark in wood-processing regions, crop residues from food-processing plants, and even old newspapers have been found to be a resource instead of a problem.

The resulting compost is in great demand by homeowners for

gardens and lawns, by commercial firms such as greenhouse operators, and by municipalities for parks, athletic fields, and golf courses. Properly processed compost has no odor; contains levels of nitrogen, phosphorus, and potash in proper proportions for good plant growth; releases those nutrients more slowly than do commercial fertilizers; and contains conditioners that enhance the tilth (consistency) of the soil and therefore the root growth of plants. Composting occurs at relatively high temperatures, which destroy the germination potential of weed seeds often found in manure and crop residues, reducing the need for herbicides. Composting helps in the control of flies and other insects by destroying the environment for the hatching of insect eggs.

ALCOHOL FUELS

Although the fuel properties of alcohol for transportation have been known from early in this century, it was not the fuel of choice for internal combustion engines as long as gasoline was available in quantity at relatively low prices. Gasoline consumption has increased at a rate of 3 to 5% per year nearly every year since 1960, except for 1973, the first of the oil crises. Americans returned to upward use trends almost immediately, until consumption reached a peak of 7,555,000 barrels of gasoline per day in 1978.

The combination of higher prices, smaller cars, reduced driving, and the legally imposed 55-mph (90-km/h) speed limit have kept consumption levels between 6,700,000 and 7,000,000 gal per day since that time. There has been little apparent detrimental effect on the lifestyle of most people.

Alcohol as a fuel has several distinct advantages as a replacement for petroleum-based fuels. Alcohol fuels can be readily used in the highway transportation sector without further refining, a quality not shared by most other fuels. Conversion of vehicles to use alcohol is relatively easy and inexpensive. The technology to produce alcohol is well known both to large producers and small-time operators (those who sometimes operate surreptitiously). Feedstocks for alcohol production can be made available in large quantities without resorting to imports. Alcohol fuels burn cleanly in existing engines and produce about the same power and mileage as gasoline, depending on the individual engine, driver, and driving conditions. Emissions are reduced for HC and CO, although NO_x levels are slightly increased.

There are some disadvantages to the use of alcohol fuels. Alcohol has a solvent property which may attack some rubber and plastic parts (elastomers). This property may harm some auto finishes and cause various gaskets and seals to decompose. Anhydrous (water-free) alcohol is expensive to produce, and if alcohol is less than 198 proof (99% pure) there may be a phase separation (water/alcohol) which could cause combustion problems.

Two forms of alcohol are most often considered for use as fuel: methanol (CH_3OH) and ethanol (C_2H_5OH). The former is commonly called wood alcohol and is poisonous to human beings and animals if ingested. The latter, grain alcohol, is the basis for beverages which are enjoyed (and abused) all over the world.

Ethanol production is based on the type of feedstock used. Plants such as sugarcane, sugar beets, and sorghum contain simple sugars which can be converted directly into ethanol. Starchy plants such as grains and potatoes must be converted first to sugars, then to alcohol. Cellulose-based materials contain glucose, which differs from starch mainly in the chemical bond between glucose units.

Three production steps are needed: formation of a solution of fermentable sugar, fermentation of the sugar to ethanol, and separation (distillation) of the ethanol from the mixture (see Figure 10-8). If grain is used as the feedstock, it must first be ground to expose the starch granules to enzyme action. Water is added to make a slurry, and the mixture is heated to 155 to 200°F (68 to 93°C) to aid in enzyme action. The starch is converted into soluble high-molecular-weight sugars called dextrins by the enzyme α-amylase.

The mash is then cooled to 135 to 140°F (57 to 60°C) and a conversion enzyme, glucoamylase, is added. After conversion of the dextrins to fermentable sugars, the mash is cooled further, to 85 to 90°F (30 to 32°C), where it is held while fermentation takes place with the aid of distillers' yeast. This takes a period of 48 to 120 hours. Alcohol and CO_2 are produced as well as heat, so cooling is necessary to maintain optimum temperature conditions.

In the final stage, heat is used to vaporize the alcohol, which is collected and cooled to convert it to liquid form. The spent yeast and residual grain mash are removed. The elimination of water from the mash produces a product known as distillers' dried grains (DDG), which is used as a feed supplement for animals. The fermentation and distillation process removes the carbohydrate fraction from the grain but leaves the protein portion intact.

The stillage, as it is often called, must be used as animal feed within 1 or 2 days, implying limited storage and transportation conditions. As with other feed supplements, DDG must be maintained as a consistent part of the animal ration. If not used for this purpose, it can be used as a fertilizer for land near the production site.

The CO_2, produced in about equal amounts with alcohol during the fermentation stage, is relatively pure if the fermenter is tightly closed. It can be dried to remove water vapor, compressed, and stored

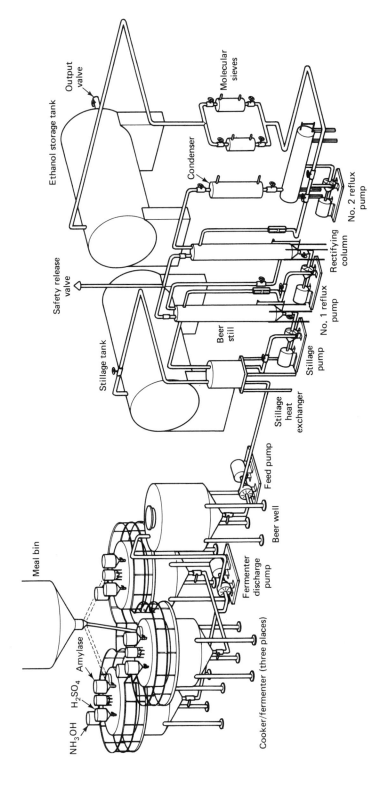

Figure 10-8 Alcohol production. (From the U.S. Department of Energy.)

or used. Since the cost of recovery equals the cost of the product in the marketplace, its use is site-specific if it is to compete with commercial production of CO_2.

One of the unresolved controversies with regard to the use of ethanol is that of the net energy balance of production. Some studies compare the total energy contents of the products and by-products with the fossil energy consumed in their production. Other studies compare the amount of crude petroleum energy required to produce a given amount of petroleum substitute. The objective is to compare the energy input with the energy output of the system. Another critical factor in these comparisons is the Btu value given to the resulting alcohol. These values range from a low of 73,000 Btu per gallon for 190-proof (95%) ethanol to a high of 88,000 Btu per gallon for 200-proof (100%) anhydrous ethanol. Gasoline is considered to have a lower limit of 117,000 Btu per gallon.

The different feedstocks vary considerably in their costs of production, depending on soil and weather conditions as well as farming practices. Since corn is the most used grain for alcohol production, most comparisons are made against it. There is one area of agreement. The amount of ethanol that can be expected to be produced from 1 bushel of corn is about 2.6 gal (9.8 liters).

As an additive to gasoline, to produce gasahol (5 to 10% by volume) each gallon of ethanol displaces (replaces) about 0.8 gal of gasoline. However, since the alcohol increases the octane rating, refineries can, if they choose, reduce the octane rating of gasoline during manufacture, thereby reducing refinery costs, and add alcohol at the end of production or at the point of consumption. This trade-off can result in an additional 0.36 gal of gasoline equivalent saved.

The crucial factor in all the calculations is the type of fuel used in the distillation process for alcohol. When oil or natural gas is used, there is apt to be a net energy loss in production. When coal, agricultural wastes, or solar energy is used as the heat source, or when "waste" heat from some other production facility is available, it is likely that a net energy gain will result.

Several crops besides corn provide a potential for feedstock material. Sugar cane is a top contender, since its sugar production can be as high as 50 tons per acre. The crop residue, known as bagasse, can be used as a fuel source, reducing petroleum-based energy inputs. Sugar cane requires tropical or subtropical conditions for growth, which limits its production areas to the extreme southern United States, Central America, and the northern parts of South America. Brazil in particular has mounted a major ethanol production program to reduce its dependence on imported oil.

Sugar beets, more versatile in terms of climate and soil conditions, remain a top choice for feedstock. Another important consideration is beet pulp, which is a valued animal feed but cannot be used effectively as a boiler fuel. Sweet sorghum can also be grown over a wide range of conditions. Its residue can be used either as animal feed or as boiler fuel.

Jerusalem artichokes, fodder beets, and fruit crops are being researched for their potential as feedstock sources. Crop residues, composed mostly of cellulose, have a good potential as well. Wastes from food processing, spoiled grain, root crops, and fruits are possible sources. A major problem with these materials is their inconsistent availability. Large-scale plants could not depend on these unreliable sources for continuous production. Small-scale (farm-size) plants, however, may be able to make use of such sources.

Current technology and law allows small-scale farm stills to produce alcohol of no more than 190 proof. This is suitable for use in internal combustion engines without blending with gasoline but would result in phase separation if it were mixed. Some modification in ignition timing and of carburetor jets is needed, as well as provision for preheating during cold weather [below 40°F (5°C)].

In spite of these restrictions, the development of small-scale units could serve individual or cooperative needs, particularly where damaged crops, culls, and spoiled grain are the feedstock materials and crop residues are used as fuel. Some of these units use open-flame systems for heating. Alcohols have a higher flashpoint than gasoline, 55°F (13°C) for ethanol, 52°F (11°C) for methanol, but their vapors are flammable in a wide range of concentrations. Small-scale operations generally will not have trained, experienced operators to monitor the operation. This could result in lower production of alcohol and potential mismanagement of systems.

Although single-batch-size and small continuous-operation plants are reasonable for a single farmer or small group to set up, medium-size units (100,000 gal per year, for example) seem too large for "farm use," yet too small to provide a net positive return for the investment. Some plants of this type, set up as demonstration units, have encountered problems in attempting to find markets for the alcohol produced in excess of their on-site needs. The infrastructure for dispersal is not available to the "marginal" producer. Large-scale (20 to 50 million gallons per year) plants are designed with markets and distribution systems similar to those of existing gasoline wholesalers and retailers, to make such large supplies economically feasible.

The use of grain for feedstock creates a "food versus fuel" controversy. The following factors should be considered. Typically at least one-third of the corn grown on farms is used where it is grown. The processing of corn into ethanol also results in the by-product of DDG, which retains the protein content of the original grain. It is estimated that between 1 and 5 billion gallons of ethanol (1 to 5% of the fuel used in transportation) could be produced without changing the market price or demand for corn. Such production levels would require between 400 million and 2 billion bushels of corn. Annual corn production levels fluctuate by at least that much each year.

Production of alcohol at this level could justify the construction of 20 large-scale plants, each capable of 50 million gallons (190 million liters) each year. This would increase employment by an estimated 1500 to 2000 persons for the plants themselves. Construction would require a minimum of 2 to 4 years per plant.

By comparison, the development of a corresponding methanol fuel structure would require 6 to 8 years per plant. Each methanol plant would employ 350 to 400 persons. The best feedstock for large-scale methanol production would most likely be coal, a nonrenewable resource. The prospects for development of this technology would be long term, on the order of 10 to 15 years. In addition to coal, many other feedstock sources for methanol production are available. These include cellulose from wood, municipal solid waste and sludge, and crop wastes.

The Fischer-Tropsch process, developed by Germany during World War II and used today in South Africa, converts coal into heavy hydrocarbons. Mobil Oil, in contracts with DOE, uses a similar process, but uses a zeolite compound as a catalyst to convert the methanol to gasoline.

Methanol has an octane rating of 106 compared to 85 to 100 for gasoline. Methanol prevents knocking and dieseling and may be mixed with gasoline in blends ranging from 5 to 30%. When it is used in this way, CO emissions are reduced by 14 to 72%, fuel economy is increased by 5 to 13%, and acceleration is increased up to 7%.

Producers of alcohol must obtain a permit from the nearest regional office of the Federal Bureau of Alcohol, Tobacco and Firearms. Additionally, the state in which alcohol is to be produced, whether experimental or production, may have additional legal requirements dealing with construction and electrical codes, storage requirements, and OSHA standards if employees are involved.

BIOMASS UTILIZATION

"Biomass usage is far greater than all other forms of solar energy put together" [Bente et al., p. 2]. Such statements seem unrealistic until one views them in the context of how such comparisons are traditionally made. Energy cost comparisons are traditionally made only in relation to sources that are purchased in the open market, as are coal, natural gas and other fuels. Since many forms of biomass (wood for heating and crop residues used as bedding for animals) and other materials are used in a "closed-cycle" system, they never compete on the open market and are not considered a part of the energy cost picture.

Typical of the use of noncompetitive fuels in closed-cycle processing is the wood-processing industry, which has reached a 45% level of energy self-sufficiency by using waste wood, chips, and bark for fuel. Biomass provided 8% of the world's energy in 1983. It is estimated there could be a five- to tenfold increase in the use of this resource using just the available biomass. Problems arise when the biomass materials are widely scattered and/or when recovery is made more difficult by rough or wet terrain or other environmental conditions.

The use of kelp as a commercial product is not a new industry. Giant kelp (*Macrocystis pyrifera*) has been harvested since the early 1930s off the coast of California for algin, a product widely used in food processing and in industrial processes. The naturally growing plants are harvested with equipment that has been tested and used for long periods of time. This led to the cultivation of kelp in an "energy farm" concept.

Wave-actuated pumps raise nutrient-rich water from depths of nearly 1000 ft to nourish the kelp. This results in an impressive 15 tons per acre per year, comparable to the best yields of biomass from terrestrial farming. Experiments at the Woods Hole Oceanographic Institution with over 50 species of seaweed showed red seaweed (*Gracilaria tikvahiae*) to be more than three times as productive as giant kelp under controlled conditions, but the yield did not carry over into full-scale production. Other efforts are being made to join these types of "energy forms" with ocean energy thermal conversion (OTEC) systems, which discharge nutrient-laded H_2O during their operating cycle. More basic research is needed before these approaches can be competitive economically.

A more recent development uses aquatic plants such as kelp or water hyacinths or terrestrial grasses and shrubs to produce the ingredients for methane production. The residue from the process is often used on nearby agricultural land for fertilizer. Water hyacinths are considered a scourge in the waterways of southern United States. They block the way for boats and are judged impossible to eradicate. Experiments have been conducted to cut these "weeds" (weeds are plants for which no beneficial use has yet been found) and spread the plant mass on fields as fertilizer. Aquatic plants, with their high water content, are better suited to anaerobic digestion processes than to gasification or other processes where low-moisture-content materials are required.

One major advantage of the presence of water hyacinths has been found: They have the ability to convert polluting sewage wastes into a protein-rich biomass which could be used for chemical production and animal feed as well as for other uses mentioned above.

A total of 540 metric tons of crop and forest residues remain in fields and forests after conventional harvesting operations. This represents a thermal energy equivalent of about 12% of the fossil fuel consumed in the United States. Only about 20% of total residues could be used for conversion because of severe environmental impact and difficulty of access and collection. Readily available residues could provide a net energy gain of (1) high-grade liquid fuel equal to 1% of gasoline consumption, (2) 4% of electrical energy now used, or (3) 1% of energy consumed as heat.

The use of marginal land not suitable for conventional crops is being explored. Certain woody shrubs and scrub plants growing in areas of marginal land, where increased land use would not interfere with established crops, forests, or rangeland, have been investigated for use as fuel producers. Several species show promise and may be converted from "weed" status to commercial crops.

SUMMARY

Nonrenewable resources are in ever-decreasing supply. With increasing population and longer life expectancy in the United States and around the world, the demand for energy and raw materials continues to increase. As production continues to increase to meet the demand for more goods and services, so do waste materials. Finding uses for waste and by-products becomes ever more essential. Finding new uses for renewable resources has been shown to be a large part of the answer to an extended supply base.

Agricultural products have many uses as renewable energy source materials, both as direct cropping projects and as a source of reuse of waste materials. Grain crops can be used to produce alcohol, and the residue that remains contains nearly the same total feed value as the original grain. The stalks, husks, and hulls from various crops can be used to produce alcohol or methane, but the loss of ground cover and subsequent topsoil loss may not warrant their use in this manner. Manure from dairy and feedlot operations is a highly potent source of methane. Both small- and large-scale digester systems have been found to be practical in economic terms.

Urban waste can be a resource rather than a troublesome disposal problem, with the use of garbage burners to produce steam and/or to generate electricity. Metal and glass discarded in the "waste stream" can be reclaimed at costs which are frequently lower than those of production from increasingly scarce raw materials. Existing landfills are potential sites for the development of methane generation located near the potential use of the gas. The removal for commercial use tends to alleviate potential odor problems and may make the surface of the landfill suitable for other productive purposes.

Some plant materials are being studied for their potential contribution to energy resources. Many show promise not only for energy but also as substitutes for animal feed and as raw material sources for medicines and pharmaceuticals. Some are able to be grown on land that is too poor in nutrients for conventional crops or in arid lands that have been suitable for agriculture only when heavily irrigated.

ACTIVITIES

1. Identify the major sources of waste materials in your community. Identify the content of those waste materials. Which have the potential for use as energy resources?
2. Calculate the cost of landfilling 1 ton of each type of "waste," and contrast this with the net cost of "recovered" energy.
3. Identify the locations of recycling centers in your community and the types and amounts of materials recycled.
4. What plant life could be introduced into the ecology of your community which would aid in energy production?

BIBLIOGRAPHY

Alternate Sources of Energy, quarterly (various issues), Milaca, Minn.

Auerbach, Les, A Homesite Power Unit: Methane Generator, 242 Copse Road, Madison, CT 06443.

Bente, Paul, et al., *Energy from Biomass and Solid Wastes: Prospects and Constraints*. Washington, D.C.: House Committee on Science and Technology, 96th Congress, June 1980.

Biocycle, Journal of Waste Recycling, July–August, 1982, The JG Press, Inc. Emmaus, Pa.

Deudney, David, and Christopher Flavin, *Renewable Energy, the Power to Choose*. New York: W.W. Norton, 1983.

Energy Information Administration, *1982 Annual Energy Review*. Washington, D.C.: U.S. Department of Energy, April 1983.

Lincoln, John Ware, *Methanol and Other Ways around the Gas Pump*. Pownal, Vt.: Garden Way Publishing, 1976.

Meynell, Peter-John, *Methane: Planning a Digester*. New York: Schocken Books, 1976.

NCRR Bulletin, The Journal of Resource Recovery (various issues), National Center for Resource Recovery, Inc., Washington, D.C.

New York State Energy Research and Development Authority, *Methane Recovery from Sanitary Landfills.* Report 78-13, Albany, N.Y., 1978.

——, *Technology Assessment of Biomass as an Energy Source in New York State.* Report 78-3, Albany, N.Y., 1978.

Nygard, Nils, "Wood Gas Generator for Vehicles." Minneapolis: Disti-Pure International, 1979.

Office of Technology Assessment, *Energy from Biological Processes.* Washington, D.C.: OTA, 1980.

——, *Energy from Biological Processes*, Vol. 2: *Technical and Environmental Analyses.* Washington, D.C.: OTA, 1980.

——, *Gasohol, A Technical Memorandum.* Washington, D.C.: OTA, September 1979.

Pimentel, David, et al., "Biomass Energy from Crop and Forest Residues," *Science* (AAAS), Vol. 212, June 5, 1981, pp. 1110–1115.

Renewable Energy News (various issues), CREN Publishing, Ottawa, Canada.

Singh, Ram Bux, Bio-Gas Plant, Gobar Gas Research Station, Ajitmal, Etawah (U.P.), India.

The State of New York, Legislative Commission on Science and Technology, *The Status of Alternative Energy Technologies.* Albany, N.Y., March 1980.

U.S. Department of Agriculture, *Small-Scale Fuel Alcohol Production*, Washington, D.C.: USDA, March 1980.

U.S. Department of Energy, Division of Solar Technology, *Biomass-Based Alcohol Fuels: The Near-Term Potential for Use with Gasoline.* Washington, D.C.: DOE, August 1978.

——, Solar Energy Research Institute, *Fuel from Farms—A Guide to Small Scale Ethanol Production.* Washington, D.C.: DOE, February 1980.

U.S. National Alcohol Fuels Commission, *Fuel Alcohol, An Energy Alternative for the 1980's.* Washington, D.C.: NAFC, 1981.

Chapter 11
Hydropower
Energy from Flowing Water

CONCEPTS

1. Historically, hydropower has been harnessed to move water, grind grain, saw lumber, generate electricity, and to perform other necessary tasks.
2. Hydropower is the most economical method known for producing electricity commercially.
3. Most of the larger, feasible sites for hydroelectric power have been developed in the United States.
4. The rising costs of fossil fuels and increased environmental degradation have created renewed interest in hydroelectric power.
5. Renovation of old, abandoned hydroelectric sites is becoming popular with individuals, small companies, municipalities, and utilities.
6. Hydropower has few adverse environmental effects.
7. Since 1978, large utility companies have been required by federal law to purchase available electrical energy from small-scale producers.

GLOSSARY

CAVITATION—wear on turbine blades caused by air bubbles in the water in contact with the blades; air in the water, resulting in reduced efficiency.
HYDRAULIC HEAD—the difference in elevation between two water surfaces.
HYDROTURBINE—a device that is rotated by the flow of water through or around a fanlike configuration.
IMPOUNDMENT—a lake or reservoir formed by a barrier such as a dam.
RETROFIT—to modernize, improve, rehabilitate, or redevelop.
RUNNER—the part of the turbine that rotates.

INTRODUCTION

The kinetic energy of flowing water was undoubtedly first used to transport materials to downstream locations by floating the materials themselves (i.e., logs) or through the use of rafts, boats, or barges. Since the energy of flowing water has limited direct application in human endeavors, efforts to divert and convert it to useful forms have been devised over the years.

Devices to extract and use the mechanical energy from streams and rivers were the first to be developed. As the need for energy forms shifted from mechanical to electrical, water-driven turbines connected to generators became the major application of hydropower.

Small independently owned hydroelectric-generating facilities were soon replaced by larger plants operated by utility companies, which could produce electricity and distribute it over a greater geographic area to meet the growing demand. This continued until most of the best hydropower sites in the United States were developed. Renewed interest in small-scale hydroelectric power has been spawned by federal and state legislation and by the interacting forces of continually increasing costs and the environmental impact of fossil-fueled electricity-generating plants.

THE DEVELOPMENT OF A TECHNOLOGY

Other than for transportation, it is conjectured that very little use was made of the energy in flowing water until after the wheel was developed. Early records show that vertical-axis waterwheels were used in Greece around 85 B.C. and that horizontal-axis units were in use beginning about 15 B.C.

When placed in moving water, an appropriately designed wheel and axle assembly can convert the fluid's lineal motion to axle rotary motion. This rotary motion can be connected via shafts, pulleys, cables, and gears to machines which consequently reduce the level of human or animal power involvement in a task. By using combinations of pulleys or gears of varying diameters, the relatively slow rotational speed of a water-powered shaft can be increased, decreased, changed in direction, or transmitted to various locations to do work.

As use of water power continued, rapid changes and improvements in the designs to harness the available energy were made. It was soon learned that the low-level kinetic energy of the flowing water was inadequate to provide sufficient power to meet the demands placed on the system. Rivers and streams also contain potential energy by virtue of elevation changes. This energy could also be gleaned if the force and weight of the falling water could be applied to the waterwheel. Subsequent designs made use of this fact by incorporating dams, diversions, and by piping water from higher elevations to the wheel.

As the limited number of ideal hydropower sites were developed and the human population expanded, ever-greater demands were placed on the energy output of the existing gristmills, sawmills, and other small hydro facilities. Demand for these services became overwhelming for many small mill owners, and employees had to be hired to meet the need for services. Because employees often desire to live near their place of employment, towns and villages soon sprang up near the mills. Before long, factories were built along the waterways and employed people to meet production needs.

Devices used to extract hydropower are categorized according to how and where the water strikes the mechanism. Waterwheels, the simplest of these devices, have provided mechanical power for nearly 2000 years. All waterwheel designs involve large-diameter slowly rotating wheels which provide the high torque necessary for driving heavy, slow-turning mechanical equipment.

The undershot wheel is the most basic or primitive of waterwheels. Water passing under the wheel strikes the paddles or blades, causing the wheel to turn. This design requires a wheel diameter of about 15 ft (4.6 m) and can operate on very low heads. A disadvantage is that it is low in efficiency because it does not take advantage of the weight of the water. (see Figure 11-2).

Figure 11-3 shows the Poncelet wheel, which is an adaptation of the undershot design. Higher efficiencies are achieved with this unit because the blades or paddles are curved and advantage is taken of the force in the water created by the restricted opening. The weight of the water also contributes energy. Care must be taken to prevent wheel damage by screening rocks and debris from entering the narrow opening under the wheel.

Additional efficiency and more energy can be acquired by directing

Figure 11-1 Hydropowered gristmill at Port Washington, New York. (Courtesy of Nassau County Museum.)

the flow of water to the upper portion of the wheel. The breast wheel (Figure 11-4), by receiving the water at a higher level, takes advantage of both the force and weight of the water. These units are more difficult to build because of the "bucket"-shaped paddles, the curved breastwork, and trash racks, which are mandatory because of the very narrow space between the wheel and the breastwork.

The most efficient and the most widely used waterwheel is the overshot wheel (Figure 11-5). With this design, water is directed to the

Figure 11-2 Undershot wheel. (Used with permission from *Micro-Hydro Power: Reviewing an Old Concept*, copyright 1979, National Center for Appropriate Technology, Butte, Mont.)

Figure 11-3 Poncelet wheel. (Used with permission from *Micro-Hydro Power: Reviewing an Old Concept*, copyright 1979, National Center for Appropriate Technology, Butte, Mont.)

top of the wheel by a flume. Some linear force, but mostly the weight of the water entering the vanes or buckets, causes the wheel to rotate. The buckets are designed to retain the water as long as possible during rotation.

The feasibility of a particular hydropower site was, and still is, a very important consideration before construction of a facility can begin. The amount of power available from flowing water is proportional to the rate at which the water flows and the vertical distance the water falls or drops, this vertical distance being the hydraulic head. The potential power obtainable from any stream can therefore be mathematically calculated by using the factors of flow volume, average cross-sectional area of the stream, velocity of the flow, and the height from which the water falls.

The world's first hydroelectric plant began operation in Appleton, Wisconsin, on September 30, 1882. The initial intent was to provide

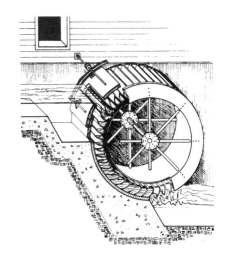

Figure 11-4 Breast wheel. (Used with permission from *Micro-Hydro Power: Reviewing an Old Concept*, copyright 1979, National Center for Appropriate Technology, Butte, Mont.)

Figure 11-5 Overshot wheel. (Used with permission from *Micro-Hydro Power: Reviewing an Old Concept*, copyright 1979, National Center for Appropriate Technology, Butte, Mont.)

illumination for the owner, but neighbors soon desired the same convenience and the first electric "utility" was born.

Unlike direct water power and the mechanical power acquired from it, electricity can be transported through wires to areas that are some distance from the generating facility. Since electricity is an extremely versatile energy form in that it can be easily transported as well as converted to light, heat, and mechanical energy via a motor, it rapidly became a very popular form of energy. No longer was it necessary to live near the river or transport raw materials to the mills at the hydropower sites in order to use the mechanical energy of the flowing water. Many more people could now benefit because electricity could be generated and brought to an area to run a gristmill or sawmill motor, illuminate, heat water and homes, or for a multitude of other uses.

Although early electricity use was limited due to the small size of generating plants and restricted to areas near the plants because direct current cannot be transformed to travel long distances, rapid developments occurred as demand for the energy form soared. New generators were designed, producing alternating current that could be transformed to high voltages for transmission to more distant consumers. Many small, individually owned, local generating plants sprang up along rivers and streams where the water flow and topography were conducive to development.

It was not long before the best sites along many of the flowing bodies of water were being utilized and the demand for electrical energy rose above their generating capacity. The dilemma was met by companies (utilities) building much larger generating plants along the rivers. Where appropriate sites were not available, rivers and streams were dammed to establish impoundments which often encompassed several square miles. By controlling the release of impounded water at the dam, water that normally would have continued on to the sea could be used to generate electricity even during dry seasons. These dams also helped to prevent flooding of downstream areas during spring runoff or after severe rainstorms. Small independent producers of electricity found that they could not compete financially with the large utility companies and were soon forced out of business.

The utility companies, with their improved efficiency and ability to generate larger quantities of electricity, were no longer using the low-speed, high-torque waterwheels of earlier years. To produce electricity, generators require high rotational speeds. Although waterwheels can be (and still are) geared or connected to pulley systems to greatly increase their 5 to 20 rpm, the approximately 1500 rpm necessary to drive a generator can better be acquired through the use of turbines.

TURBINE DESIGN AND USE

Turbines can be classified according to their design. There are basically two types: impulse and reaction. Impulse turbines are used in high-head applications and include the Pelton, Turgo, and cross-flow designs. There are two major types of reaction turbines, the Francis and the Kaplan.

Patented in 1880 by Lester Pelton, the Pelton turbine is the most common of the impulse units. It has been used in hydraulic heads as low as 50 ft (15.2 m) and as high as 3000 ft (912 m) in South America. Most applications are with heads of 100 ft (30.4 m) and above. The Pelton design involves the use of one to four nozzles with restricted openings which produce a jet of highly pressurized water directed toward a disk with cups attached to the outside edge. The cups are designed to extract the greatest amount of energy possible from the water (see Figure 11-6). Depending on the head available, rotational speeds upward of 1300 rpm can be acquired and no gearing is needed. Where lower heads and therefore lower pressures are present, gearing to 1500 rpm is required.

The Turgo impulse turbine is an improved variation of the Pelton unit. Developed in 1920 by Eric Crawdson, this unit is manufactured exclusively by Gilkes of England. The runner is a cast wheel which has a shape similar to a fan blade that is closed on the outer edge. A water jet is applied at an angle to the blades rather than perpendicular and strikes the front, goes through the blades, and exits on the opposite side. As with the Pelton, more than one nozzle may be used. Hydraulic heads of 40 ft (12.2 m) and above are required. This unit can produce

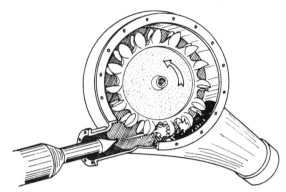

Figure 11-6 Pelton Turbine. (Used with permission from *Micro-Hydro Power: Reviewing an Old Concept*, copyright 1979, National Center for Appropriate Technology, Butte, Mont.)

power equal to the Pelton with a smaller-diameter runner rotating at a higher speed.

Produced commercially by Ossberger of West Germany, the crossflow turbine incorporates an advanced Poncelet design, resulting in a hybrid impulse-reaction turbine. Applicable to both high heads with low-volume water flow and low heads with high-volume flow, the unit was patented in 1903 by A. G. M. Mitchell. The runner of this unit resembles a squirrel-cage blower which is driven by a flow of water from a rectangular opening. The water is forced through the blades, across the inside of the unit, and out the other side, thus extracting most of the kinetic energy available (see Figure 11-7). These units have been successfully installed in heads as low as 39 in. (1 m). They may also have application for use in ocean tidal and current energy systems.

Although providing the same function as that of impulse turbines, reaction turbines work on a different concept. These turbines are placed directly in the water stream and use the water pressure combined with its velocity, rather than its velocity alone, to acquire high-speed rotation.

Francis turbines are generally installed in very large hydroelectric facilities with hydraulic heads often exceeding 1000 ft (305 m). As you may have guessed, Francis (James B.) is responsible for its design. With this unit, water enters openings around the entire top and falls through the squirrel-cage-type fins, causing rotation. In addition, water discharging into the draft tube at the bottom provides suction and thereby makes most efficient use of the head between the turbine and the tail water (see Figure 11-8).

An Austrian engineer by the name of Viktor Kaplan recorded patents in 1913 and 1915 for his reaction-type turbine. As an improvement of the propeller-type design, which looks much like a ship's propeller enclosed in a housing, Kaplan's unit has variable-pitch turbine

Turbine Design and Use

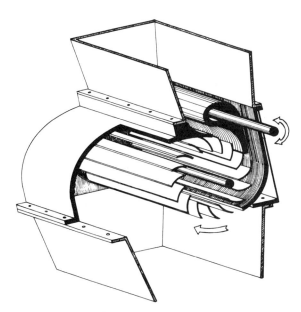

Figure 11-7 Cross-flow turbine. (Used with permission from *Micro-Hydro Power: Reviewing an Old Concept*, copyright 1979, National Center for Appropriate Technology, Butte, Mont.)

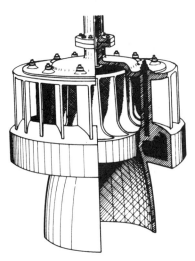

Figure 11-8 Francis turbine. (Used with permission from *Micro-Hydro Power: Reviewing an Old Concept*, copyright 1979, National Center for Appropriate Technology, Butte, Mont.)

blades which are used to control the speed of rotation under seasonal flow conditions. The fixed-pitch propeller reaction turbines available before Kaplan's innovation required a constant flow rate to operate efficiently.

Axial-flow turbines have been found to be economical in extracting power from medium-head sites. In these units, the direction of the water flow to and from the turbine is parallel to the axis of the unit. In the bulb-type turbine, the turbine is connected directly to the generator encased in a steel bulb and the bulb centered in the water flow. The tubular type has an extended shaft which connects the turbine with the generator, which is installed above the water passage.

Other water-driven mechanisms used to generate electricity from the energy available in flowing water have been devised and are in various stages of experimentation and development. More detail on these units is given in the section on tidal energy in Chapter 12.

OTHER CONSIDERATIONS

Care must be used with all turbine designs to eliminate as much foreign matter from the water flow as possible. Trash racks must be kept free of debris, and grit and stones must be eliminated as much as possible because of their abrasive effect on the turbine blades. Cavitation is also a potential problem, especially with the Francis design.

The relative efficiencies of the various water wheel and turbine designs are given in Table 11-1. From this display it is easy to see why commercial efforts to produce electricity have utilized turbines.

TABLE 11-1 Relative Efficiency of Water Wheels and Turbines (%)

Runner	Efficiency Range of the Runner	End-Use Efficiency After Gearing and Generator Losses
Water wheels		
Undershot	25–45	19–34
Poncelet	40–60	30–45
Breast	35–65	26–49
Overshot	60–75	45–56
Turbines		
Reaction	80	60
Impulse	80–85	60–64
Cross-flow	60–80	45–60

Source: Adapted from *Micro-Hydro Power: Reviewing an Old Concept*, copyright 1979, National Center for Appropriate Technology, Butte, Mont. Used with permission.

As the demand for electrical energy increased, utility companies, groups of utility companies, and even federal agencies became involved in developing larger and larger hydroelectric facilities until most of the sites in the United States feasible for building large units were utilized.

One example of a large hydroelectric facility is Hoover Dam (Figure 11-9). The U.S. government became involved in its development to alleviate a serious flooding problem on the Colorado River, and uses the impounded water to generate electricity. Sale of this electrical energy has nearly paid for the development of the facility.

Located in Nevada, Hoover Dam is 726 ft (221 m) high, 1244 ft (379 m) long, and impounds Lake Mead, which is America's largest human-made reservoir. The dam was completed in 1935 and generators were added unit by unit until 1961, when the seventeenth (and last) brought its total electrical generating capacity to 1345 MW. This electricity is distributed to customers in Nevada, Arizona, and California.

Hoover Dam was the largest hydroelectric facility in the world until 1949. Since then dams have been built as high as 5800 ft (1768 m) at the Reisseck site in Austria, and generating capacity has reached 4500 MW at the Bratsk Dam facility in the USSR. The highest dam in the United States, with a height of 2558 ft (780 m), and the greatest producer of electricity is the Grande Coulee Dam, at nearly 2000 MW. Figure 11-10 provides a schematic illustration of a typical plant, much of which is underground and hidden from view.

Figure 11-9   Hoover Dam. (From the U.S. Bureau of Reclamation.)

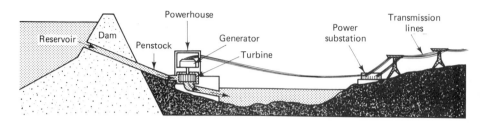

Figure 11-10 Schematic of a typical hydroelectric system. [From *The Encyclopedia Americana* (Danbury, Conn.: Grolier, Inc., 1982).]

In areas where the volume of available water is inadequate to turn a turbine to meet electricity demands on a daily basis, water may be recycled. After flowing through the turbines, the water is retained in an impoundment below the generating plant. During periods of low electrical demand (nights, weekends, and holidays), water is pumped to another impoundment above the dam for reuse when demand for electricity increases (see Figure 11-11). One way to do this is to use very large motor-driven pumps. Another way to pump water is to reverse the entire system. An interesting point is that a generator produces electricity when mechanical energy is used to rotate the armature

Figure 11-11 Pumped-storage hydroelectric facility. (Courtesy of the New York Power Authority.)

in the magnetic field; the generator becomes a motor, producing mechanical energy when electrical energy is directed to rotate the armature. The motor-generator is used to drive the turbine in reverse, thus pumping water from the lower to the upper impoundment. Kaplan turbines are specifically applicable to this use and approximately two-thirds of the energy expended to pump the water to the upper impoundment can be recaptured during the next high-demand period.

EXPANDING ELECTRICITY DEMAND

As the population expanded and dispersed across the United States, hydropowered electricity-generating plants were built to meet the energy demands of people in the various regions. Where flowing rivers were not available and impoundments impossible to construct, utility companies were forced to seek other energy sources to rotate their turbines.

Steam became the turbine driving force because of the great kinetic energy available when it is under pressure. Easily transportable chemical energy in the forms of wood, coal, fuel oil, natural gas, and in recent times nuclear, release their high levels of thermal energy to boil water, creating steam which is directed to rotate a turbine.

As the U.S. demand for electrical energy continued to increase, hydropower contributed less and less to the total electrical energy production. Soon after 1900, hydropower contributed 57% of total U.S. electricity. In 1969 it provided 17%; in 1970 this figure dropped to 15%, and it went down to 12% in 1983. This downward trend has now begun to reverse as costs associated with fossil-fueled electricity-generating plants continue to escalate.

Using the power of flowing water has proven to be the most economical way of producing electrical energy. Another advantage is that water flow through a turbine can be regulated to increase or decrease electrical output rapidly according to consumer demand, or to interface with other methods of producing electricity.

POTENTIAL FOR HYDROELECTRIC FACILITIES

Hydropowered systems provide approximately one-third of the total worldwide electrical energy. Although there are some countries, for example Norway, which obtain nearly all their electricity in this manner, many others rely on fossil-fueled facilities. It is estimated that only 10% of the world's hydro potential has been developed.

The countries that have the greatest potential for hydropower development are in Africa, Asia, and South America. Interestingly, most of these countries lack significant coal deposits, which provides impetus to use these undeveloped assets. China, for example, has over 60,000 installations and continues to build small, 35-kW units for local use since they do not have an extensive interconnecting grid. Most European countries have already built plants on the available sites.

Depending on the reference you select, the United States still has between 38 and 90% of its potential hydroelectric sites undeveloped. If your position is for total possible development (no matter what), the upper figure may be accurate. If, however, you consider such things as economical feasibility, the quality of life, and question the need for excessive electrical energy, then the lower figure is appropriate. Total possible development, of course, would mean the impounding of all potential sites, including wild rivers, wilderness areas, scenic areas, and state and national park sites. Just think of the quantity of electricity that would be available to people in the southwestern part of the United States if the Grand Canyon were fully developed (dammed) as a hydroelectric facility!

Realistically, new, large-scale development of hydropower in North America is generally restricted to the mountainous areas of the Pacific Northwest and Alaska, although there are still some potential sites in the Rockies and the eastern United States, especially the Northeast. In 1979, the Army Corps of Engineers reported that there are approximately 50,000 sites that could be developed.

The Pacific Northwest currently produces over 90% of its total electrical energy through waterpower. As of 1983, there were 1288 facilities contributing to total U.S. electrical demands, and many smaller-scale sites in various stages of completion or renovation.

RESURGENCE OF SMALL-SCALE HYDROELECTRIC POWER

A number of events have caused renewed interest in the development or redevelopment of small hydro facilities. Most of the feasible large-scale sites are already being utilized, the high cost of land makes large impoundments financially impractical, and continued increases in fossil fuel prices have escalated electricity costs. In addition to new technology which permits the efficient use of small-scale, low-head sites, several other incentives are attracting attention.

There is currently a struggle over who has the legal rights to the energy available from hydropower, and this has required the federal court system and the Federal Energy Regulatory Commission (FERC) to become involved. The FERC is the regulatory agency for hydroelectric projects. Specifically, the struggle is over the consumer cost of electricity and the inequitable distribution of this cost. There are several areas in the nation where urban electricity customers are required to pay for energy generated in fossil-fuel-fired plants, while customers in the less densely populated areas enjoy the lower cost associated with locally generated hydropower. In New York State, for example, consumers in the New York City area are required to pay 6.5 times as much as consumers in the northern part of the state for equal kilowatt-hour usage.

As a result of studies and cost analyses, the National Energy Plan of 1977 emphasized the expansion of hydroelectric facilities at new or existing sites because of reduced costs compared with new coal or nuclear generating facilities of equal capacity. An added asset is that many small hydro facilities already exist near high-demand centers.

One of the larger incentives for hydroelectric development as well as other renewable energy electrical producers is the Public Utilities Regulatory Policies Act (PURPA) of 1978. This act requires large utility companies to purchase electrical power from small, privately owned generators at rates equal to the utility's savings from not generating that electricity themselves. This savings is called the "avoided cost" and is high where expensive, imported oil is burned at low efficiency to produce electricity. Examples in 1983 values are 5 cents per kilowatt-hour in Massachusetts, 6 cents in New York, and 8 cents in New Hampshire.

As might be expected, many utility companies have resisted implementation of PURPA because there are several complications associated with interconnecting smaller systems with the grid. Included in these problems are such concerns as the type of electricity being generated (ac or dc), phase and cycle matching, safety of both equipment and personnel, and payback rates, not to mention the loss of previous consumers who are now providing electricity for themselves. This resistance has taken the form of foot-dragging delays, inflexibility, added unclear requirements, insurance coverage demands, and other obstacles. A federal court case upheld the act as constitutional in the spring of 1983. A similar law passed in 1979, called the Northwest Regional Power Act, mandates the purchase of electricity from renewable sources when the cost is less than or equal to power from new conventionally fueled electricity-generating plants.

In addition to the PURPA incentive, the DOE has provided funding for feasibility studies and demonstration grants, and has supported various hydroelectric projects. The Crude Oil Windfall Profits Tax Act (COWPTA) of 1980 provides a federal tax credit for capital improvements on facilities up to 25,000 kW of generating capacity. The FERC provides exemptions from licensing requirements for units under 5000 kW and shortcuts through these requirements for 5000 to 80,000-kW facilities. The Economic Recovery Tax Act (ERTA) provides for the 5-year amortization of capital investments in new equipment and dam reconstruction and repair. The U.S. Supreme Court has ruled that when utility company licenses expire on hydroelectric plants, public utilities have an opportunity to purchase them for the original cost less depreciation. Many states have also encouraged small hydro development through exemptions from taxes on property improvement, providing corporate income tax credits, issuing low-interest loans, establishing payback rates under PURPA and reducing environmental requirements. Of all these incentives, the assurance of selling electrical energy to the utility and the tax advantages are the prime movers of the resurgence.

Small-scale hydroelectric units can be categorized in two different size ranges. Facilities that generate between 100 and 20,000 kW are considered small scale, while those producing up to 100 kW are termed microhydro units.

Many of the small and microhydro sites that are receiving attention are those which were operable in the past and fell into disuse when the large, centrally located utility companies became established (see Figure 11-12). At many of these sites, the dam and impoundment already exist and retrofitting efforts are necessary to place the unit back in service (see Figure 11-13). Many redevelopers are making repairs

Figure 11-12 Small, old hydroelectric facility: interior view. (Courtesy of James River Corporation.)

Figure 11-13 The old facility; remodeled interior view. (Courtesy of James River Corporation.)

to the dams but are purchasing new turbines and generators because of the efficiency of the newer units. Establishing new sites complete with dam and impoundments is also being done, but the costs related to totally new facilities is expensive.

Small-scale facilities are being developed by municipalities as well as industrial and commercial entities. Their primary objective is to produce electrical energy and sell it at a profit to local consumers or the regional utility. Several municipalities have retrofitted older sites and are providing their citizens with low-cost electrical energy. New England currently has 23 sites in various stages of operation, including a 3000-kW unit in Berlin, New Hampshire, and a 15,000-kW unit in Lawrence, Massachusetts. In addition, New York has 19 municipally owned/operated units in the upstate area, and cities in several other states are also becoming involved.

New microhydro units are being installed and older ones redeveloped by individuals as well as corporations. The primary objective of these units is to provide low-cost electrical energy for on-site use. The average residential demand for electrical energy is about 6000 kWh per year, which could be provided by a 1-kW production unit, but most installations exceed on-site maximum needs. Whether intended for personal or corporate use, excess energy generated can be sold to the local utility.

Technological developments in the design and efficiency of tur-

Figure 11-14 Aerial view of an ultra-low-head hydroelectric plant under construction. (Courtesy of Energy Research & Applications, Inc.)

Figure 11-15 Microhydropower system. (Courtesy of Small Hydroelectric Systems & Equipment.)

bines and generators have been a contributing factor to the resurgence of small-scale hydropower. Standard turbine designs are available for low-head applications, low-cost computer control systems are available, and induction and synchronous generators with governors automatically adjust to 60-cycle ac to provide compatible current interface with the commercial utility.

For those systems that generate dc rather than ac, electrical energy produced beyond that which is needed can be "stored" by converting it to chemical energy in batteries. This energy would then be available to meet peak demands. For developers who do not wish to purchase an expensive bank of batteries, a synchronous inverter can be installed which changes the dc to ac and permits utility interface at a constant 120-V 60-Hz rating. Electricity to meet peak demands or needs due to system failures can be purchased back from the utility, thus using the utility as the "battery."

Costs of producing electricity using the energy of flowing water vary greatly. For those persons desiring to do much of their own work and to adapt used equipment, a micro system can be constructed for about $5600. Prices in 1984 for a new turbine and generator varied from $8000 to $28,000. This, of course, did not include pipe, wire, generator house, or dam. Complete-15 kW units, except for the dam,

Figure 11-16 Gemini synchrononous inverter. (Courtesy of Windworks, Incorporated, a Subsidiary of Wisconsin Power & Light Company.)

sold for about $36,000, with expected payback periods of 12 to 14 years.

ENVIRONMENTAL ISSUES

Many of the current hydroelectric plants were constructed prior to the era of strong public concern for the environment. The most economical, easily accessible sites were consequently developed without such concerns. Now, as interest is regained in hydropower, utility companies, municipalities, and small independent producers must be cognizant of all environmental interrelationships.

Large hydroelectric facilities require great pressures to drive the turbines, and this necessitates high dams. These dams create large impoundments, which result in a number of environmental impacts, some positive, some negative. Lake Mead, which is 110 miles (177 km) long, Roosevelt Lake (created by the Grande Coulee Dam), which is 151 miles (243 km) long, and other large reservoirs provide prime boating, fishing, and recreational areas which did not exist prior to construction. In addition, devastating floods previously caused by local storms and spring runoff are no longer a hazard to people living in the immediate area. Instead, this water is now available for irrigation purposes during dry periods, and soil erosion has been controlled.

Large impoundments, on the other hand, flood many acres of land which may previously have been residential areas, fertile farmland, wildlife areas, or scenic vistas. They drastically raise the water table of the surrounding area and change the soil mineral levels. Earthquakes are also a possibility due to the weight of the water concentrated in one area. Over a period of many years, impoundments have a tendency to fill with silt and therefore become less effective as generating facilities due to the reduction of the hydraulic head. Although the loss of wildlife habitat is often a strong point against the development of impoundments, the loss of naturally existing scenic areas is irreplaceable.

Efforts are being made to review hydropower site potential on a state-by-state basis and many are finding that some new site development as well as redesigning existing sites for improved generating capacity holds minimum potential for environmental damage. For example, the state of Maine has determined that it could increase its hydroelectric production by 12% without significantly harming the environment. Others are finding that site development is unacceptable because of the loss of agricultural lands, wildlife habitat, fishing and boating opportunities, fish spawning areas, and wilderness areas.

FUTURE OF HYDROELECTRIC POWER

With world demand for electrical energy increasing at an 8% annual rate, new, expanded, and retrofitted hydroelectric installations are expected to play a significant role in meeting these demands.

Although the small and micro facilities individually contribute comparatively little to the total supply picture, continued upgrading of the many old sites and the development of new sites will add significantly to our needs. Nationally, our hydroelectric output could increase from the current 12% of the total electricity produced to 23% if the old, previously used sites were renovated. Utility companies, especially those using large quantities of fossil fuels, can be expected to continue their resistance to cooperate as the "little guys" continue to "make waves" and cut into "their" profits.

Utility companies, in attempting to reduce their generating costs, will also build new and renovate old hydro sites wherever possible. The development of pumped-storage hydro facilities will increase in the industrialized nations, especially in those where the feasible, naturally occurring, extensive impoundments have already been developed.

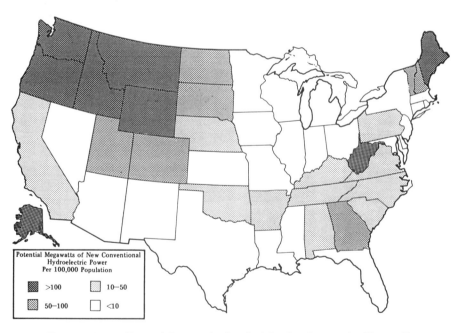

Figure 11-17 Potential new hydroelectric development. (From the Federal Power Commission.)

Remote locations that have large-scale hydro potential will be impounded and new electricity transmission techniques will be used to transport the power to the demand centers. Where possible, hydropower will be used as the supply system for the base load to reduce fossil fuel consumption.

Other projections for the future include using wind power to pump the water at pumped-storage projects, municipally developed facilities extracting the energy from water supply and water treatment plants, and perhaps hydroelectric-powered electrolyzers, which create hydrogen gas from water by dissociation. The resulting hydrogen and oxygen gases produced during low-demand periods could be sold.

SUMMARY

Our society has been increasingly dependent on electricity since its earliest days of development as an energy form. As clean, easily transported, and easily converted energy, it has become a virtual necessity in our daily lives. The quest not only to meet our needs but to reduce costs has led us from hydropower to chemical fuels and on to alternative energy sources to turn the turbines. As the availability of economical hydro sites diminished, chemical fuels to provide thermal energy to turn the turbines were employed. Now, as the fossil/chemical fuels are becoming depleted and more expensive, alternatives to this energy must be developed and made to play a larger role in generating electricity.

Harnessing the power of flowing fresh water is the most economical method of producing electrical energy. Although most of the large ideal sites in the United States have been exploited, there are still many sites that could provide adequate energy for small and intermediate-size generating units.

There has been a resurgence of interest in hydroelectric power in recent years due to the escalating costs and negative environmental impact associated with using fossil fuels. Acid rain, fuel oil spills, air pollution, water pollution, radiation leaks, radioactive spent-fuel storage, and decommissioning of electrical power plants contribute greatly to the direct and indirect costs of using fossil fuels to generate electricity.

ACTIVITIES

1. Investigate the historical development of the various designs of waterwheels and turbines used to convert hydropower to more usable forms of energy.
2. Visit and read about hydropower sites that have contributed historically to your geographic area.

3. Tour a modern hydroelectric plant. Prepare a sketch of how it is designed and how it functions. Discuss the hydraulic head, turbine design, water flow rate, electrical power output, the amount of energy it contributes to the total electrical demands of the area, the consumer cost per kilowatt-hour, and the number of hydroelectric plants in the area.
4. Explore the requirements necessary for the installation of a microhydroelectric system. Research the various units that are available and "select" one to meet your needs.

BIBLIOGRAPHY

Alward, R., S. Eisenbart, and J. Volkman, *Micro-hydro Power: Reviewing an Old Concept.* Butte, Mont.: The National Center for Appropriate Technology, 1979.

"Hydroelectric Power," *The Encyclopedia Americana.* Danbury, Conn.: Grolier, 1982.

Kohler, Joe, "Home-Size Hydro Power—A Practical Case Study," *Solar Age,* September 1983.

Leckie, Jim, G. Masters, H. Whitehouse, and L. Young, *More Other Homes and Garbage.* San Francisco: Sierra Club Books, 1981.

Loeb, William A., "How Small Hydro Is Growing Big," *Technology Review,* August–September 1983.

Marier, Donald, "Hydropower's Renaissance: Technical and Financial Innovations Pave the Way," *Alternative Sources of Energy,* July–August 1983.

Meador, Roy, *Future Energy Alternatives.* Ann Arbor, Mich.: Ann Arbor Science, 1978.

U.S. Department of the Interior, Bureau of Reclamation, *Hoover Dam.* Washington, D.C.: U.S. Government Printing Office, 1982.

Chapter 12
Ocean Energy Systems
A Sea of Energy

CONCEPTS

1. Ocean energy forms are inexhaustible.
2. Each form is applicable to on-site and regional use.
3. Each system has its own special characteristics.
4. Each system involves conversion to electrical energy.
5. Ocean energy systems have few negative environmental effects.

GLOSSARY

DELTA T (ΔT)—the difference between two temperatures.
DOLDRUM—a region of relatively calm winds near the equator.
ESTUARY—an area where a river meets a sea and both river current and ocean tides are present.
GYRE—a circular course or pattern.
HYDRAULIC RAM—a pump driven by water.
INTERFACE—the common boundary of two bodies or spaces.
PERMEABLE—the ability to allow passage through a material.
PNEUMATIC—operated by air or air pressure.
SALINITY—pertaining to or containing salt.

INTRODUCTION

Water covers approximately 71% of the earth's surface. As a plentiful, inexhaustible resource, large bodies of water possess both thermal and motion energy, but the problem throughout history has been how to extract and apply these energies economically. The motion energy of flowing fresh water was the easiest to exploit by conversion to mechanical energy, and consequently many of the ideal naturally occurring sites have been developed (see Chapter 11). Although there is strong interest in smaller freshwater hydropower applications, additional attention has been focused on the use of ocean waters as a source of inexhaustible energy. Research and development efforts continue as we constantly strive toward an understanding not only of the characteristics of flow but also of tides, currents, waves, thermal differences, salinity differences, and the plants that live in our ocean waters.

TIDAL ENERGY

Tidal energy, the first ocean energy to be harnessed, has been used throughout history to supply needs for power. During the eleventh century, tidal mills were used to grind grain in England, France, and various other sites along the European coast. Hundreds of mills were built in the seventeenth and eighteenth centuries for grinding grain and sawing wood in England, and similar mills were built later along the New England coast (see Figure 12-1). Remnants of tidal dams are still evident today along the Maine coast.

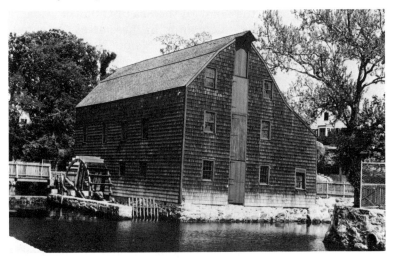

Figure 12-1 Tidal gristmill still operating at Saddle Rock, New York. (Courtesy of Nassau County Museum.)

The invention of the steam engine rapidly replaced tidal-powered facilities because the new device was not site-specific. The mechanical power for operating machinery could now be supplied near the site where raw materials were acquired, thereby eliminating considerable transportation-related time, costs, and efforts. With the reduction of business, operators of tidal-powered mills soon abandoned their facilities.

In more recent times, large-scale tidal energy projects were not proposed until 1928, when the French designed a pilot electric generating plant for construction at l'Aber Vrach. These efforts were followed by those in the United States, England, and the USSR, but none came to fruition because they were not economically competitive with hydropower for generating electricity. France was the only nation that actually followed through by building a plant. Located on the Rance River estuary, having a tidal range of over 42 ft (13 m), the plant continues to produce approximately 240 MW of electrical energy annually.

The interaction of gravitational forces of the earth, sun, and moon cause the tides to rise and fall. Tidal flow, however, depends on the shape of the cove, channel, estuary, or bay bottom and the configuration of the adjacent landmasses. Where large quantities of water are moved through constricted openings by rising and falling tides, the potential for energy exploitation exists. Worldwide, there are relatively few places where this phenomenon occurs at the tidal range of 16 ft (5 m), considered to be the minimum elevation change necessary for generating electricity using traditional technology. The best sites include Argentina's Patagonian coast, the English Channel, the Bay of Fundy, the Murmansk coast, and the coast of the Sea of Okhotsk. Tidal ranges of 16 to 36 ft (5 to 10 m) are common in these areas. Potential sites along the Bay of Fundy, located between Maine, New Brunswick, and Nova Scotia, possess nearly one-half of the world's theoretical tidal energy.

Two high tides and two low tides occur every 24 hours 50 minutes. This corresponds with one full rotation of the moon around the earth. Highest tides occur at full and new moons and the lowest occur during the first and third quarters of the 29½-day lunar cycle. Seasonal distances between the earth and sun also affect tidal levels. All of these variables are cyclical, therefore predictable, and must be considered when designing facilities that use tidal energy.

A number of different designs have been proposed for harnessing tidal energy, and nearly all require a bay or estuary that can be dammed economically to form a tidal pool or permit diversion of the water flow. Most designs are similar to the conventional low-head hydropower units used on rivers. The Rance estuary system incorporates a reversible turbine electric generator built into a dam impounding a single tidal pool (see Figure 12-2). As the ocean tide rises, water flows through the turbine on its way to the tidal pool. When the tide recedes, water from the pool again flows through the turbine on its return to the sea. Maximum generation occurs if the flows through the turbine are delayed approximately six hours to acquire maximum hydraulic head. Electricity is consequently generated four times in each tidal cycle. When the water levels on each side of the dam are equal, there is no hydraulic head (no potential energy) and no electricity can be generated.

Worldwide, there are only two tidal energy plants generating electricity commercially, the French Rance project and Kislaya Guba in the USSR, which is a 400-kW unit built in 1967.

As might be expected, Canada is a leader in tidal energy development due to its lengthy shoreline on the Bay of Fundy. Although several projects are being contemplated, three hold the greatest potential. The first to be completed is a 200-MW electricity-generating plant scheduled for operation in 1984. Located on the Annapolis River at Annapolis Royal, Nova Scotia, this demonstration unit is to be North America's first tidal energy project (Figure 12-3). A second project, proposed by the Nova Scotia Tidal Power Corporation, involves building a tidal dike between Economy Point and Cape Tenny across the Minas basin. The 49.2-ft (15-m) hydraulic head would be used to generate electricity for export to New England.

Canada's third proposed project, a cooperative effort between the United States and Canada, is for the Passamaquoddy Bay on the Bay of Fundy and would require two tidal pools. Passamaquoddy Bay, the upper pool, would be filled during high tide and the lower pool,

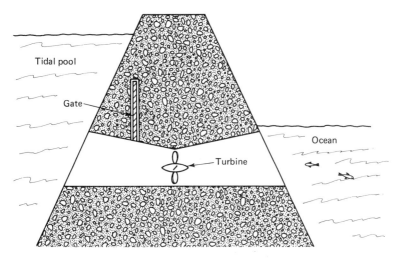

Figure 12-2 Tidal power dam with turbine (schematic).

Figure 12-3 Annapolis tidal power project. (Courtesy of Nova Scotia Power Corporation.)

Cobscook Bay, would be emptied during low tide. A dam between the two would contain the electricity-generating plant.

Eight more tidal projects are actively being planned by other countries since the high costs of conventional fossil-fueled plants have now made these systems economically competitive. There are 27 additional sites around the world with tidal ranges adequate for the installation of proven, conventional tidal energy systems and a multitude more that are appropriate for the newer, experimental designs. Two new designs have recently been advanced which represent a different approach to converting tidal energy to electrical energy. Both have received grants from the DOE and developmental modeling is continuing. Daniel Schneider has patented a "lift translator" which employs fluid dynamics and looks like a giant venetian blind (see Figure 12-4). Based on the principle that makes aircraft wings lift, an array of curved foils is mounted on an endless-belt mechanism which changes lineal flow to rotary motion. As the unit is rotated by fluid power, the foils revolve around the two axles, which can be connected to generators, compressors, or pumps. As the foils go over the axles they flop over to extract additional energy from the flowing water as it exits the translator. Fixed guide vanes can be moved in front of the unit on either side to help divert the fluid to an optimum angle, thus permitting flow from 180° directions (incoming and outgoing tides). The device is proposed for use not only in tidal power projects but also for use with river or ocean currents and wind energy systems. The initial design has been

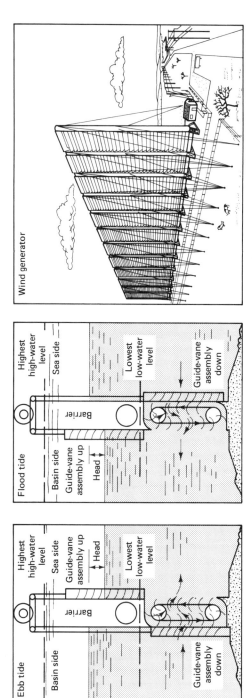

Figure 12-4 Lift translator—Schneider. (Reprinted from *Popular Science*, 1983, with permission of Times Mirror Magazine, Inc., and Charles Berger, Illustrator.)

tested in South Korea. The possibility of using the unit in low-head [2 to 25 ft (0.6 to 7.6 m)] applications is encouraging, especially since it could eliminate the need to construct large, expensive dams or locate only at sites possessing large tidal ranges.

The second design, developed by Alexander Gorlov, involves the use of a movable hollow plastic dam (Figure 12-5). Operable in tidal ranges of 6.5 to 16 ft (2 to 5 m), the hollow dam admits and expels water which compresses air inside to drive pneumatic turbines and generate electricity. In addition to the advantage of low-head application, this dam can be moved to permit the passage of ships or migrating fish. Estimates are that this design could be up to 30 times more economical then conventional tidal power systems. The lift translator and plastic dam designs both hold considerable promise because they eliminate most of the environmental concerns associated with tidal power utilization.

Three additional designs are being proposed for use in ultra-low-head tidal and freshwater river applications. One is of Canadian design, called the Vertical-Axis Water Turbine (VAWT), which looks much like an underwater windmill. The design is said to be functional in hydraulic heads as low as 1.64 ft (0.5 m) and is expected to produce up to 10 MW. The other two designs are being developed by the British firm AUR Hydropower Ltd. of London. One is named the AUR Water Engine; the other, the Salford Transverse Oscillator (STO). Both are designed for simple, economical, reliable operation in Third World countries. They use hydraulic rams (pumps) to extract the tidal energy and require

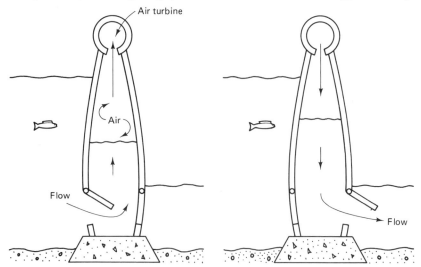

Figure 12-5 Movable plastic dam—Garlov. (Adapted from *Renewable Energy News*, June 1983.)

heads of 1.64 to 9.84 ft (0.5 to 3 m). More reports will be seen about these and other designs in publications that announce current events in our use of inexhaustible energy sources. Watch for them.

A major problem with tidal-driven electricity-generating plants is that maximum output may not occur at times of peak demand and the cyclical nature of the output is difficult to integrate with the overall electrical power generation and distribution system. Since electricity cannot be stored or saved, it must be produced when that energy form is needed and little variation of output is permissible. The only conventional method for generating electricity that is feasibly compatible to balance the cycles of tidal electricity generation is hydroelectric, since output is quickly variable and automatic interface is possible. Thermal electric plants do not respond as easily to rapid fluctuations in demand, owing to the time required to increase or decrease temperatures and steam volume. Another problem is that tidal power sites are frequently long distances from major cities or demand centers.

The power available from ocean tides is proportional to the square of the tidal range. When designers and engineers analyze a potential site for harnessing the available energy, they must not only measure the hydraulic head but also calculate the volume of water flowing past a given point. Since the power is proportional to the square of the height, a 13-ft (4-m) head can provide 16 times the power of a 3.28-ft (1-m) head.

Environmental impacts of tidal energy extraction can adversely influence shipping, and the impounding of estuaries can affect fish, wildlife, and traditional human activities in the area. Since dams or diversion walls are necessary to implement most systems, ship and boat access may require the construction of locks or may be restricted to specific times during the tidal cycle when the surface elevations are equal on both sides of the obstruction. Fish and wildlife reproduction, feeding, access, resting areas, and even life itself could be affected due to the tidal pool alteration and whirling turbines. Other ocean systems are believed to be less costly to the environment.

OCEAN CURRENT ENERGY

Ocean currents of the world can be thought of as rivers flowing within the seas. Just as we have harnessed energy from flowing freshwater rivers inland, proposals are being made to take advantage of ocean currents.

Often categorized as a form of solar energy, current systems are caused by prevailing global winds. Winds are created by the uneven heating of the earth by the sun. In blowing across water, winds create a surface stress causing both waves and currents. In 1835, Gaspard Gustave de Coriolis presented a paper describing the distortion of fluid motions attributed to the earth's rotation. When this fluid motion, termed the Coriolis effect, is combined with prevailing winds, currents are established within the oceans.

Figure 12-6 shows the world's ocean current systems. In the northern hemisphere, ocean current gyres move clockwise and because of the Coriolis effect are generally strongest along the western shores of landmasses.

Major ocean currents considered to have flow rates sufficient for electrical power generation include the Benguela (western coast of southern Africa), Agulhas (eastern coast of southern Africa), Torres Straits (between Australia and New Guinea), Kuroshio (Japan), and the Florida Current (southeastern United States). The Kuroshio Current possesses an average flow rate of 6.5 feet (2 m) per second with a volume of 777 million cubic feet per second, while the best rate, present in the Florida Current, averages 6.2 ft (1.9 m) per second but has a volume of 1.1 billion cubic feet per second. The Florida Current exhibits a peak summer rate of 7.5 feet (2.3 m) per second and drops to 4.9 feet (1.5 m) per second in winter. Use of this energy source is projected to be ideal, since electrical energy in Florida reaches its highest demand level in summer.

Although there are presently no operating current-driven electricity-generating plants, designs for various mechanisms are being developed, modeled, tested, and patented. Meetings on use of the Florida Current have been held and an installation projected for construction off the Florida coast should be cost-competitive by 1985.

Since the energy delivered is proportional to the cube of the ocean current speed, as with wind energy, proposals tend to favor several smaller hydroturbines rather than a very few large ones. Smaller units would be easier to fund, manufacture, move into place, and moor to the ocean bottom. Dams or diversions are not necessary as in conventional hydroelectric plants because the hydroturbine is placed directly in the ocean current flow. Electricity generated can be sent to shore over high-voltage cables or a factory ship producing energy-intensive products or goods could be moored near the site.

A recently proposed design for a unit involves a huge 6000-ton (26 million kg) ducted turbine called a Coriolis turbine. Rotors are housed in the duct which, through hydrodynamics, increases the current flow speed two- to threefold. Similar designs are being used in generating electricity from the wind.

"The central mechanism of the ocean turbine is a two-stage, axial-flow turbine consisting of a pair of counterrotating rotors, which are driven by the ocean current. In the large units, the turbine blade tips

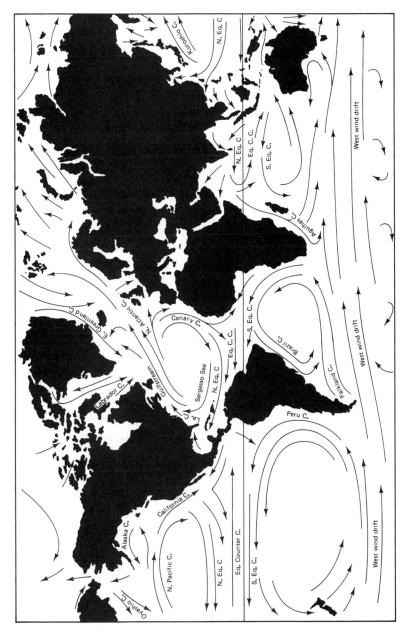

Figure 12-6 World ocean current systems.

Figure 12-7 Coriolis turbine. (Courtesy of AeroVironment Corporation.)

are mounted to stiff circular rims. The rotating rims turn the friction wheels, which drive electrical generators mounted inside the hollow cylindrical duct that houses the rotor" [Lissaman, p. 75].

Figure 12-7 illustrates the Coriolis turbine, which if installed in current flow rates found in the Florida Current would be comparable to or less in cost than an equal-output nuclear-, coal-, or oil-fired electricity-generating plant, and constant fueling would not be necessary.

Other designs that have been advanced include the lift translator, discussed in the ocean tidal energy section, and a method consisting of parachutes attached to a closed-loop drive cable. The cable drives a pulley connected to a generator (Figure 12-8). Laboratory tests have proven successful. Both of these designs are projected to be more economical to construct than turbine driven systems.

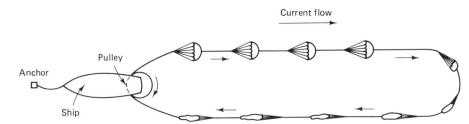

Figure 12-8 Parachute design for acquiring ocean current energy. [Adapted from Joseph P. Riva et al., *Energy from the Ocean*, prepared by the Science Policy Research Division, Library of Congress, for the House Committee on Science and Technology (Washington, D.C.: U.S. Government Printing Office, 1978.)]

As an alternative to generating electricity, one designer has proposed mounting vertical-axis turbines in the ocean floor. Each unit would be a submerged hydrogen generator driven by the current. The hydrogen produced would be piped to shore to fuel automobiles, aircraft, and electricity-generating plants.

Environmental concerns for areas near units extracting ocean current energy are philosophical at this time. Since the turbines would be located below the surface, shipping would be relatively unaffected except for surface factories or ships located on site. Whether a large array of units would have a significant effect on the current, adjacent landmasses, people, animals, or fish remains to be studied.

OCEAN THERMAL ENERGY CONVERSION

Exploitation of the potential energy in naturally occurring thermal gradients was proposed in 1881 by Jacques d'Arsonval. The concept uses the temperature difference between warm surface water and cold water from the ocean depths to run a heat engine, which, in turn, operates an electrical generator. Unlike tidal or wave energy, ocean thermal energy conversion (OTEC) systems using the massive energy storage capacity of the ocean can provide a steady output appropriate for base-load electricity supply.

The efficiency of OTEC systems is directly proportional to the thermal differential or delta T (ΔT) between the warm- and cold-water supplies. As may be expected, only tropical regions are suitable, where cold currents are found below warm surface waters. A ΔT of 31°F (17°C) is considered minimum for driving a system, thereby restricting potential plant location to generally within 20° latitude of the equator. There are exceptions, such as the Gulf of Mexico. Temperature differences of 36 to 40°F (20 to 22°C) are common over large areas of the world's tropical oceans. Some areas in the western Pacific have ΔT values as high as 44°F (24.5°C). Figure 12-9 illustrates ocean regions having OTEC feasibility. Winter surface temperatures within these regions average 77°F (25°C).

Equatorial doldrums will probably prove to be the best locations, since in these areas hurricanes are rare, winds seldom exceed 25 knots, and ocean currents are under 0.62 miles (1 km) per hour.

The first OTEC pilot plant was built onshore in 1930 at Mantanzas Bay, Cuba, by French engineer Georges Claude, This was an open-cycle system in which warm seawater was flashed to steam in a vacuum or low-pressure chamber, expanded through a turbine, and condensed when cooled by the flow from the deep water pipe. This and a succeeding effort were successful in generating 22 kW of electricity, but both failed due to structural problems caused by waves and currents acting

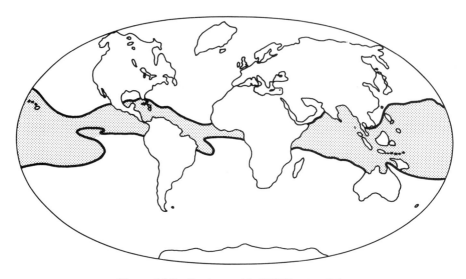

Figure 12-9 Regions with OTEC potential.

on the 6.5-ft (2-m)-diameter iron cold-water pipe. Claude's efforts were terminated in the mid-1930s due to the pipe problem and the fact that there was no net gain of energy (i.e., the plant used more energy than it produced). A third plant was planned by Claude and the French government for installation in Guadeloupe in 1958 but was terminated in favor of a fossil-fueled plant.

Although both open- and closed-cycle systems have been proposed for OTEC plants, the closed cycle or Rankine cycle originally proposed by d'Arsonval appears to offer the most promise for commercialization. Figure 12-10 shows a schematic diagram of an OTEC closed system. Warm surface water is used via a heat exchanger to expand and vaporize a pressurized working fluid such as liquid sulfur dioxide, isobutane, propane, ammonia, Freon, or other fluorocarbons. This release of energy is directed to a low-pressure turbine connected to an electrical generator. After expending its energy, the working fluid is condensed by the cool water from the ocean depths and recycled to the warm-water section of the system.

The U.S. government began sponsoring research and development on OTEC systems in 1972. The state of Hawaii, Dillingham Corporation, Lockheed, and Alfa-Laval joined forces in 1978 to produce a closed-system model. In August 1979, Mini-OTEC (Figure 12-11), anchored off Keahole Point on the coast of the island of Hawaii, became the first successful unit to produce a net energy gain from the OTEC process. Built on an old Navy barge, Mini-OTEC used the Rankine cycle with ammonia as the working fluid, driven by a temperature of $76°F$ ($24.4°C$) at the ocean surface and cooled by $40°F$ ($4.4°C$) water

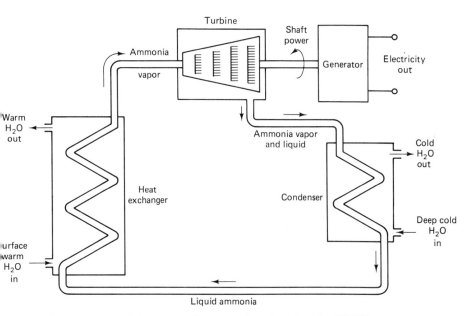

Figure 12-10 Schematic diagram of a closed-cycle OTEC system—Rankine cycle.

from a depth of 2165 ft (660 m) (a ΔT value of 36°F or 20°C). The unit generated 12 kW of electricity, but many engineering problems remain to be resolved. Japan commissioned another small plant in the fall of 1979.

The first large-scale at-sea system was Project OTEC-1 sponsored by the DOE. Constructed on a converted World War II Navy T-2 tanker, the SS *Chepachet*, and renamed SS *Ocean Energy Converter*, OTEC-1 was located in the 4000-ft (1219-m)-deep Alenuihaha Channel 18 miles west of Hawaii. This 1980 venture was a closed-cycle 1-MW experimental unit with the purposes of evaluating heat exchanger performance, development of processes and control mechanisms to reduce biological growth fouling of the system, solving other engineering problems, and studying environmental effects. The 40°F (22.2°C) thermal differential at this site permitted a 2 to 3% conversion efficiency.

There are still a great number of problems to be solved and conditions to be overcome. D'Arsonval's energy conversion system has been proven successful many times, while the ocean has remained a challenging, often forbidding environment. The obvious problems relate to the ocean itself with its hurricanes and storms, wind and waves, and ocean currents. How do you appropriately attach a fixed platform to the ocean bottom or anchor a floating OTEC plant in these ever-changing conditions in depths often greatly exceeding 2000 ft (610 m)? What technique do you use to attach power cables from this rotating

Figure 12-11 Mini-OTEC. (Courtesy of Dillingham Corporation.)

and translating plant to shore so that the energy can be utilized? Alternating-current electricity cannot be transmitted economically at distances exceeding 30 miles (48.3 km) without additional transformers. How can you keep the nutrient-rich ocean water from biofouling the heat exchangers? How do you attach, support, and retain a 2000-ft (610-m)-long cold water pipe 33 ft (10 m) in diameter weighing 110,000 lb (49,896 kg) to a pitching and rolling ship? What environmental impact is likely from discharging large volumes of cold water at the normally warm surface?

Technological progress toward the solutions of these problems has been evident from Claude's iron cold-water pipe to Mini-OTEC's reinforced rubber pipe and on to OTEC-1's cluster of three polyethylene pipes. Some designs incorporate the cold-water pipe, mooring, and power cables in one unit. Biofouling problems of the heat exchangers have been addressed on OTEC-1 by using both mechanical means to scrub the pipes and chemicals (chlorine from the sea) to retard rapid accumulation. Titanium metals have proven strong and corrosion resistant but are very expensive due to the limited availability of the base metal. Ammonia has been shown to hold advantages over the other working fluids because it has the highest heat of evaporation, greatest thermal conductivity, is readily available and inexpensive, and leaks are easily detected.

Research and development efforts directed toward system improvement and eventual commercialization are continuing. Since bottom-mounted plants are impractical in deep waters, designers have focused on floating platforms. Such platforms can be categorized according to shape, one type incorporating a rectangular ship or barge-type hull, while the other is of spar bouy design and generally cylindrical in shape. Figure 12-12 illustrates the spar bouy configuration, which considerably reduces the vertical and angular forces imparted by the sea on floating objects. This design should greatly reduce the problems associated with the connections of the cold-water pipe and power cable to the platform.

In addition to the open- and closed-cycle systems already discussed, vapor lift and direct thermoelectric conversion systems have been advanced. The vapor lift design is an open-cycle system recently proposed by Beck, Zener, and the team of Charwat and Ridgeway. The concept involves converting the warm water to mist or foam in evacuated chambers, raising it to a higher elevation, where it is condensed with cold water, and using the head (height) of the lifted water to drive a hydraulic turbine. Advantages of this design are that no heat exchangers are required and fresh water is a by-product—an additional boon for plants located near islands or desert regions.

The direct thermoelectric conversion concept is the least complicated of all the designs. With this system, large numbers of semiconductors would be wired in series across the temperature differential to produce direct-current (dc) electricity. Advantages cited include no moving parts, low maintenance, no boiler or vacuum system, less biofouling than with the closed cycle, and the semiconductor materials are said to be plentiful and economical. Disadvantages are limited use for

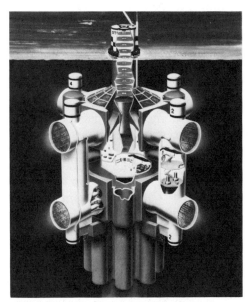

Figure 12-12 Floating OTEC plant. (Courtesy of Lockheed Corporation.)

dc and the inability to transmit it for long distances unless it is converted to ac. Resultant efficiencies and economic comparisons with the other systems are currently unknown.

Although most efforts to commercialize OTEC involve the conversion of thermal energy to electrical energy for use at major load centers on shore, on-site use would eliminate the problems associated with the power cable connector and often excessive distances between the plant and end-use applications. Schemes for electricity utilization on site involve the production of a marketable product or synthesis of a storable, transportable fuel or energy form. Products would appropriately be those which are energy intensive to produce, such as aluminum or the smelting and refining of other metals. Synthesis from seawater of industrial chemicals such as caustic soda and agricultural-grade ammonia has also been proposed. Additionally, the dissociation of seawater through electrolysis produces hydrogen and fuel-grade anhydrous ammonia, which could be synthesized and shipped to shore. An advantage of floating manufacturing or converting plants is that they could be free to move on a seasonal basis to locations possessing the greatest thermal differential or closer to the source of new materials or product markets. Aquaculture, taking advantage of the nutrient-rich deep ocean water, has strong potential as a companion for all OTEC systems. Experiments have been under way on St. Croix Island in the Caribbean since 1972.

The construction of a 40-MW experimental plant is planned for location off the coast of Oahu, Hawaii, in 1986. The site was selected due to the presence of ideal OTEC conditions located within a few miles of shore. Several plants floating in the waters near tropical developing countries are expected in the near future, since many of these nations lack fossil fuels and would have to import them for electrical power generation. Potential for trade exists where OTEC plants and technology could be exchanged for minerals or other raw materials needed in the industrialized nations.

Environmental impacts related to OTEC units have proven negligible to date, with no serious problems foreseen in the future. The most popular working fluid for closed systems, ammonia, is toxic, but since it is easily detectable and biodegradable there have been few concerns. The production of titanium-alloy heat exchangers holds some concern because it is not an abundant mineral and the metallurgical production process involves high-temperature reactions with magnesium or sodium metals which are potentially explosive.

Extensive numbers of OTEC units in a small geographic area could affect the climate. Fossil- and nuclear-powered plants have been found to cause changes by adding to the earth's heat load, producing CO_2, and expelling large quantities of water vapor and acids into the atmosphere.

Figure 12-13 OTEC ammonia factory ship. (From the U.S. Department of Energy, Office of Public Affairs.)

Many OTEC plants discharging cold water could lower the ocean's surface temperature, affecting the evaporation and CO_2 exchange rates in an area. This could also lead to a situation in which plants operated by one nation influence the thermal structures of another nation's territorial waters. The volumes of cold water estimated to be required are 500,000 gal/min for a 10-MW plant and 6 million gallons per minute for a 100-MW plant, so some of these concerns are valid.

Commercial fishing is expected to benefit. Cool water from the depths of the ocean is rich in nutrients and living organisms (plankton) and when brought to the surface, would attract a wide range of sea life. The effect on deep-water marine life is not expected to be significant. Territorial rights for OTEC sites are expected to be analogous to those governing the international fishing industry. In total, the environmental impact is expected to be benign or subtle even when units are concentrated in a small area.

The utilities' acceptance of OTEC systems has been minimal. In 1981 the federal government moved to eliminate all funding. Without adequate funding or tax credits, the goals of OTEC research may not be achieved rapidly. The basic technology has been proven and further testing is necessary, but few companies are willing to take unsubsidized financial risks. Projections are that in the next 10 years there will be several commercially operating plants in the prime areas near Hawaii, Puerto Rico, the Gulf coast of Florida, and in waters of the tropical developing nations.

WAVE ENERGY

Ideas to extract the mechanical energy from ocean waves have been proposed for the past two centuries, with patents for system designs recorded since 1856 in Great Britain. The rapid growth in need for energy during the Industrial Revolution spurred innovations that attempted to acquire usable mechanical energy and later generate electrical energy from this indirect solar energy source.

As the sun heats the atmosphere at differing rates and temperatures, a flow of air (wind) is produced which moves from areas of cooler, more dense air to those with warmer, rising, less dense air. Where this occurs across large expanses of open water, waves are produced as some of the wind energy is transferred to the water surface. The velocity and direction of wind varies according to season and geographic location, with winter winds producing the greatest waves—at the time when energy demands are often greatest.

Island countries or nations located east of oceans possess the best wave-energy resource in areas of prevailing westerly winds. For this reason the British Isles, Norway, and the west coast of North America are the recipients of the world's most energetic waves. Other locations with high potential for wave energy development include the Falkland Islands, Fiji, Japan, Sweden, and the São Paulo coast of Brazil. As with other hydro-related energy sources, the energy content in the waves increases as the square of the height. A 6-ft (1.83-m)-high wave would provide four times as much energy as a 3-ft (0.914-m) wave. Geographic areas with average wave resources are not currently, and may never be, feasible for wave-energy development.

As an industrialized island nation importing much of its energy, Great Britain has been the world leader in the development of wave-energy technology. More than 350 patents for devices have been issued to inventors who hope to extract the Atlantic Ocean's wave energy to reduce dependency on imported energy supplies. The United States has issued only one-sixth this number of patents, even though much of the Washington/Oregon/northern California coast has significant wave-power potential.

Under the leadership of various political or industrial sponsors, Great Britain, the United States, Norway, and Sweden have initiated and sponsored wave-power research. Great Britain's Department of

Energy withdrew from program sponsorship in 1982, claiming that it no longer had the financial resources for continued involvement and that it was responsible only for the developmental stages of the technology. Since then the Department of Industry has taken over the country's leadership role and is sponsoring the world's first full-scale wave-energy electricity-generating plant near the Isle of Lewis off the coast of Scotland. This will be a 4-MW plant with completion expected in 1987. Future projections indicate that Great Britain may be able to acquire half of its electrical energy needs from wave-energy conversion.

Numerous schemes have been advanced over the years for converting the multidirectional, constantly varying energy of the ocean's waves to usable energy forms. Mechanical energy, until recently, has been the goal for many designs, since the generally low-level energy of waves was determined to be inadequate to rotate conventional turbines for generating electricity. Since high speed is necessary for conventional hydroturbine rotation, energy from the waves, producing mechanical energy of relative motion in solid-body mechanisms, can then be converted by nonconventional turbines to rotate generators. Nonelectricity uses are similar to those available from OTEC systems: water desalination; production of hydrogen, ammonia, and uranium from seawater; catalytic conversions of hydrogen and carbon oxides to fuels; and at-sea manufacturing processes. Few other nonelectrical concepts involve hydraulic (mechanical) and thermal energy. Hydraulic systems would be used to pressurize oil or seawater, which could be used on site or transferred to shore in a single pipe for running devices such as hydraulic motors. Thermal conversion would involve using the wave energy to drive a function brake. The brake in turn would be used to heat water or oil, which could be used for direct heating or preheating in a manufacturing process or onshore electricity-generating plant.

The first patent for a wave-energy conversion device was issued in 1799 for a design involving a ship attached to a long lever. The pitching of the ship in the waves caused an oscillating motion of the lever which could be used to drive machinery or pumps on shore. Many other variations of floating structures have been proposed through the years which transferred the rocking motion of the waves through ropes, cables, levers, or gears to perform needed tasks requiring mechanical energy.

Several designs of wave-driven water pumps or rams have met with various levels of success. When used to raise water to a height suitable to turn conventional hydroturbines when released, this method has proven to have relatively low efficiency and has been abandoned by most researchers in favor of other designs. Applications to pump water to onshore reservoirs in pumped-storage hydropower systems is being considered.

An early application of conversion to electricity was used by M. Bochaux-Pracëïque in 1910 at Bordeaux, France. This 1-kW residential system involved a turbine driven by air which was pressurized by waves acting in a vertically bored hole in a cliff adjacent to the sea. This concept was used in 1965 by Y. Masuda in his patent and subsequent Japanese manufacture of self-contained air-turbine bouys which generated 60 W of electricity to illuminate navigation buoys. There are still approximately 500 of these small electricity-generating units in Japanese waters and they are the only wave-powered devices in regular use today. Continued application of this basic pneumatic concept in research and development efforts in both Japan and Great Britain has led to progress in wave energy to electricity conversion technology. Figure 12-14 illustrates the concepts of the oscillating water column used to drive an air turbine.

Masuda also designed a floating raft system which Sir Christopher Cockrell of Great Britain continued to develop. Called contour rafts, these units were hinged together to extract energy from the waves (see Figure 12-15). Newer developments have included double-action pumps activated by the sequential pitching of the rafts, providing fluid pressure that could be used in hydraulic motors coupled to electrical generators. Tests at the Massachusetts Institute of Technology project that this

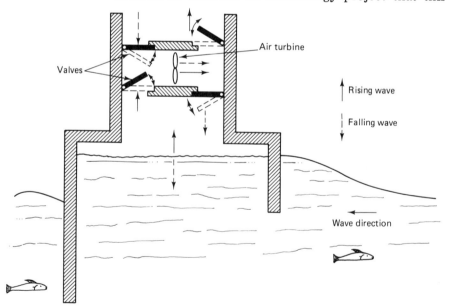

Figure 12-14 Oscillating water column with air turbine. (Adapted from Robert A. Meyers, ed., *Handbook of Energy Technology and Economics*. Copyright © 1983. Reprinted by permission of John Wiley & Sons, Inc.)

Wave Energy

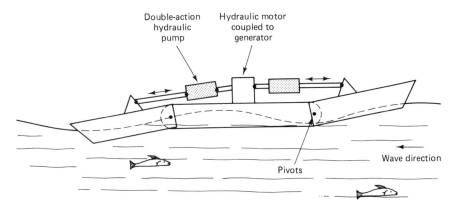

Figure 12-15 Cockrell contour raft concept. (Adapted from Robert A. Meyers, ed., *Handbook of Energy Technology and Economics*. Copyright © 1983. Reprinted by permission of John Wiley & Sons, Inc.)

system may be more economical to build and easier to anchor in the sea than other systems.

A floating cam-shaped device developed by S. H. Salter was tested in 1978 off the Massachusetts coast. Called the Salter Cam, Salter Duck, or nodding duck, several of these devices aligned along a tubular connector parallel to the waves would force a fluid through the tubing to drive hydraulic motors, which would then be connected to electricity generators. Figure 12-16 illustrates the Salter Duck, which extracts up to 90% of the energy from waves. The tests demonstrated the need to move offshore to deeper water and that a more efficient support mechanism was needed to prevent heaving and pitching of the ducks. Additional experiments and developments have been conducted in 1/50 scale on the Draycote Reservoir in England and 1/10 scale on Lock Ness

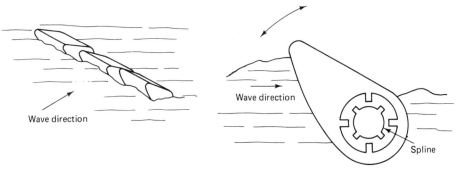

Figure 12-16 Salter Duck. (Adapted from Robert A. Meyers, ed., *Handbook of Energy Technology and Economics*. Copyright © 1983. Reprinted by permission of John Wiley & Sons, Inc.)

in Scotland. More recent designs incorporate gyroscopes in the "beak" of the duck to supply the energy for pumping the fluid.

The experimental ship *Kaimei* (Figure 12-17), built by the Japan Marine Science and Technology Center, contributed greatly to wave-conversion technology during its 1978–1980 operations. This floating laboratory, 262 ft (80 m) long by 59 ft (12 m) wide, with 22 air-filled chambers open to the sea at the bottom and accommodating air-driven turbines at the top, was an application of Masuda's original design. The rising water of each incoming wave forced the air in the chambers through the turbines (Figure 12-14). One-way valves allow air to refill the chambers as the waves pass.

The *Kaimei*'s many separate chambers provided the opportunity to test and compare several different design concepts and their respective efficiencies. Through the joint efforts of Japan, Great Britain, Ireland, Canada, and the United States, the test program under the International Energy Agency has made impressive gains. One Japanese unit exceeded all others tested when it produced 291 kW of electricity.

Various buoy concepts have been designed and tested using both hydraulic and pneumatic mechanisms. In addition to the small air-turbine navigational buoys, larger units moored at both surface and subsurface levels have proven successful. One hydraulic design involved taut-line mooring. An internal diaphram operating a hydraulic piston was deflected by the fluctuating pressure within the waves. The Bristol

Figure 12-17 The Kaimei—built for wave-energy experiments. [Courtesy of Japan Science and Technology Center (JAMSTEC).]

cylinder is also a hydraulic design. Within this air-filled cylinder, moored just under the ocean surface, is a fluid-pumping piston attached to the anchor line. As each wave raises and lowers the cylinder, the piston provides hydraulic pressure. This unit has proven very efficient, but mooring and fluid cooling have been difficult. A pneumatic design which also incorporated a flexible rubber diaphragm in a cylindrical buoy concentrates and constricts the movement of air inside with passages leading to an air-driven turbine.

The second generation of oscillating water columns (as used on Kaimei) developed by the National Engineering Laboratory of Glasgow, Scotland, continued with experimentation toward the optimum size and shape of both fixed and floating systems. The 4-MW demonstration plant scheduled for completion in 1987 off Scotland's coast utilizes this design. A concrete breakwater 197 ft (60 m) long will contain pneumatic turbines driven by air forced through vertical columns in the structure by the waves breaking against the breakwater. The current projection, even before completion of the project, is that this design is still not economically competitive with present fossil-fueled electricity-generating plants. However, large arrays of mass-produced units could successfully rival present systems costs.

Several other designs possessing great potential have been proposed within the last few years. Among them are the Lanchester Clam, the Lancaster Flexible Bag, a Dam-Atoll design, and a submerged resonant duct. The Lanchester Clam uses wave-operated flaps to force air through a turbine (see Figure 12-18), while the Flexible Bag uses wave pressure to move air through a turbine and back to the bag via a one-way valve system. The last two designs are submerged designs which involve "bending" or directing the energy of the waves toward a central area or focal point where the concentrated flow of water can be harnessed. Still other designs involve linear generator designs, gas concentration cells, and piezoelectric applications.

The three basic categories of wave-energy-conversion designs are the buoy type, oscillating water column, and the submerged oscillating water column. Both hydraulic (fluid)- and pneumatic (air)-activated systems function to provide the adequate energy level and speed increase necessary to rotate turbines connected to electricity generators. The numbers of designs indicate that the pneumatic systems are the most popular, perhaps because of the simplicity of the mechanisms.

There are currently no large-scale, commercial wave-energy-conversion devices in operation anywhere in the world, even though they are technically feasible. Of the generating systems available, preference appears to be toward the use of alternating current synchronous machines due to their operating efficiency and lower comparative cost. Coupled together via transformers, high-voltage electricity can be

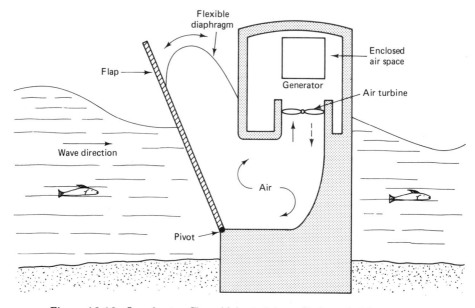

Figure 12-18 Lanchester Clam (Adapted from Robert A. Meyers, ed., *Handbook of Energy Technology and Economics.* Copyright © 1983. Reprinted with permission of John Wiley & Sons, Inc.)

transmitted to onshore switching stations. Operating costs are projected to be relatively low, but initial capital investments are going to be high.

Each energy-conversion device has technically related problems which require continued research and development to determine its acceptance or rejection for large-scale, commercial adoption. Additionally, there are technical problems and lack of data concerning near-shore waves; the effect of wave forces on the structure of the devices; materials that will endure the harsh ocean environment; methods of mooring the devices appropriate to wave, current, and wind direction; electrical power generation from the slow, irregular action of the waves; best methods of using or transmitting the converted energy to shore; and effects on the ocean and shoreline environment.

The environmental impact of wave-energy-conversion systems is projected to be minimal—definitely much less severe than many of the conventional fossil-fueled energy-conversion systems. Concerns that have been identified involve potential interference with fishing and offshore oil exploitation efforts, shipping, the physical state of the adjacent coastline, spawning of fish, and the human visual and social amenities.

Commercialization of wave energy is very near. The technology is approaching readiness, with research and development efforts continuing in Scotland, Japan, the United States, Norway, and other countries. Costs are nearly competitive with conventional generating plants and when wave devices can benefit from mass-production techniques, the costs may even be lower, especially when considering the environmental impacts. Within 20 years, the energy needs of countries with ocean coastline and possessing adequate wave strength will be using wave-energy conversion systems backed up by fossil-fueled systems for those tranquil days at sea.

SALINITY GRADIENT POWER

Energy may be extracted from water by using the pressure potential that exists across a membrane between two solutions of differing salinity levels. This potential is created by the process of osmosis, which is the tendency of a fluid to pass through a semipermeable membrane into a solution on the opposite side where its concentration is higher (Figure 12-19). Hydrostatic or osmotic pressure is present until equilizing conditions have been established on both sides of the membrane.

An example of this process occurs when dried peas, beans, or raisins are placed in a pan of water. Since the material inside cannot escape into the water, swelling of the dried materials occurs due to

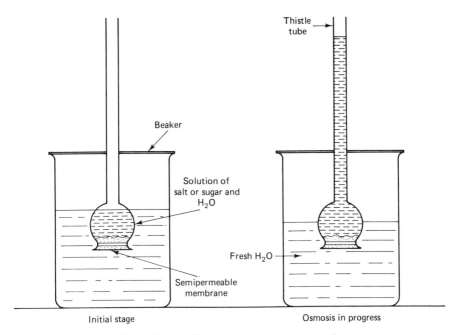

Figure 12-19 The phenomenon of osmosis.

inward passage of water through the semipermeable skins. This phenomenon will continue until equilibrium is reached—a state in which the vapor pressure of the water inside the cell is equal to the vapor pressure of the external water or until the concentration of the two solutions is balanced. The passage of water molecules through a semipermeable membrane is always in the direction from the dilute to the more concentrated solution.

This same process can be used where salt water and fresh water are separated by a manufactured semipermeable membrane (see Figure 12-20). Fresh water passes through the membrane, resulting in a head of water on the saline side which can be used in conventional hydroturbines to generate electricity. Since raising the level of the entire ocean, or even large landlocked lakes, is totally out of the question, saltwater impoundments would be necessary so that the hydraulic head can be contained for use. The Dead Sea and Great Salt Lake, due to their very high salt content, hold great potential for this technology.

Another method for salinity gradient energy conversion is through the use of a dialytic battery. The process is basically reverse electrolysis and would require a large number of specially designed membranes of two types. One membrane would permit passage of negatively charged ions, as in chlorides, while the other would allow positively charged ions, as in sodium, to go through. Electrodes would be connected to facilitate electrical energy acquisition.

A third technique involves the differing rates of evaporation of fresh and salt water. Established pressure differentials would cause vaporization of the water, and the vapor would be directed through turbines.

Each of these systems requires additional research and develop-

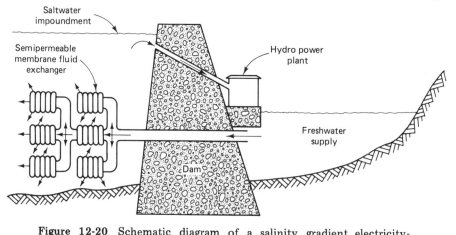

Figure 12-20 Schematic diagram of a salinity gradient electricity-generating plant.

ment efforts in the laboratory and small-scale experimentation. The projected costs are exceedingly high compared to those for conventional electrical energy systems. Estimates range from 5 times to 50 times more expensive for the osmotic pressure system and the dialytic battery, respectively. The major stumbling block is the design and development of an economical manufacturing process that can produce the semipermeable membranes in appropriate sizes and with adequate strength to endure the pressures and water-environment interface. Research is continuing, but commercialization is not expected until well after the turn of the century.

At this time the environmental impact is projected to be minimal and very similar to that associated with tidal energy systems. Impoundments created in landlocked lakes are anticipated to cause fewer problems than those involving ocean estuarial ecosystems.

OCEAN BIOCONVERSION

Energy from the plant sector of the biological environment in the ocean holds great potential because it is renewable and does not compete with land-based forestry or agriculture. Commercial harvesting of naturally occurring giant kelp has been conducted off the coast of southern California since the early 1930s and efficient equipment for harvest has been developed to provide kelp for the food-processing industry.

The establishment of ocean plantations or farms would provide high yields of quality plant life from concentrated, specific locations. These farms would use the nutrient-rich water from the depths to nourish the plants toward rapid, prolific production. The growth would then be harvested and converted to fuels such as methane. Other products, such as fertilizers, fish, shellfish, and other human and animal foods, provide additional forms of chemical energy for meeting society's needs and reducing fossil fuel consumption. For example, the production of fertilizers from ocean plants would reduce the use of natural gas currently required for conversion to fertilizers. Refer to Chapter 10 for details concerning the conversion of plant materials to fuels.

Integration with other ocean energy technologies appears appropriate. Wave-driven pumps could be used to acquire the nutrient-rich waters required for rapid plant growth. Coordination with OTEC systems would seem logical since one of the ecological concerns of their use is the discharge of nutrient-rich water at surface levels.

The establishment of ocean plant farms and the conversion of plants to fuel is not currently economically competitive. As new uses and products are developed, the importance of ocean plants and their contributions to society will increase.

SUMMARY

The oceans of the world hold considerable promise for contribution to society's energy needs. Energy from the ocean, if used where it is available, would save conventional fuels for geographic areas with fewer alternative energy forms. For this reason it is expected that tropical developing countries will use ocean energy systems and could export transportable energy forms through international trade agreements.

Each of the ocean energy forms requires additional research and development to be commercially available on an economically competitive basis with conventional energy sources. Some forms require extensive efforts (e.g., salinity gradient). Ocean current systems require moderate development, while OTEC, tidal, wave, and bioconversion methods will begin contributing to the world's energy supply in the mid-1980s and 1990s.

Since electrical energy is not transportable for long distances and there are few major markets for electricity located in the best ocean energy areas, conversion to transportable fuels or product manufacturing at on-site locations is expected.

ACTIVITIES

1. Construct a model or illustrate and describe how tides interact with the ocean bottom and adjacent landmasses to provide exploitable energy.
2. Explain why ocean thermal energy systems must be located in tropical geographic areas.
3. Describe the potential environmental impact if OTEC plants were used extensively in a small geographic area.
4. Describe the technical and engineering problems associated with OTEC plants.
5. Conduct an experiment to demonstrate the phenomenon of osmosis as found in the use of ocean salinity gradient energy. Figure 12-19 illustrates how the experiment can be set up. Vary the concentration of salt in the water solutions and measure the height of the column of the solution displaced. Try various sugar solutions. Compare results using equal volumes of sugar and salt.

BIBLIOGRAPHY

Beck, E. J., "Ocean Thermal Gradient Hydraulic Power Plant," *Science*, Vol. 189, July 25, 1975.

Charwot, A. F., and S. L. Ridgeway, "The Mist-Lift OTEC Cycle," in *Energy*, Vol. 5. Oxford: Pergamon Press, 1980, pp. 511-524.

Foy, James A., "Harnessing the Tides," *Technology Review*, July 1983.

Johnson, Tom, "Electricity from the Sea," *Popular Science*, May 1981.
Kocivar, Ben, "Lifting Foils," *Popular Science*, February 1978.
Krenz, Jerrold H., *Energy: Conversion and Utilization*. Boston: Allyn and Bacon, 1976.
Lissaman, P. B. S., "Undersea Turbines," *Popular Science*, September 1980.
Meador, Roy, *Future Energy Alternatives*. Ann Arbor, Mich.: Ann Arbor Science, 1978.
Meyers, Robert A., ed., *Handbook of Energy Technology and Economics*, New York: Wiley, 1983.
Newman, J. N., "Power from Ocean Waves," *Technology Review*, July 1983.
Pryde, Philip R., *Nonconventional Energy Resources*. New York: Wiley, 1983.
Riva, Joseph P., et al., *Energy from the Ocean*. (Prepared by the Science Policy Research Div., Library of Congress, for the House Committee on Science and Technology.) Washington, D.C.: U.S. Government Printing Office, 1978.
Shaw, Ronald, *Wave Energy: A Design Challenge*. New York: Halsted Press, 1982.
Wilson, M. N., "Slow Speed Generators with Superconducting Windings," *Proceedings, First Symposium on Wave Energy Utilization*, Gothenburg, Sweden, 1979.
Winer, B. M., "Electrical Energy Transmission from Ocean Current Power Plants," *Proceedings, Third Workshop on Ocean Thermal Energy Conversion*, Houston, Tex., 1975.
Wood, Chris, "British, Canadian Devices Offer Unique Low-Head Tidal Approach," *Renewable Energy News*, July 1982.
Wood, Chris, "Tidal Conference Endorces Fundy Tidal $23 B Project," *Renewable Energy News*, July 1982.
Zener, C., "Foam Solar Sea Power Plant," *Science*, Vol. 189, July 25, 1975, pp. 294-295.

Chapter 13

Conservation and Legislation
The Wise Use of Energy

CONCEPTS

1. Energy conservation depends primarily on the attitudes and values of the energy user.
2. Reduction of costs and financial return on the amount invested are the primary incentives for conservation.
3. Many conservation efforts can be implemented quickly and with small to moderate expense.
4. Conservation (efficient use of energy) can be accomplished with minimal effects on one's life style.
5. Conservation is the quickest, cleanest, and least expensive way to counteract the effects of higher energy costs and shortages.

GLOSSARY

ASHRAE—American Society of Heating, Refrigeration and Air Conditioning Engineers.
CONSERVATION—the act of saving or preserving.
DEGREE-DAYS (DD)—temperature differential between a reference temperature, for example 65°F (18°C), and the average of high and low temperatures during a 24-hour period.

GIGAWATT—1 million kilowatts.
HVAC—heating, ventilation, and air conditioning.
INSULATION—a material, or the use of a material, which reduces the rate of thermal energy transfer.
PAYBACK PERIOD—the length of time required for the cumulative net savings from an energy investment to equal the original investment.
PPM—parts per million.
VAPOR BARRIER—any surface or material that restricts or stops the flow of water vapor (low permeability).

INTRODUCTION

Americans waste nearly 50% of the energy they use. Until 1973 we believed that energy would always be available, and at reasonable cost, so why be concerned with conserving it? Conservation is strongly dependent on attitudes and values, but it is also financially related. To see how conservation can make a difference, we must first identify energy use in the various components of society.

Installation of more efficient equipment, reduction of the amount of waste materials through recycling, better production controls, and the refitting of buildings for better efficiency in heating and cooling have all helped to reduce energy costs in the U.S. industrial sector. Denis Hayes makes a strong argument for energy conservation through increased efficiency. He has calculated that the United States could, in the next 25 years, "meet all its new energy needs simply by improving efficiency in energy use" [Grossman and Daneker, p. 47]. Lovins and Lovins have shown conclusively that it is possible to reduce energy consumption and still maintain production levels. "The ten most energy-intensive industries during 1972-1979 decreased their energy consumption per unit of product by an average of more than fifteen percent" [Lovins and Lovins, p. 247].

In spite of these efforts, much more can be done. Industry spends eight times as much on advertising as it does on research and development of energy-efficient production of consumer goods [Grossman and Daneker, p. 111].

ENERGY CONSUMPTION

Energy use levels are usually compiled for four sectors: commercial, industry, transportation, and residential. Energy use for industrial purposes is the largest of the four, as shown in Figure 13-1. Transportation is second, with residential and commercial following in that order.

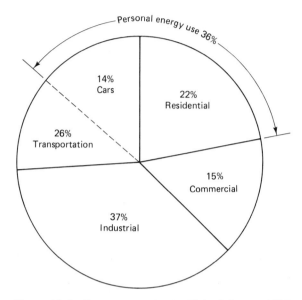

Figure 13-1   Energy usage in the United States, 1980.

Of these four sectors, industry has done the most to reduce the amount of energy wasted in its operations. Energy consumption is one of the major cost factors in production. Industry generally uses two strategies to conserve: reduction of wasted energy and improvements in efficiency.

Industrial energy use reached a peak consumption level of 26,234 trillion Btu in 1973, the year of the first oil embargo (see Figure 13-2). Energy consumption in industry dropped by nearly 13% in the next two years but then rose again to 25,739 trillion Btu in 1979, almost the level of the previous high point. The second "oil crunch" made everyone aware (at least for a time) of the need to conserve. Industry-instituted measures to reduce energy use have resulted in a decline to 24% of the 1979 energy use level [*Energy User News*, January 23, 1984].

Just as impressive as the reduction in consumption is the amount of energy consumed per unit of output (dollar of energy purchased per unit of production output). In 1970 the ratio was 65,920 Btu per dollar of production (see Figure 13-3), but that figure continually declined until it reached a level of 44,900 Btu per dollar of production in 1982, a 31.9% reduction in the ratio of energy use to production [*Energy User News*, January 23, 1984].

The Alliance to Save Energy conducted a study which estimated that commercial and industrial firms using 60% of the nation's electrical energy have invested in only 15% of the electricity conservation projects that are economically attractive. . . . The most capital-intensive industries—steel, chemicals, paper, aluminum, and petroleum refining could reduce their energy costs by 10% with waste-heat recovery, computerized process control and upgrading and reuse of by-product gas. The rate of return on these conservation investments frequently exceeds 30%. ["Energy and Education," February 1984, p. 1]

The commercial sector has also reduced its energy demands since 1973. The energy consumed for each unit of activity (8 hours of effort by one employee) was reduced by an amount corresponding to that of the industrial sector, from a peak of 549,000 Btu per employee per day

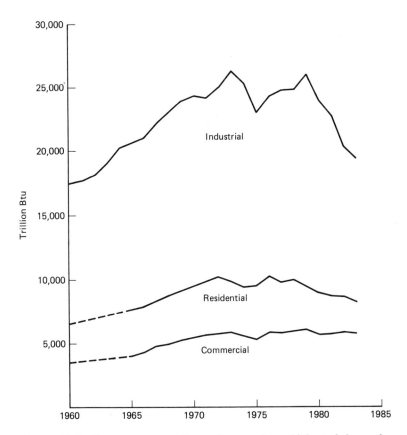

Figure 13-2 Energy consumption: three sectors. (Adapted from the U.S. Bureau of Labor Statistics.)

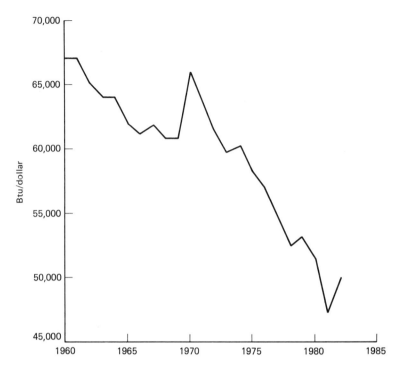

Figure 13-3 Energy use per unit of output: industrial sector. (Adapted from the U.S. Bureau of Labor Statistics.)

in 1971 to 418,000 Btu in 1983 (Figure 13-4). This represents an improvement of nearly 25% [*Energy User News*, January 23, 1984].

Some gains have been made in more energy efficient operations of office buildings through the use of energy management systems that control the times when lighting and heating demands occur. Electronics control equipment costing from a few thousand dollars to systems costing 10 to 100 times that much often save that cost, plus that of installation, within a few months. Other factors include the reduction in the level of lighting by removing "excess" lights, changing to more efficient lighting systems, and replacing equipment with models using less energy.

The transportation sector accounts for about one-fourth of the energy consumed in the United States. Improvements have been made in several areas, such as smaller, lighter, more fuel-efficient vehicles. Internal combustion engines rank very low in efficiency, about 25%. Since most travel uses automobiles, they are a prime target for conservation efforts.

The number of vehicle-miles traveled has been lowered somewhat, from a high average per car of 10,180 miles in 1972 to just over 9000 miles per vehicle in 1981. Higher fuel prices and the enforcement of

Energy Consumption

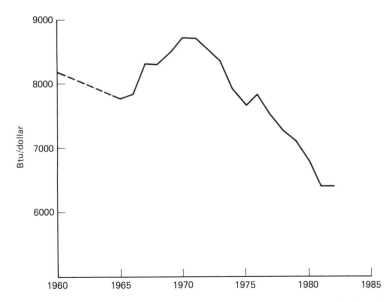

Figure 13-4 Energy use per unit of output: commercial sector. (Adapted from the U.S. Bureau of Labor Statistics.)

the 55-mph speed limit as well as an increase in car pooling were factors that helped to increase the passenger miles per vehicle.

Prior to the first oil embargo of the early 1970s the demand for larger cars with more comfort and appearance features increased, and the fuel rate for passenger cars declined each year from 1960 through 1973, from 14.3 miles per gallon (mpg) to 13.1 mpg. An even more important "conservation" aspect has been the reduction in highway deaths—nearly 10,000 per year—since the lower speed limit was imposed.

Since 1973 the trend in consumption has been reversed, with the result that fuel efficiencies increased to 15.5 mpg per passenger car in 1981. The use of alternative fuels has made a small impact in the consumption of petroleum products, but much more remains to be done. It is estimated that the "improvement" could reach 26 mpg by the mid-1980s. This is poor by European standards. Fuel efficiency is a vital area of concern, since automobiles are still the major consumers of petroleum-based fuels in this country, burning almost one-third of the petroleum we use. The automobile is the most energy-intensive means of land transport that we use by a factor of 15 compared to bus and by a factor of 25 compared to rail transit.

In 1960, each person in the United States consumed 100,000 Btu of energy (mostly from fossil fuels) during one 24-hour period (Figure 13-5). Heating and cooling costs were comparatively inexpensive and the demand for more "creature comforts" increased that consumption

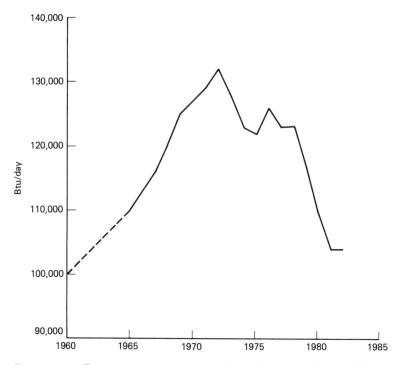

Figure 13-5 Energy use per person: residential sector. (Adapted from the U.S. Bureau of Labor Statistics.)

by nearly one-third, to 132,000 Btu per person in 1972. Nearly every year since then has shown a decline, to the present level of about 104,000 Btu per person per day. Consumption has decreased primarily as the result of the introduction of better insulation and "tightening" of buildings through caulking, door and window repair and replacement, the installation of more efficient gas and oil furnaces, and the use of more efficient appliances.

PERSONAL ATTITUDES TOWARD CONSERVATION

The attitudes toward energy conservation reflected by individuals in their private actions tend to be centered first on immediate, short-term savings and second on a fear of lowered standards of living. Long-term savings (payback over several years) is not considered a major factor by many persons, particularly those who are poor and/or those who rent their housing from others. Those who have little or no savings do not consider the purchase of energy-saving materials or services when the basic costs of food, clothing, and shelter take most or all of their income.

Cost factors seem to be the major immediate consideration in whether people invest in energy-saving measures. Reports from several sources indicate similar results; families cut back on the use of appliances but do not give them up, nor do they replace them unless they are beyond repair. Nationally, only a small number of homeowners, frequently less than 20% of those surveyed, have spent as much as $100 for weather stripping or storm windows. The average expenditure per household in a Minnesota study was $266. Since fuel prices have risen rapidly during the past few years, any real savings through these measures were lost due to increased costs of heating.

Owners of rental property have little or no incentive to improve the energy efficiency of their holdings since the lessee is most often the one who pays the energy bill. Renters are rarely inclined to put improvements into the property of others; their "payback" is seldom as much as a "thank you" and may be in the form of an increase in rent for the "improved" property. Incentives in the form of compensation to renters and owners for energy-saving measures could aid in changing this attitude.

Such measures have been introduced at the federal level and some in state legislatures, but little real progress has been made. There is still an attitude of "I don't care," a feeling that there is an "oil company plot to rip off the public," and that "there really isn't an energy shortage." Until the public at large is convinced that saving energy is in their best long-range interest, much will remain to be done in the conservation effort.

RECYCLING USED MATERIALS

Some consideration is being given to "life-cycle costs" of production. Instead of calculating only the initial cost of building and materials, together with the immediate production and shipping costs, some companies are including the costs of waste disposal and recycling in their financial considerations—a life-cycle financial consideration.

Easily recoverable materials such as scrap metal, plastics, and wood scraps have long been considered as usable by-products of the manufacturing process. Some materials are not as easily recovered.

The recovery of spent industrial chemicals is increasing as it is recognized that dumping this material is not only hazardous to the environment but damaging to the financial structure and public relations of any company or corporation that violates the law. Some used chemical compounds have been found to be useful "raw materials" for other uses. Industrial waste exchanges (see Chapter 10) attest to increased awareness and use of this technique.

Every pound of aluminum that is recycled saves about 90% of the energy required to replace it from bauxite ore. The energy used in the refining process is almost entirely electrical energy. Recycling aluminum could save as much as 200 million Btu/ton.

Other metals consume substantial amounts of energy during the refining process. Recycling steel would save 12 million Btu/ton and similar effort would save 42 million Btu/ton for copper. The EPA estimates that 90% of the steel and 60 to 75% of the aluminum used in this country could be recovered and reprocessed.

Recycling of glass and other container materials does not save quite as much energy, but other savings are found in the reduction of litter, which not only is unsightly but can provide a breeding place for vermin and insects. Any such reduction will result in a lowered potential for the spread of disease. The dollar amount of the cost cannot be easily measured, but any reduction in the amount of lost time, production, and services is a positive factor. From a personal standpoint, avoidance of illness is a definite advantage.

Waste recovery studies (see Chapter 10) have shown that most of the solid waste in the United States is composed of paper. If it were to be recycled, it would require only 25% of the energy used to process paper from wood pulp. Paper manufacturing plants have made more progress in reducing waste and in using waste than have most other industries.

Waste from the manufacture of plastics can be used in one of two ways: either ground up and reprocessed, or burned under carefully controlled conditions as a substitute or supplementary fuel. In one plant, paint solvents removed from a spraying operation were rerouted to the ovens used to dry the paint, reducing the natural gas normally used by two-thirds.

In the steelmaking industry, replacement of open-hearth furnaces with "basic oxygen" furnaces has saved a similar amount of energy. Manufacturers of portland cement have reduced energy consumption in the high-temperature steam process just by lowering the temperature of the steam. Europeans mix fly ash (a by-product of coal- and oil-burning steam plants) with portland cement as an "extender," with little change in the characteristics of the resulting road material.

ENERGY-SAVING TECHNOLOGIES FOR INDUSTRY

Industrial production equipment wears out and becomes obsolete. In particular, electric motors and controls are leading candidates for replacement by more efficient units. Lovins and Lovins have shown that if industry were to systematically replace all electric motors with new, more efficient, properly sized motors, this country could save 75 gigawatts of electricity. This amount of electricity would be enough to replace the entire generated output of the nuclear power program now in existence.

Several factors must be taken into account when planning for energy conservation in the industrial sector. Among these are initial energy costs as a base from which to begin calculating payback costs. Also included are the age and condition of the buildings; whether the buildings are used daily, or for more than one shift; the severity of the weather; and obviously, the cost of the energy conservation measures themselves.

Heating or cooling functions that take place during production (and might provide a source of "free" energy) should not be overlooked. It is frequently advantageous to investigate energy sources with the view toward extracting more usable energy from "waste heat," which may be discharged as steam, hot air, or hot water.

Five technologies that have been shown to return their initial cost quickly are infrared heaters, personnel sensors, fluorescent bulbs in place of incandescent, flue dampers, and portable heat scanners.

Infrared units can be used to heat small, localized areas. Infrared rays are absorbed by objects, which in turn radiate heat to workers. The air is not heated directly, resulting in a much more efficient transfer of heat and less loss due to rapid air changes.

Personnel sensors work most efficiently for light control when the areas they control receive at least 2 watts per square foot (22 watts per square meter). Two types of systems are available: a passive type which uses infrared sensors to detect heat changes caused by physical activity, and an active system using ultrasonic transmitters. The latter fill the area with 40-kilohertz transmissions which are disturbed when someone moves in the room. When the pattern is disrupted, a signal is sent to turn on the lights. This application is useful in warehouses with intermittent activity by workers. Both types of systems have been shown to reduce energy usage by an amount equal to or greater than their installed cost within one year.

It has long been recognized that a fluorescent lamp can provide several times the light of an incandescent bulb of the same wattage. Fluorescent bulbs convert electricity to light energy with an efficiency of 20%, four times better than that of incandescent bulbs. Furthermore, fluorescent bulbs have a longer life—as much as 10 times longer, according to manufacturers' ratings. The development of screw-in fluorescent bulbs has made it possible to replace many less efficient lamps with no change in wiring. Low-pressure and high-pressure sodium lamps and mercury vapor and metal halide bulbs can be used in many applications.

They are even more efficient than fluorescent bulbs (low-pressure sodium is rated at about 55% efficiency).

Flue dampers limit the heat loss from flues when boilers are not operating. Savings of 8 to 30% have been realized within one year. Not only does the damper cut off the upward flow of heated air, but when the heating system calls for more heat, the system first circulates the heat from the stack to warm the boiler before turning on the boiler unit. This process saves from 80 to 90% of the heat that would normally be lost. Cold air is also stopped from entering the building via the chimney.

Portable hand-held heat scanners measure the amount of heat being emitted from a surface. The warmer the object, the more heat, in the form of infrared rays, is emitted. Small units of this type range in cost from $500 to $2500. A small heat loss amounting to a Btu cost of less than $1 per hour could, if undetected, cost as much as the price of the heat scanner in a matter of 2 or 3 months.

Another very effective technique for salvaging heat energy is the use of cogeneration equipment and heat exchangers. They are described in Chapter 14.

FACTORS AFFECTING HEATING AND COOLING

No building can be said to be completely airtight; some infiltration is always taking place. Some air exchange is necessary to maintain sufficient oxygen levels for human beings, animals, and plants as well as for any combustion that may be taking place as in furnaces and boilers. Gaps around doors and windows are not always completely sealed. Joints where walls and floors meet and where ceilings and roof surfaces connect can be expected to be less than perfectly sealed. Buildings that were constructed prior to the rise in fuel prices were not as tightly sealed and contained less insulation than those of recent construction. Concrete structures and those with massive amounts of glass on all sides are also notorious wasters of energy for both heating and cooling.

Commercial buildings often have heavy traffic through entrance ways, and many have loading or delivery entrances that provide ample opportunity for heat to escape in winter and for heat and humidity to enter in summer. Strips of plastic or rubber material, similar to those found in front of open-shelf freezer sections in grocery stores, are used as a barrier to the flow of warm air in and out of dock and warehouse areas. The strips inhibit the airflow but do not impede normal loading/ unloading operations to any great extent.

A small gap of 1/8 in. (3 mm) would, if extended around a standard door, create an opening equivalent to that of a 3 in. by 5 in. (7.65 by

12.75 cm) opening. Openings this small, as well as more noticeable ones, all contribute to the heating and cooling load.

Heat transfer is always occurring. It is desirable to limit the flow to areas where it is wanted and to restrict it from going where it is not wanted—to the outside in cold climates, to the inside in warm climates. Heat always flows from areas of high temperature to areas of low temperature, so it is important to restrict its flow with barriers with appropriate insulating value.

Heat is produced from several sources. All human and animal occupants of a building contribute to the heat load. Lighting, in particular incandescent bulbs, helps raise interior temperatures. Mechanical devices give off heat as they operate. Each of these should be taken into consideration when determining the HVAC needs of the building. The design of many buildings does not take into consideration these sources of heat. In some instances there is a need for air-conditioning equipment to operate even during cold weather. Standard Oil has been able to eliminate the boiler units in its office building in Chicago and now uses recovery equipment to transfer heat generated by computers to other parts of the building. The retrofit has resulted in a 44% reduction in energy consumption since 1977, with a payback period of less than 2 years [*Energy User News*, January 23, 1984].

PAYBACK (OR RETURN ON INVESTMENT)

Profit, without which free enterprise would not function, can hinge on a few percentage points of the wholesale price of a product. As other countries become more competitive, with lower wages and with modern, efficient plants, U.S. manufacturers must seek ways to reduce production costs. One of the best ways is to reduce the energy-related costs of production.

Grossman and Daneker report that between 1977 and 1979 there was a 3.9% expansion in the economy with an energy use increase of only 1.9% [p. 36]. The Bonneville Power Administration has shown that a unit of energy saved by conservation costs one-sixth of the equivalent unit of energy produced from a new thermal power plant [p. 46].

The Alliance to Save Energy made an analysis of utility conservation programs, specifically that of Arkansas Power and Light. Their results showed that only 15% of AP&L's customers had made investments in energy-efficient equipment or practices. The cost of the conservation programs was $22 million and achieved peak savings of 34 MW per year, a cost of about $650 per kilowatt. The cost of a new generating plant of this size would be at least several hundred dollars per kilowatt more and would take several years to build. The Congressional Research Service estimates that conservation efforts will have an impact equal to 161,000 MW by 1990 and 214,000 MW by the turn of the century. Load management programs would save 104,000 MW by 1990 and 124,000 MW by 2000. Passive solar would add 3000 MW and 10,000 MW, respectively, and cogeneration and small power producers are likely to contribute an additional 11,000 MW by the year 2000.

RESIDENTIAL CONSERVATION

Energy consumption for residential purposes makes up 22% of the total energy pattern. About 70% of residential energy use is for space heating and cooling. Domestic water heating takes an additional 20%, and the remainder is used for lighting, preparing meals, and operating appliances.

Reducing residential energy consumption can be accomplished in several ways. The least expensive in both initial cost and in maintenance are energy conservation methods. Slightly more costly is the designing of features into construction which will increase energy gain and prevent energy loss. Solar systems, as described in Chapter 7, range from passive to active systems and increase in cost from the former to the latter.

The best "payback" is in the lowering of the thermostat setting in winter for heating and increasing the setting in summer for cooling. A 6°F lowering in winter, from 74°F (23°C) to 68°F (20°C) would result in a saving of 15% in energy costs. This could amount to a saving of more than 570,000 barrels of oil per day in the United States. Raising the air-conditioner setting by a similar amount in summer, from 72°F (22°C) to 78°F (26°C), would save almost 50% as much because of the reduction in running time for the compressor in the air-conditioning unit. This could save the equivalent of 190,000 barrels of oil per day.

About 10% of the annual energy supply goes into building construction. Design features should be more carefully calculated since most buildings are routinely overbuilt, with safety factors calculated to custom-built standards rather than efficient use of materials. If major changes were made in standard sizes of materials used in construction, there would be some initial confusion among architects, contractors, and builders. Standard material sizes can be used with more careful attention to sizing and types used. Instead of using plywood as the outside sheathing on a wood-frame structure, builders could use it only on the corners of the structure, with insulating sheathing or other materials for the majority of the wall. The result is a loss of only about 5% in structural strength but a substantial saving in both weight of materials and cost. Increased insulating value would also be realized.

In some instances, vertical wall supports (studs) made of aluminum or other metal are being used in new construction and remodeling. Aluminum is lighter in weight, but the energy cost and the need to adapt different construction techniques must be considered as part of any energy savings. Aluminum conducts heat rapidly, and perhaps more important, aluminum is not a renewable resource.

CAULKING AND WEATHER STRIPPING

The word *infiltration* has an insidious sound. It is used militarily to indicate the process of sneaking into the enemy position. Used in construction, it literally means the same thing. If unwanted air infiltrates a structure, it can upset the most careful calculations with regard to heating and cooling loads. Infiltration accounts for 38% of the heat loss in the average U.S. home. Reducing infiltration is the easiest, least costly, and most effective way to save energy in the home. Properly caulking and weather stripping those homes now heated with gas could save enough gas to heat another 4 million homes.

To reduce or eliminate the uncontrolled movement of air currents, it is necessary to carefully seal all openings in a structure. Closing gaps of small dimensions is fortunately quite easy with the use of a variety of readily available, inexpensive materials. The most common sealing material is caulk, which is available as a thick paste in either self-propelling cartridges or tubes designed to be used with caulking "guns." The latter are reusable holders with a hand-operated lever which forces the caulk out of the tube through its nozzle. A bead or seal can be secured in almost any location, sealing cracks or joints between various building materials or features.

The first step is to seal all points where joints have been created in the construction process. Building materials expand and contract during temperature changes, creating tiny openings. Places to begin are around door and window frames, where the structure rests on the foundation wall, and where chimneys, pipes, and wires penetrate the structural envelope. In new construction, caulking should be done as the building is being erected or assembled. A small bead of caulk on inside and/or outside accessible surfaces can bring rapid dividends, often making up the financial expenditure in less than one heating season.

There are several types of caulk available: elastomeric and silicone compounds, which are relatively costly but have a long life; acrylic latex; butyl rubber; and oil-based caulks, in descending order of cost and life span. Table 13-1 details the types, their characteristics, and

TABLE 13-1 Caulking and Weather-Stripping Materials

	Advantages	Disadvantages	Other names
Caulking compounds			
Elastomeric	Adheres to most materials, long lasting	Most expensive, hard to smooth, cannot be painted	Polysulfide, polyurethane, silicone
Latex and butyl rubber	Adheres to damp surfaces, can be painted	Not for metal surfaces, will shrink slightly	Latex, acrylic latex, butyl rubber
Oil-based caulks	Lowest price, indoor use	Shortest life	
Special purpose			
Polyurethane	Expands to fills holes and cracks	Expensive	
Asphalt base	Seals roof cracks, vents, and chimney flashing	Black color, sticky when hot	
Rope caulk	For large cracks, pliable	Temporary, does not adhere well	
Weather-stripping materials			
Door sills			
Interlocking	Very effective, durable	Expensive, hard to install	
Insulated	Less heat loss than through metal	Shorter life	
Automatic	Adjusts to changes in door or sill	May wear rapidly	
Jambs			
Spring metal	Compresses between door and jamb	Not easily moved, may vibrate in wind	
Gasket (rolls and strips)	Flexible, less costly, self-adhesive	Wears out quickly, absorbs water	Felt, foam tape, vinyl bead, rigid-backed strip

estimated life. Many factors affect the latter, including the type of surface(s) to which they are applied, temperature changes they must endure, and expansion and contraction problems. If caulk has previously been applied and it shows gaps, peeling, or breaking away, it should be completely removed and new material applied.

Weatherizing also takes the form of closing air gaps around surfaces that are often opened and closed. Seals for this application include foam strips, some backed with wood or plastic; wool felt, also found with a supporting backing; vinyl or rubber strips; and flat metal strips, usually bent into a narrow V shape which can spring out against a door or window edge to form a seal. Each type has a range of purposes and can often be interchanged according to availability, price, personal preference, and the sealing effectiveness desired.

A most important consideration is the sill area of doors. Care should be taken to seal the bottom of doors, since cold air that enters will remain near the floor and cause more discomfort than air entering at a higher level in the room. Hardwood sills are very attractive but can warp. Metal does not warp but rapidly conducts the heat outward. Several combination metal-plastic or wood-plastic configurations are available to maintain appearance and control heat flow. Information about each of these materials is available from manufacturers and retailers that handle the products.

INSULATION: TYPES AND USES

Increasing the amounts and types of insulation has made substantial energy savings possible. With energy costs escalating, even greater savings are expected. If attic insulation were increased in the estimated 15 million under-insulated single-family homes in the United States, about 8% of the heating oil consumed could be saved each winter.

The life of a building is measured in decades; houses are not replaced as quickly as are other energy-consuming devices such as appliances and automobiles. The best alternative is to retrofit existing structures with energy-reducing features, including insulation, weather stripping and caulking; storm windows and doors; fluorescent fixtures where appropriate; energy-efficient appliances; and solar collection features.

The need for insulation is not a consideration solely for persons who live in cold climates. Insulation and other techniques can reduce the influx of sun and hot air into a structure and make the work of air conditioners much easier. Many people in the Sun Belt pay more for cooling their homes and offices than those who live in northern climates pay for heating. Insulation can be just as important in warm climates as it is in colder areas. Heat transfer can be restricted with the use of either thermal barriers or air-movement barriers. Thermal barriers are most often insulative or heat-reflective materials, while air movement is best restricted by sealing or "encapsulating" an area to reduce air infiltration.

Insulation is made from many materials and is produced in several forms. The most common are mineral or rock wool, cellulose fiber, vermiculite and perlite, foam board materials (polyurethane and polystyrene), foamed-in-place materials (urea formaldehyde and urethane), fiberglass, and reflective foils. Table 13-2 shows the types and features for each of the more commonly used materials.

The types and amounts of insulation used are based on several factors. The most important of these is the amount of heating or cooling needed to maintain the desired temperature for activities in the structure. A major factor in the determination of heating needs is the heating "degree days" (DD) value. The values are closely related to the latitude and altitude at which the building is located. The extreme southern part of the United States may have less than 1500 DD, while the northern part of the United States and southern Canada will be in the range of 7000 to 10,000 DD. Proximity to large bodies of water can change these numbers, as can the presence of major ocean currents. The western coastline of North America is warmed by the currents flowing northward, creating a narrow strip of less severe weather, requiring less need for insulation. Determination of the DD values for a particular location can be found in many texts, references, and newspapers or by contacting the nearest office of the National Weather Service.

Wind patterns can have the effect of increasing the DD for specific sites. The protection offered by structures, trees, or natural formations such as hills is of particular importance. The appropriate location of trees, both coniferous and deciduous, can help in deflecting away the cold winds in winter and routing cooling breezes nearer in warm

TABLE 13-2 Insulation Chart

Types	Forms	Advantages	R-value per inch
Rock (mineral) wool	Loose	Can be blown or poured	3.12
Glass fiber (fiberglass)	Batts and rolls	Easily fastened	3.17
Cellulose fiber	Loose	Can be blown in	3.7
Vermiculite–perlite	Loose	Granular: easily poured	2+
Urea formaldehyde	Boards	(Exterior finished sidewalls only)	5.1 (new); 3.4–4.2
Polystyrene	Boards	High R-value for thickness	4.0
Polyurethane	Foam boards	High R-value for thickness	6.5
Polyisocyanurate	Foam boards	High R-value for thickness	5.0

TABLE 13-3 Recommended R-values

Heating Degree Days	Ceilings	Walls	Floors Over Unheated Areas
Up to 2000	19-26	11	11
2000-4000	26-38	13	13
4000-6000	30-44	19	19
6000-8000	33-49	19	22
8000-10,000	38-60	19	22

weather. Savings of 8 to 12% can be realized by appropriate selection and placement of wind buffers.

The R-value ratings for each of the insulative materials are also shown in Table 13-2. The R-value is a material's resistance to heat flow. Normal insulation needs are determined by the recommended R-values for ceilings, walls, and floors, shown in Table 13-3.

Heating and cooling calculations should take into consideration solar gain and the heat-producing value of the contents of the structure: the people, appliances, and their effect on the amount of supplemental heat or cooling needed.

ENERGY AUDITS AND APPLIANCES

Several methods are available to determine an energy-saving program for buildings. Energy audits can be performed by independent contractors, utility companies, or the occupants themselves. The most thorough audits will be performed by contractors or utility employees at a modest cost. Some utility companies conduct free audits. For people who wish to conduct their own energy analysis, information and forms are available from a number of sources, including books, magazines, and local Cooperative Extension offices. Appendix E may be used if you wish to conduct an audit as an energy activity. No matter who conducts the audit it will provide valuable information as to which features of the structure are not energy efficient and require attention.

The items most often identified as needing attention are the building insulation, if any, the condition of caulking and weather-stripping material, and the need for storm windows or double glazing. The age and condition of the heating and/or cooling system is very important. Be sure that any energy audit includes energy gain from appliances and south-facing windows, as well as the losses from the structure.

Lighting levels and fixture locations are another energy use category to check. If lighting levels are too high, lower-wattage bulbs can be used, or fewer bulbs can be mounted in multiple socket fixtures.

(*Do not* leave a socket without a bulb. Screw in a burned-out bulb.) Localized or task lighting can be used instead of total room lighting. Timed lighting can be an energy saver. Fluorescent bulbs can be used in place of incandescent bulbs. Appliances should be sized to needs. Appliances that are too large use more energy than needed to do the job, and those that are too small may have to operate too long.

Does the water heater need extra insulation? All but the newest ones do. To find out, place a layer of insulation on the top or tape it on the side of your water heater (be sure the vent is not obstructed). Wait 24 hours. Place your hand under the insulation to determine if heat is being retained with the extra insulation. *Be careful*—in some instances, the heat retained may be enough to cause burns. A more sophisticated way is to measure the temperature with a thermometer. If a second layer of insulation—a heater "blanket"—is installed, it frequently saves enough energy to pay for itself in less than a year.

Most water heaters are set for 140°F (60°C) when shipped from the factory. Homes and offices rarely need water this hot, and it can be dangerous to children and those with sensitive skin. Turning down the temperature to 110 to 120°F (43 to 50°C) can result in a savings of up to 22% in energy use.

Are heat ducts and hot-water pipes insulated? This is easily accomplished during new construction and for existing open systems but is obviously difficult or impossible to do in finished walls without major reconstruction. Some contractors cover both hot- and cold-water pipes. There are several specialized insulating materials available for insulating pipes and ducts. Just a thin layer or covering on the cold-water pipes can reduce or eliminate condensation of moisture during the warm humid months and prevent dripping onto floors, ceiling tiles, or sheet rock.

One of the provisions of the Energy Policy and Conservation Act of 1975 was the requirement that manufacturers place labels on most major appliances showing their estimated annual operating costs. When purchasing new appliances you should make close comparisons of models, makes, and energy use. The cost of operating an appliance can often exceed the purchase price in 2 or 3 years. (See Appendix D for examples of appliances and the energy they use.)

Appropriate use of appliances must also be considered. For example, baking a few potatoes alone in a full-sized oven is very wasteful of energy. Appropriate meal planning would call for an "oven-prepared meal" where not only the potatoes but also the meat, vegetables, dessert, and perhaps even an appetizer could be cooked at the same time.

AIR POLLUTION

With increased energy costs and the desire to use energy more efficiently, more insulation and weatherization brings a serious disadvantage —indoor air pollution. For houses and buildings with conventional insulation and weatherization, outdoor air replaces indoor air about once every hour. As buildings are made tighter, air exchangers become fewer, until they reach as little as one air exchange every 5 hours in so-called "superinsulated, airtight" houses.

Urban residents normally spend 90% of their time indoors. As the number of air exchanges per unit of time decreases, the potential for retention of harmful pollutants increases. Studies have shown that the concentration of pollutants varies in direct proportion to the ventilation rate. Unfortunately, little has been done by either governmental agencies or environmental groups to seek solutions for corrective measures or new standards.

Gaseous emissions, particulates, and organisms from literally thousands of origins have the potential to create pollution. These may come from such diverse sources as combustion by-products, radon and radon decay products, formaldehyde, asbestos, chemical fumes and particulates, tobacco smoke, aerosols, and microorganisms found naturally in the environment.

Combustion

Major contributors to indoor pollution levels are the products of combustion that occur during heating and cooking. Fuels produce several by-products when burned: carbon monoxide (CO) and carbon dioxide (CO_2), oxides of nitrogen (NO_x), formaldehyde, hydrocarbons, and various particulates. If unvented heating units such as kerosene heaters are used, elevated levels of oxides of both carbon and nitrogen are present. In homes where gas stoves are used for cooking and where no venting is used, the peak hourly concentrations of these pollutants may exceed EPA standards by two to seven times.

The National Academy of Sciences and ASHRAE have both recommended that the occupational standards established by EPA for evaluating health risks caused by pollutants be those used for all indoor applications including residential structures.

Oxides of nitrogen, both NO and NO_2, as well as sulfur dioxide (SO_2) can cause greater susceptibility to respiratory infections, particularly among the very young and the elderly. The EPA standards call for limits of 0.05 ppm for NO_2 and 0.14 for SO_2. Those who use un-

TABLE 13-4 Air Quality Standards

	Standard Concentration		
	ug/m^3	ppm	Averaging Time
Suspended particulate matter (TSP)	75	—	Annual geometric mean
	260	—	24 hours
Sulfur dioxide (SO_2)	80	0.03	Annual mean
	365	0.14	24 hours
Carbon monoxide (CO)	10,000	9	8 hours no more than once per year
	40,000	35	1 hour no more than once per year
Nitrogen oxides (NO_2)	100	0.05	Annual mean
Ozone (O_3)	235	0.12	1 hour daily maximum, once per year
Nonmethane hydrocarbons	160	0.24	6-9 annual mean no more than once per year
Lead (Pb)	1.5	—	3-month average

Source: U.S. Environmental Protection Agency.

vented kerosene stoves may be exposed to as much as 0.10 ppm in a small room. Both types of oxides bind to the hemoglobin much like CO, causing impaired breathing, bronchitis, and emphysema. Other effects may include increased reaction time and depression.

Combustion of solid fuels results in the presence of suspended particulate matter composed of airbone solids and low-vapor-presssure liquid particles less than a few hundred micrometers in diameter. Wood burning can also cause increased levels of CO, especially if there is improper venting. Burning of paper with colored inks, such as Sunday comic pages, can release traces of arsenic vapor into the room. Wood that has been treated with preservatives emits toxic fumes when burned.

Radon

The presence of earth, rock, and mineral products exposes all of us to another pollutant—radon. It is an odorless, colorless, radioactive gas, the natural decay product of radium 226, which occurs in the earth's crust. Radon 222 has a half-life of 3.8 days. It decays further into polonium, several types of lead, and bismuth. Radon is a major portion of the background radiation occurring naturally in the environment. It is believed to be a major cause of lung cancer.

The preferred construction materials for passive solar structures (granite, brick, concrete, and native rock) are also the materials highest in radon. Buildings with several tons of rock-type materials which also happen to be constructed with heat-saving techniques serve to trap and concentrate radon gas rather than allowing it to mix with outside air and be diluted. Effective heat exchangers can reduce this hazard while providing energy savings.

Formaldehyde

Particle board, plywood and some insulation materials, textiles, carpets, and carpet pads are manufactured with chemicals containing formaldehyde (HCHO). This colorless, water-soluble, strong-smelling gas can cause upper-respiratory irritation. Concentrations of as much as 0.5 ppm have been measured in some homes. A level of 0.1 ppm can cause irritation and at 0.25 ppm asthmatics and children are at risk.

Formaldehyde is a component of urea-formaldehyde foam insulation and in some instances can outgas over long periods of time. Mobile homes frequently have high concentrations, since one of the major components is particle board. Measurements have shown levels as high as 2.4 ppm.

Physiological effects include eye irritation and tearing at low levels from 0.1 to 0.4 ppm. Concentrations of 10 to 100 ppm can cause serious lower-respiratory irritation and pneumonia symptoms. At levels as low as 6 ppm, formaldehyde is a carcinogen.

Asbestos

One of the most widely used fireproofing materials has been found to be the cause of skin irritation, lung and abdominal cancer, and a lung disease called asbestosis. Since the mid-1970s other materials have been substituted for asbestos for fire-retardant coatings. There remain many places where asbestos is still in place, providing a source of air pollution.

The primary preventative is the removal of asbestos from each place it can be found. As long as asbestos is not disturbed (i.e., the surface is not broken or abraded), the fibers will remain in their proper place. Where removal is impossible or expensive, it can be sealed so that the fibers cannot be released.

Tobacco Smoke

The smoke from a cigarette can contain as many as 2000 different chemicals, including CO, formaldehyde, NO_2, phenols, hydrogen cyanide, ammonia, and radioactive polonium 210. Concentrations of pollutants have been measured in offices at several times the levels allowable from chemicals in industrial workplaces. Particulate levels of 35 ppm and CO concentrations as high as 2 ppm have been found in bars, restaurants, conference rooms, and the smoking section of airplanes. Many of the substances identified in tobacco smoke are known carcinogens and can cause irritation of the mucous membranes of the respiratory tract, as well as changes in heart rate and blood pressure. Children of smokers are reported to have respiratory illness more often and more severely than those not exposed. Conservation efforts are increasing the exposure to these pollutants.

Microorganisms and Allergies

All living organisms emit various types of substances. Airborne microorganisms are one of the primary means of disease transmission; direct contact is another. Other organic pollutants include pollen, molds, fungi, and algae, which have all been known to cause allergies. Tightly sealed, energy-efficient buildings create prime conditions if precautions are not taken to change temperature and humidity conditions on a regular basis.

More insidious is the slow, often years-long buildup of allergic reactions. They are sometimes rapid and severe but can sometimes appear only after prolonged exposure to chemicals and other products. Combinations of materials are sometimes the causative factors. Elimination of a single item may reduce the reaction or end it completely.

LEGISLATION AND REGULATIONS

The federal government has developed several pieces of legislation to deal with the complex situation of air quality and the control of contaminants found to be harmful or toxic. The first of these to have a major impact on the lives of people in the United States was the Clean Air Act of 1970 (PL 91-604). In addition to environmental provisions, two regulatory standards address the matter of indoor pollutants. They are the National Ambient Air Quality Standards and the National Emission Standards for Hazardous Air Pollutants. Standards exist only for combustion pollutants, CO, NO_2, and particulates.

The Toxic Substances Control Act provides broad powers to EPA to control chemical substances that "present an unreasonable risk of injury to health or the environment" (15 U.S.C. Sect. 2601-2629, 1976 & Supp. IV 1980). This act regulates the manufacture, processing, distribution, and commercial use of chemicals, their labeling, instructions for use, and notice of potential health risks to purchasers or the general public. It is difficult to enforce this act with the "manufacture" of urea formaldehyde foam insulation since its manufacturer is the installer and control would require an inspector on site at every home or commercial site where the foam is injected. Additionally, attempting to regulate all the uses of particle board would be impossible, short of banning its manufacture entirely.

The Consumer Product Safety Act gives the Consumer Product Safety Commission (CPSC) the authority to control consumer products that release pollutants. The control generally regulates appliances, not construction materials. The commission must also rely on voluntary consumer product safety standards to reduce or eliminate the risk of a product. In the absence of such a voluntary standard, the CPSC could ban the production of a product where an "unreasonable risk of injury" exists.

Standards have been established by ASHRAE for minimum ventilation required to achieve acceptable indoor air quality. ASHRAE standard 62-73R has been proposed to revise previous standards which are now incorporated in the building codes of 45 states. The proposed standard specifies both short-term (30 minutes to 1 hour) and long-term (24 hour) exposure for industrial chemicals as well as for substances such as carbon dioxide.

The rate of ventilation is also based on the level of activity of the persons expected to be in the area. ASHRAE recommends that the minimum outdoor air requirement be 5 ft^3/min per person for sedentary work. As the level of activity increases, from shopping, for example, to heavy factory work, the ventilation demands increase by a factor ranging from 2 to 4.5.

LEGISLATION AND TAX INCENTIVES FOR CONSERVATION

When the DOE was established in 1977, among its tasks were the promotion of conservation of available energy resources, to reduce the overall energy demand (as well as the demand for imported energy), and to promote the development of nonconventional energy sources. The levels of funding and policy decisions about priorities have varied with each administration. Much emphasis has been placed on incentives for fossil fuel development and for nuclear-fueled electricity generation.

The National Energy Conservation Policy Act of 1978 (PL 95-619) was passed in an effort to reduce the rate of consumption in nonrenewable energy resources yet maintain economic growth. Included in the Act were provisions for weatherization grants for low-income families, a loan program for solar heating and cooling for residences, grants to schools and hospitals to improve energy efficiency, and efficiency standards for new appliances. It also provided penalities for violating fuel economy standards for cars and promoted the use of recycled materials.

Electricity-producing plants using natural gas or petroleum as their primary fuel were urged to convert to coal under the Powerplant and Industrial Fuel Use Act of 1978 (PL 95-620). This Act sought to promote the use of a nonrenewable fossil fuel with reserves far in excess of those of oil and natural gas. Loans for the purchase of air pollution equipment were also included.

The law that provides the most incentive for the energy consumer is the federal Energy Tax Act of 1978 (PL 95-618). It provides tax credits for residential insulation and weatherization, the purchase of solar and other renewable energy equipment, and the development of geothermal energy; provides tax incentives for "van pooling"; and exempts alcohol fuels from the gasoline excise tax.

Small and medium-size WECS and hydroelectric facilities received much incentive from the Public Utility Regulatory Policies Act of 1978 (PL 95-617) (PURPA). It requires public utilities to interconnect with private producers of electricity, encourages cogeneration and other small-scale power production, and establishes rate-adjustment standards to reflect the "avoided cost" of fuel for public utilities. Funding is provided for state and consumer representation at proceedings which establish those rates. The latter provision has been useful in several states and with some public utilities which were less than eager to interconnect and pay equitable rates for power purchased from private producers.

Another "incentive" to use less natural gas was provided with the passage of the Natural Gas Policy Act of 1978 (PL 95-621). The major provision is the deregulation of the price of new gas and certain intrastate shipments, to establish an incremental pricing system for large consumers (to encourage the use of other fuels, conservation, and cogeneration), and the curtailing of gas use in emergencies coupled with presidential authority to allocate natural gas use in an emergency.

The Crude Oil Windfall Profits Tax Act of 1980 (PL 96-223) provides a 40% federal tax credit for residential WECS and a 15% tax credit over and above the normal 10% investment tax credit for nonresidential wind systems.

Further tax credits are provided in the Economic Recovery Tax Act of 1981 (PL 97-34). Maximum tax brackets were reduced from 70% to 50%, making tax shelters less valuable and investment in profit-making ventures more so. Depreciation rates were changed to provide an "Accelerated Cost Recovery System" (ACRS) based on four classes:

3-, 5-, 10-, and 15-year public utilities. Investment tax credits now apply to property with an ACRS of 5 years rather than 7.

Under PL 97-34, conventional and renewable energy systems are both depreciated under the same accelerated cost recovery schedule. This has eliminated the competitive edge that had aided the developing renewable energy industry. Several reports have indicated the need to continue support for renewable energy incentives until the industry is well established. Any losses in federal and state revenues as a result of these incentives are more than made up in increased energy production, increased industrial production (often of existing plants and technologies), and increased jobs.

Individuals and businesses may enjoy a tax credit of 40% of the first $10,000 spent on solar, geothermal, or wind-powered equipment, to a maximum of $4000. Installation costs are included. Home energy conservation costs may account for a tax credit of 15% of the first $2000 spent. A new $300 limit applies if you move into another home at a later date. The details and limitations of the materials and applications that can be used to meet the regulations for the tax credit are found in Department of the Treasury publications at your local library or can be requested through district offices of the Internal Revenue Service.

In addition to the federal tax credits, loans, and grants, most states have encouraged the development of nonconventional energy resources and conservation with additional incentives. These incentives are most often in the form of income tax credits for the installation of solar systems, some for wind and hydro production, and a few for woodburning space-heating units, insulation, storm windows, weather stripping, and other fuel-saving measures. In addition, refunds are granted in some states for the installation of energy-producing equipment, including waste conversion or alcohol-producing plants. Legislation for the various states changes rapidly; it is important to check your state's current provisions for the current status of tax incentives. New York and other states provide a property tax exemption for the assessed value of the solar system or WECS portion of the property, as well as a 15% state tax credit. Massachusetts is more generous, offering a state tax credit of 35% after the federal amount.

The DOE has recently announced plans to improve its Conservation and Renewable Energy Inquiry Referral Service. Another service, the National Appropriate Technology Assistance Service, has been funded to provide technical service for anyone with problems in the design, operation, or maintenance of renewable energy technology.

SUMMARY

Energy conservation and economic growth are compatible. The per capita use of energy in Sweden and West Germany is one-half that of the United States, yet their standard of living is comparable to ours. The opportunity to effect tremendous energy savings exists in America—but the desire to do so is questionable. Our national economy, security, and ability to compete in foreign markets make conservation imperative.

Why conserve? Conservation is the cheapest, safest, fastest, and most economical way of "developing" energy. The most compelling reason for conservation is the cost, to ourselves and to future generations, of not conserving. Many natural resources are becoming less available and thus more expensive. Some renewable resources have been "tapped" for a limited amount of energy, but more remains to be done. A majority of that which is discarded as "waste" in our society has the potential for recycling and/or energy extraction. Less waste disposal means less pollution and less environmental degradation. In many instances, inappropriate energy sources are used simply because of their convenience rather than their appropriateness to perform a task.

Transportation offers many opportunities for conservation, such as driving less, combining trips, car pooling, and using alternative transportation, including mass transit, bicycles, and walking. Vehicles that are lighter weight, have smaller engines, and are driven at lower speeds could save enough energy to cut in half our need for imported oil.

Matching the transport mode to the need would save large quantities of fuel. Using rail instead of trucks for long-haul runs would "cost" 700 Btu/ton versus 3000 Btu/ton. Using rail instead of air transport would reduce energy consumption by 60%.

Shelter, in the form of private residences as well as workplaces, could be made more energy efficient with weatherizing and insulation techniques. The resulting savings could be as much as 50% in energy costs. The purchase of more energy-efficient appliances would reduce energy consumption 10 to 20% in homes and would be a welcome relief from higher electricity costs.

Although not considered conservation—but a substitution of energy sources—the addition of passive, hybrid, and active solar systems can be accomplished with little or no change in the outward appearance of a home and will result in tremendous savings in energy costs over the life of the building. Similar savings can be accomplished with commercial and industrial buildings.

The industrial sector consumes nearly 40% of the energy produced in the United States. Many energy- and cost-saving approaches have been developed which are applicable across the country. Only about 15% of the potential applications have actually been put in place. Heat-recovery equipment, supplying external combustion air and preheating of incoming air, could reduce fuel consumption by 30%.

The value of the materials found in the waste stream has been well established. "Resource recovery," not waste disposal, is the concept that must be developed! Whether waste is burned to produce heat for process steam, or is shredded and separated by various separating techniques, the result is the same—waste reduced in volume, energy produced and/or saved, and costs at all levels reduced. In addition, environmental conditions improve and employment opportunities increase.

ACTIVITIES

1. Select one industry in your community. Request a field trip or tour of the facility with emphasis toward the energy conservation measures that industry has taken (or is planning to take). What other use(s) could be made of the energy released after each process?
2. Check for and measure the size of gaps around doors and windows of a home, school, or other building. Calculate the equivalent open area through which heat is lost.
3. Conduct a home energy audit (see Appendix E).
4. Request an energy audit for your home or office from your local utility. Develop a plan to reduce or eliminate the energy losses identified. Calculate the costs involved and the payback time in terms of energy savings.
5. Maintain a log of the daily heating degree-days over a period of time. Calculate the volume of energy needed and the cost to heat your home during this period.
6. Check the laws of your state or province to identify the tax incentives currently in force. Identify those measures you could take in your home or business to take advantage of the laws.

BIBLIOGRAPHY

Consumer Reports, October 1982, pp. 504-507.

Energy Information Administration, *1982 Annual Energy Review*. Washington, D.C.: DOE, April 1983.

Energy User News, "Efficiency Trends by User Type," *Energy User News*, January 23, 1984, p. 5.

Frieden, Bernard J., and Kermit Baker, "The Record of Home Energy Conservation: Saving Bucks, Not Btu's," *Technology Review*, October 1983, pp. 23-31.

Gabel, Medard, *Energy, Earth and Everyone*. Garden City, N.Y.: Anchor Press, 1980.

Grossman, Richard, and Gail Daneker, *Energy, Jobs and the Economy*. Boston: Alyson, 1979.

Heinz, John, *Energy and Education*. National Science Teachers Association, February 1984, Vol. 7, No. 3, p. 1.

Kirsch, Laurence S., "Indoor Air Pollution and Government Policy, Part I," *Environment*, March 1983, Vol. 25, pp. 17-20, 37-41.

——, "Indoor Air Pollution and Government Policy, Part II," *Environment*, April 1983, Vol. 25, No. 3, pp. 27-35.

Lotker, Michael, "Making the Most of Federal Tax Laws," *Alternative Sources of Energy*, No. 63, September-October 1983, pp. 38-43.

Lovins, Amory B., and L. Hunter Lovins, *Brittle Power: Energy Strategy for National Security*. Andover, Mass.: Brick House, 1982.

National Audubon Society, *Audubon Energy Plan, Technical Report*. New York: National Audubon Society, April 1981.

Renewable Energy News, November 1983.

Spengler, John D., and Ken Sexton, "Indoor Air Pollution: A Public Health Perspective, *Science*, Vol. 221, No. 4605, July 1983, pp. 9-17.

U.S. Department of Energy, *Energy Projections to the Year 2010*. Washington, D.C.: DOE, October 1983.

Wadden, Richard A., and Peter A. Scheff, *Indoor Air Pollution*. New York: Wiley, 1983.

Chapter 14

Cogeneration and Heat Reclamation
Two Energy Forms for the Price of One?

CONCEPTS

1. Cogeneration increases the total efficiency of the fuel conversion process.
2. Cogeneration is applicable to many businesses and industries.
3. The use of cogeneration facilities and heat reclamation systems will become widely used in the future.
4. Efficient use of our resources will have less of a negative environmental impact.
5. Heat reclamation equipment can recover only a portion of the waste heat rejected from an energy conversion system.
6. Heat pumps can greatly reduce the consumption of the balance of our fossil fuel reserves.

GLOSSARY

COEFFICIENT OF PERFORMANCE (COP)—the efficiency of a unit based on energy output ÷ energy input.
COGENERATION—the utilization of a single energy source to produce electrical energy and thermal energy (heat).

HEAT PUMP—a device for transferring heat from a substance at one temperature to another substance at a higher temperature.

HIGH-GRADE THERMAL ENERGY—temperatures above 200°F (93°C).

LATENT HEAT—heat that causes a change of state without a change in temperature.

LOW-GRADE THERMAL ENERGY—temperatures below 200°F (93°C).

PROCESS HEAT—the thermal energy required in an industrial plant (i.e., steam, hot water, hot air, etc.).

SENSIBLE HEAT—heat that changes the measurable temperature of a material but does not change its state.

INTRODUCTION

Using energy that is readily available rather than wasting it would seem to be a wise thing to do, especially when fuels are expensive and the supply limited. Cogeneration systems, which provide two or more usable energy forms from a single source, and heat reclamation systems, which salvage heat that is normally wasted, are becoming popular as the price of conventional energy sources escalates.

Heat pumps are units that also use energy wisely. These units extract from the environment low-grade heat that would normally be lost or disregarded and concentrate it to usable levels. This heat can then be used instead of consuming additional quantities of fossil fuels.

Wise use of the resources at hand will help save fossil fuels for future generations, reduce the costs of products, provide employment opportunities, and provide a better standard of living through less environmental degradation.

BACKGROUND

A characteristic of the affluent American economy has been to use a product and then throw it away. This "disposable" attitude is easily evident as you observe the content of the trash that we dispose of daily, which ranges from glass bottles to automobiles. Nearly all of this material required energy during forming or production, and much of this material can return a portion of this energy, directly or indirectly, through wise utilization.

Not so easily observable are the vast quantities of thermal energy that we have permitted to "escape" into the environment after converting a fossil fuel source to meet our energy needs. This waste of thermal energy represents a large resource which, if used instead of wasted, would conserve considerable energy for future generations. As might be expected, many of the larger users of energy have the greatest potential to be massive wasters as well as great conservers.

Traditionally, large industries and utility companies have converted fossil fuels to meet their needs and exhausted the "excess" low- and high-grade heat into the atmosphere or adjacent waterways. As the economics of this practice become more binding due to the ever-increasing cost of fuels, many of these users will be looking at ways to make the most efficient use of their energy sources.

COGENERATION SYSTEMS

Much of industry's energy use involves steam. Paper manufacturing, milk processing, oil refining, and cement and chemical production are examples of industries requiring large volumes of steam in the production of products. Each company must therefore burn fossil fuel to produce the necessary steam. Many of these industries also use considerable quantities of electrical energy which they purchase from a utility. An appropriate procedure would be to use the same quantity of fossil fuel to supply both needs.

Cogeneration energy systems are dual (or multiple) energy conversion systems which provide electrical energy and thermal energy from a single fuel source. The thermal energy may be in the form of industrial process steam or hot water, hot water or air for space heating, potable hot water, or other necessary heat.

The technology for cogeneration is not new. In the early twentieth century it was common practice for factories to generate their own electrical power from the same fuel source that produced the process steam required in the plant. Nearly 50% of all U.S. electricity at that time was generated in many small power plants, which also provided hot water for space heating in the adjacent community. As large, centralized utility companies became established and organized complex electrical distribution systems, small producers found that they could no longer compete. By 1920, 30% of the country's electricity came from cogeneration sources. Economical and easily accessible fuels continued to make centralized generating facilities more efficient and economical, and the use of cogeneration by small producers continued to fall, to today's level of about 4%.

Internationally, the use of cogeneration systems has been more widely used. More than half of the urban homes in the USSR are heated with "waste" heat from industrial and utility plants. Denmark heats 33% of its city homes and Sweden 25% (*Soft Energy Notes*). Sweden, a long-time user of urban district heating, is installing electricity-generating units in its heating plants. They are now able to generate electricity at no appreciable increase in costs, and still circulate 210°F (99°C) water through insulated steel pipes to heat the city's dwellings and domestic water.

Nationally, cogeneration, heat reclamation, and "waste" heat distribution for district heating have not been popular. Although some systems can be found, past practice has been simply to dispose of "excess" heat as economically as possible. Inconvenience, unreasonable distances to potential "partners," and the lack of need, desire, necessity, or encouragement to be efficient were the reasons for gross energy waste in the past. Worldwide, today's higher fuel costs and the need to conserve fossil fuels have helped to establish renewed interest in cogeneration systems. Utility companies are finding that selling normally discharged thermal energy can be an additional source of income.

Most electricity is produced using a steam turbine generator, gas turbine generator, or diesel generator. Each system usually burns fossil fuel and gives off large quantities of heat which is generally wasted. Large industries which provide their own electrical power frequently employ a diesel generator system in which the generator is rotated by a multicylinder piston-type diesel engine which burns fuel oil. Gas turbine units work similarly to a jet engine. These are rotary engines in which hot, combusting, expanding gases drive a turbine connected to a generator.

Normally rejected heat energy from the exhaust gases from either diesel or gas turbine generating units can be used to boil water, producing steam. This steam can be used as the thermal energy required in process and/or plant heating operations. The fuel-use efficiency in these cogeneration systems is generally increased to 55 to 60% for standard high-speed diesel generation units and 75 to 80% for gas turbine units. Even greater efficiencies can be obtained if the thermal energy stripped from the exhaust gases is used to drive a low-pressure steam turbine generator. The sequential production of electricity from two different generators, directly and indirectly fueled by a single source, is called combined-cycle power technology (see Figure 14-1). In addition, process heat can still be recovered from the steam turbine generator, resulting in three energy outputs (two electrical, one heat) from a single fuel source.

Figures 14-2 through 14-4 show the Hoffman–La Roche diesel-powered cogeneration system. Believed to be the largest slow-speed diesel cogeneration engine in the world [85 ft long, 35 ft high, and 22 ft wide (25.9 m long, 10.7 m high, and 6.7 m wide)], the unit began operation in February 1983 at the Hoffman–La Roche vitamin-manufacturing plant in Belvidere, New Jersey. The 33,200-hp 10-cylinder slow-speed (120-rpm) diesel engine turns a turbine which produces 23,000 kW of electricity, 160,000 lb of process steam, and 262,000 lb

Cogeneration Systems

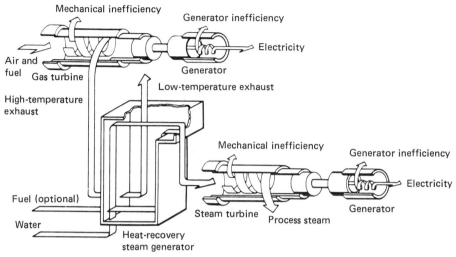

Conventional combined cycle system.

Figure 14-1 Conventional combined-cycle system. (From Robert A. Meyers, ed., *Handbook of Energy Technology and Economics*, 1983. Reprinted by permission of Charles E. Hickman and John Wiley & Sons, Inc.)

Figure 14-2 Diesel cogeneration system: plant exterior view. (Reprinted with permission of Hoffman–La Roche Inc.)

Figure 14-3 Diesel cogeneration system: The 10-cylinder slow-speed diesel engine. (Reprinted with permission of Hoffman–La Roche Inc.)

of 140°F (76.7°C) water per hour. It provides all the electrical needs for the plant, and the excess is sold to the local public utility. Waste heat from the exhaust is recovered and used to make process steam required in the vitamin-manufacturing process. The project, expected to operate at an overall energy-efficiency rate of 87%, is a DOE cost-shared demonstration project. This appreciable increase in efficiency over the normal 40% for a diesel-generator is attributed to the slow-speed operation of the diesel unit.

Utilities and large companies that generate electricity from nuclear or fossil fuels generally use high-pressure steam turbine generators, resulting in a fuel-use efficiency of approximately 30% and 35%, respectively. Condenser losses alone account for 48%. Rather than warming lakes or the atmosphere, if the recovery of rejected heat were incorporated with these units following electricity generation, an overall efficiency of 84% could be obtained even though higher initial steam temperatures and pressures are required. This system is called the topping cycle because electricity is produced at the beginning or "top" of the fuel conversion sequence (Figure 14-5).

Cogeneration Systems   349

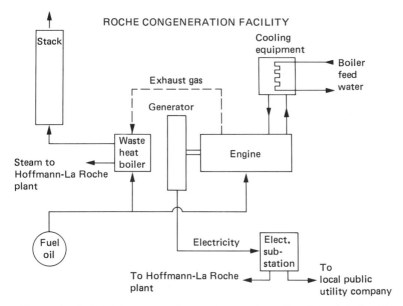

Figure 14-4 Diesel cogeneration system: schematic. (Reprinted with permission of Hoffman–La Roche Inc.)

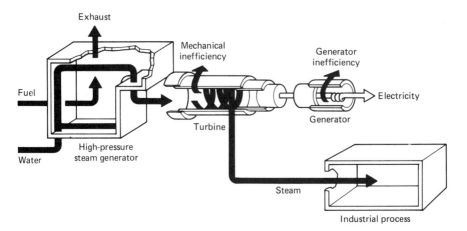

Figure 14-5 Cogeneration topping-cycle system. (From Robert A. Meyers, ed., *Handbook of Energy Technology and Economics*, 1983. Reprinted by permission of Charles E. Hickman and John Wiley & Sons, Inc.)

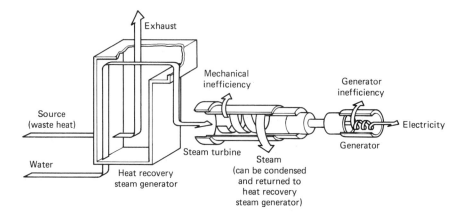

Figure 14-6 Cogeneration bottoming-cycle system. (From Robert A. Meyers, ed., *Handbook of Energy Technology and Economics*, 1983. Reprinted by permission of Charles E. Hickman and John Wiley & Sons, Inc.)

Bottoming-cycle cogeneration systems (see Figure 14-6) may be attractive in the future as energy costs escalate. These systems generate electrical energy from lower-temperature "waste" heat from used process heat, exhaust from steam turbine units, heat from exothermic chemical processes, or exhaust from ovens or kilns used in various industries. The units used to convert this lower-temperature heat require heat engines similar to the Rankine cycle engines used with OTEC and geothermal energy sources. You will recall that these systems require a binary fluid with a lower boiling point, which is heated by the source temperature and expanded through a low-pressure turbine. These systems, currently available in smaller sizes, are expensive and consequently are not attracting the level of attention that the topping-cycle systems are enjoying. Six units were successfully combined-cycle tested for the first time in 1979 on 500 to 600-kW units using waste heat from stationary diesel electric generators.

A study by Research Planning Associates for the DOE concluded that electricity cogeneration is ideally suited to five industries that constitute 75% of the industrial steam demand. These industries are pulp and paper, chemicals, petroleum refining, steel, and food processing. According to the study, these industries require about 20% of the total energy used in this country.

HIGH-GRADE HEAT RECOVERY

Fossil fuels have a combustion temperature of approximately 3000°F (1650°C). Since steam can attain a practical working temperature at only 1000°F (540°C), two-thirds of the fuel energy is lost due to conventional technology, which converts chemical energy to mechanical energy via steam. There are many industrial processes today which dispose of high-grade heat from boilers, furnaces, and ovens in the range of 2000 to 3000°F (1090 to 1650°C). High-grade heat recovery is presently being approached through cogeneration systems because durable heat exchangers capable of efficient recovery of high temperatures are simply not available.

LOW-GRADE AND "WASTE" HEAT RECOVERY

When energy is converted from one form to another there is a net loss of usable energy due to the second law of thermodynamics. The only exceptions involve solar cells and fuel cells because they do not convert the primary energy form to heat before generating electricity.

Conservative estimates place the quantity of energy wasted by all sectors of American society at about 70% [Angrist] and by individuals at nearly 50%. One solution to the problem of our dwindling fossil energy reserves is simply to reduce this waste, thereby permitting our supplies to last longer. These figures could be reduced considerably through conservation efforts, but there are practical levels beyond which efforts to conserve are no longer feasible. For example, automobile engines require a cooling system that dissipates heat created by fuel combustion. Energy could be saved if this thermal energy were recovered and used for other purposes, but how can this be done in a practical manner at 55 mph (88 km/h)? The systems to recover this heat would be more expensive than the energy they could conserve, or they would create levels of inconvenience for the user such that the systems would be disregarded. Table 14-1 shows the thermal efficiencies of some typical energy conversion systems.

Low-temperature heat for space, water heating, and industrial processes is regularly acquired from the high-temperature combustion of fossil fuels. Since combustion produces temperatures far in excess of those required, heat produced by combustion may be considered wasteful.

Considerable effort and money is currently being expended in reclaiming heat that would normally be wasted or lost to the environment. Although not practical for some energy conversion systems, such as automobiles and trucks, or at extremely low-grade temperatures, stationary systems do provide opportunity for thermal energy recovery via heat exchangers, heat pipes, and heat pumps. Some heat-recovery systems, using both latent and sensible heat, claim a potential energy savings as high as 20% of the total Btu input depending on the efficiency of the unit producing the waste heat. The problem has been to

TABLE 14-1 Overall (In-Field) Thermal Efficiencies
of Energy Conversion Systems and Devices

System	Efficiency (%)[a]
Electricity generating plant	
Fossil-fueled	35
Fossil-fueled with cogeneration	84
Nuclear-fueled	30
Home heating systems	
Heat pump (solar-assisted)	125[b]
Electric resistance	100 (30 net)[b]
Heat pump (ground or H_2O)	85
Natural gas (high efficiency)	80
Natural gas (normal efficiency)	60-80
Liquefied petroleum gas	75
Heat pump (air source)	62
Fuel oil	55-65
Solar collector	35-85
Solar photovoltaic cells	12-15
Wood stove	35-65
Stove with catalytic combuster	70-80
Fireplace	−10-50
Automobile engine	25 (5 net)[b]

[a] Variations are attributed to differences in manufacturers, models/designs, size of units, and degree of maintenance.
[b] Total efficiency from energy value of source to the converted end-use efficiency.

design and develop recovery systems that can provide a net financial gain or return on investment, thereby encouraging use resulting in energy conservation.

Heat may be recovered from many different sources and used to heat or preheat incoming water, air, or production materials. Specifically, heat may be reclaimed from:

Building ventilation systems
Cooling fluids: liquids, gases
Wastewater
Exhaust systems: chimneys, stacks, pipes
Production equipment: motors, machines, ovens, dryers, boilers
Products: metals, ceramics, foods
Refrigeration condensing units
Furnaces
Incineration of waste materials

Analysis of these sources must be made before use to determine chemical composition, temperatures, pressures, particulates, flow rate, and heat of vaporization and condensation.

Heat-recovery equipment is expensive. Government tax incentives, highly competitive markets, and the necessity to reduce expenses have prompted many industries to invest in reclamation units and systems. James H. Mueller of Regenerative Environmental Equipment Company, Inc., has said, "Equipment can be financed from energy savings with pollution control as a bonus" [Hill, p. 409].

Heat Exchangers

Heat exchangers are generally located in or around the waste heat source and may be of air/air, air/liquid, liquid/air, or liquid/liquid type. In each case the heat exchanger conducts the heat away from the source so that it may be used rather than wasted. Heat-carrying media often have fouling and corrosive characteristics. For these reasons, heat exchangers are manufactured from many materials, including copper, brass, aluminum, stainless steel, and ceramics. Many can be disassembled for mechanical cleaning. Where potable water is being heated, double-wall units help to ensure that contaminants in the heat source medium will not enter the exchange system. Figure 14-7 shows various types and sizes of heat exchangers for small residential and large commercial applications.

Large facilities that have centralized heating plants often use heat exchangers called economizers in their stacks or chimneys (see Figure 14-8). These units strip considerable heat from the exhaust gases of these fossil-fueled plants and preheat incoming water or air so that less total fuel is required to maintain normal operating temperatures. An example would be where flue gases from a boiler are cooled from 600°F (316°C) to 350°F (147°C) in the stack economizer. The returning boiler feedwater from the process or heating system is heated in the stack economizer from 220°F (104°C) to 300°F (149°C) before entering the boiler. Heat remaining in the flue gas is exhausted up the stack.

The New York Power Authority recently opened a 1-acre hydroponics greenhouse in Queens to use waste heat from an adjacent subway electricity power project to grow vegetables for food markets in the metropolitan area (see Figure 14-9). The principal energy source for heating the greenhouse is water, which is used to cool the lubricating oil of the turbine. Temperatures of 95 to 105°F (35 to 41°C) are being extracted. An aquaculture application of discharge heat by the Long Island Lighting Company is expelling warmed seawater into a lagoon used to raise oysters.

Other interesting applications of heat exchangers can be found in

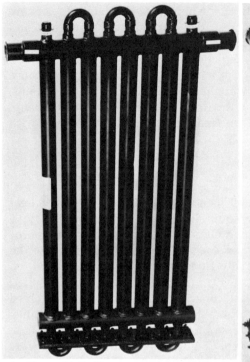

Figure 14-7 Heat exchangers (Courtesy of Doucette Industries, Inc., York, Pa.)

Figure 14-8 Stack heat reclaimer. (Courtesy of Eldon Corp., P.O. Box 10513, Jefferson, La. 70181.)

Figure 14-9 Hydroponics greenhouse heated by waste heat from an electricity generating plant. (Courtesy of the New York Power Authority.)

use on several energy-efficient dairy farms. Heat from the milk cooler's refrigeration system is recovered and used to heat water for cleaning the milking equipment. At one central New York State farm, methane from the cow manure is produced in a digester and powers an engine connected to a 19-kWh generator. The heat given off by the generator engine is recovered and provides 70% of the heat for the farmer's house and maintains the methane digester at a constant 95°F (35°C) throughout the year.

Heat Pipes

Heat pipes are superconductors compared with devices of the past. They are low in cost, simple to design, have no moving mechanical parts, and are efficient in the transfer of heat. The principle of the heat pipe was used by Robert S. Gaugler of the General Motors Corporation in 1942, but practical application did not begin until 1963. George M. Grover of the Los Alamos Scientific Laboratory worked with a similar device and established the term "heat pipe" to describe it (see Figure 14-10).

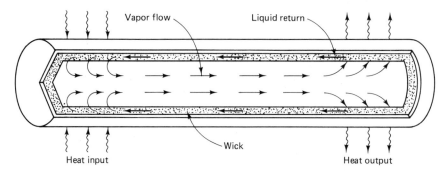

Figure 14-10 Heat pipe.

According to Roth:

> Heat pipes are closed, evacuated chambers, which inside are lined with a capillary wick, which is saturated with a volatile working fluid.
> In operation, heat pipes combine two familiar principles of physics, vapor heat transfer and capillary action. Upon the application of heat at the evaporator section, the fluid in the wick is transformed to a vapor and driven into the center cavity of the pipe. When the fluid vaporizes, a large amount of energy is absorbed, which is called the latent heat of vaporization. . . . This heat energy is stored and will not be released until the vapor is condensed to a liquid. [Roth, p. 4]

To understand more about the heat pipe function, Roth explains:

> The heat input causes a rise in pressure in the evaporator section which moves the vapor to rapidly fill the entire length of the heat pipe. The inner walls, being slightly cooler, cause condensation upon contact with the vapor. This releases the stored heat energy to the walls of the heat pipe and the resulting liquid is collected in the wick. The capillary pumping action of the wick returns the liquid to the evaporator section, where it is, again, vaporized to continue the cycle. [Roth, p. 5]

Heat pipes would be ideal for many heat reclamation and cooling applications because they can transport heat with a very small temperature loss, move large amounts of heat with small units, conduct many thousands of times better than metals, are self-contained, eliminate cross-contamination, and require no controls or power. In addition, they have a long life, no moving parts, operate automatically, require no maintenance, are noiseless, and are low in cost. The problem is that large-scale heat pipes are not yet commercially available.

Heat Pumps

Heat pumps are "mystical" devices that perform a function few other machines can perform. By using a small amount of electrical energy, heat pumps take normally unusable low-temperature heat and "pump" it up to produce higher-temperature useful heat which is many times greater than the heat equivalent of the electricity required for operation. Basically, it multiplies the primary energy by capturing normally lost, unusable, or wasted "free" heat.

A heat pump is a device that functions similar to your refrigerator. Heat removed from inside the refrigerator by the refrigerant system is discharged into room air. You can feel this heat being emitted from the condensing coils in the back or bottom of the unit while it is running. The heat pump removes heat from an external source such as air or water, concentrates it, and discharges the heat inside a structure where it can be utilized. In the summer, the unit can be reversed—extracting heat and humidity from inside and discharging it outdoors.

The concept of the heat pump is not new. Lord Kelvin produced a "heat multiplier" in 1852, and Jacob Perkins' vapor-compression refrigerator is 18 years older. Air-compression refrigerators were proposed by Oliver Evans in 1805 and built by John Gorrie in 1849. The first laboratory-made refrigerator produced by the evaporation of ether into a partial vacuum is credited to William Cullen in 1448. Earlier references to refrigeration go back to the early fifth century. The first heat pump used to heat a house was built in 1926 by Haldane in Scotland.

Like air conditioners, heat pumps are rated in tons, based on the amount of heat necessary to melt 1 ton of ice. A 1-ton rating means that the system will provide heat at the rate of 12,000 Btu per hour.

Air source heat pumps. Outdoor air in winter may seem cold, but it actually contains much heat. Only at absolute zero [$-460°F$ ($-273°C$)] is there no heat in air. Scientifically (and for advertising purposes) this sounds terrific, but to be economically competitive with fossil fuel heat sources, use of standard air source heat pumps should be restricted to temperatures above $28°F$ ($-2.2°C$). Most manufacturers honestly state that air source units work less efficiently below $40°F$ ($4.4°C$). With this limitation, air source heat pumps are not highly recommended for use in the northern one-half of the United States or where air temperatures may be below $40°F$ for extended periods of time.

Ground-Coupled heat pumps. There is also considerable heat energy available in the earth. It is not even necessary to tap into a geothermal source, although it would be nice to have one available. In a ground-coupled system, pipes are buried below the frostline and water or other fluid is circulated through this "flow loop," extracting heat from the earth. The heat pump in turn extracts this heat from the circulating fluid. Moist soil, which will conduct the thermal energy in the ground to the flow loop, is preferred (see Figure 14-11).

Approximately 400 ft of flow loop is required for each ton of heating capacity. Where the building lot is very small or other buildings are near, the pipe can be laid in a trench that encircles the house rather than in a closed-loop pattern in the yard. For retrofit applications where the existing structure has plantings, walks, driveways, and underground connections, a vertical loop can be established by drilling a deep hole and inserting two pipes linked by a U-fitting. As might be expected, this coupling most often becomes a combination ground-coupled/water source system. Completely installed units ranged in cost from $7000 to $10,000 in 1984.

Water source heat pump. More low-grade heat can be extracted from water source heat pumps (Figure 14-12) than from air source or ground-coupled units. Due to the specific heat and conductivity of water, higher efficiency is possible. The major concern when installing these units is that an adequate supply of good-quality water be available from which to extract the heat. This requires a nearby river, lake, or aquifer that may legally be used. A minimum flow of 2½ to 3 gal/min is necessary for an average home-size unit.

Excluding geothermal resource areas, groundwater temperatures vary according to geographic location and depth. The nationwide average ground temperature is around 47°F (8.3°C). It ranges from a high of 80°F (26.7°C) in Texas to a low of 44°F (6.7°C) in North

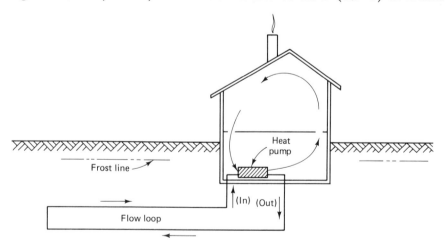

Figure 14-11 Ground-coupled heat pump.

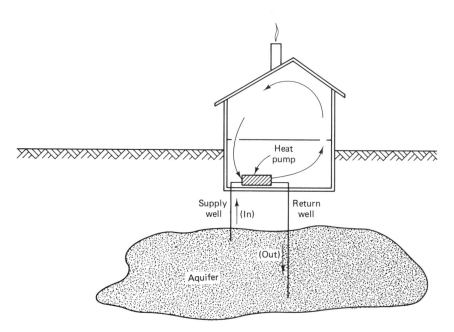

Figure 14-12 Water source heat pump.

Dakota (Figure 14-13). Where temperatures below 40°F (4.4°C) are encountered (no matter what the heat source), larger, more expensive units are required to maintain efficiency.

How heat pumps work. "It sounds great, but how do I heat my house to 68°F (20°C) with air, earth, or water that is colder and maybe even freezing"? Response: The same way that your refrigerator makes cool things colder.

It has long been known that the compression of a gas causes its temperature to rise, and if released quickly, this compressed gas causes significant cooling. In explaining the function of the water source heat pump (Figure 14-14), Gannon states:

> In a typical heating cycle, liquid refrigerant flows through a capillary tube that lowers its pressure and boiling point. Passing through the ground-water heat exchanger, the refrigerant extracts heat from the circulating water, boils and vaporizes. The cooled water returns to the ground while the warm, low-pressure gas travels to the compressor. There it is squeezed to form a high-pressure, hot gas. Pumped through a reversing valve to the air heat exchanger, it condenses, releasing heat to the circulating air. The warm air is ducted throughout the house, and the liquid refrigerant flows back through the capillary tube to repeat the cycle. For cooling, the process reverses. The compressor sends the hot gas directly to the water heat exchanger to release heat

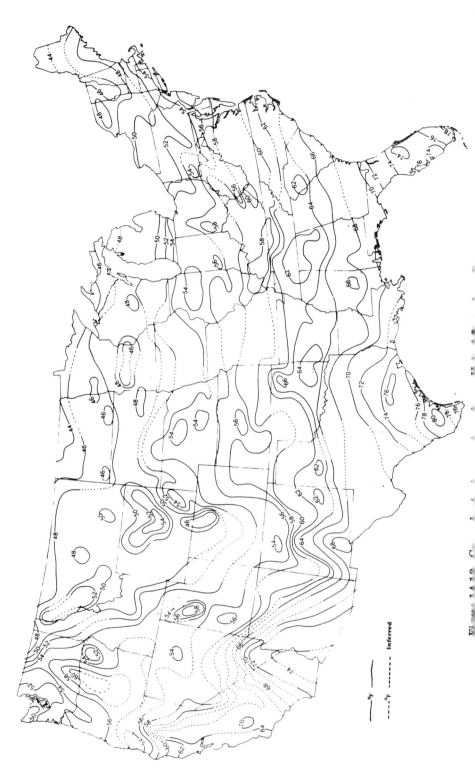

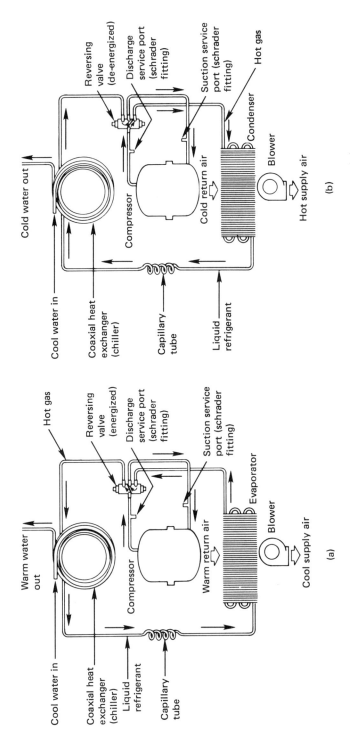

Figure 14-14 Schematic of a water source heat pump system: (a) summer cooling; (b) winter heating. (Courtesy of Carrier Corporation.)

collected from the house. The warmed water returns to the ground while the cooled, liquefied refrigerant flows through the capillary tube to the air heat exchanger. There, it again absorbs heat from the house and vaporizes. The gas returns to the compressor, which pumps it back to the water heat exchanger to renew the cycle. [Gannon, p. 79]

Less external energy is necessary to provide heat using a heat pump than is required by conventional heating methods. The "free" heat gleaned from the source medium is what makes heat pumps highly efficient. Temperatures of 190 to 200°F (88 to 93°C) are provided by the compressor in liquid Freon systems. There are other refrigerants which have good capacity for transferring heat. These, classified as inorganic compounds, include saturated aliphatic hydrocarbons, unsaturated hydrocarbons, and halogenated hydrocarbons. The latter group are most widely used due to their chemical stability and comparative safety from toxicity, flammability, and explosion.

Heat pump units are rated by their coefficient of performance (COP). The COP is based on electric resistance heating, in which 1 kW provides 3413 Btu in 1 hour for a COP of 1. Stated another way, with electric resistance heating you get 1 watt (W) of heat energy out for

Figure 14-15 Water source heat pump unit: external view. (Courtesy of Carrier Corporation.)

Figure 14-16 Air source heat pump unit: cutaway view. (Courtesy of Trane, Tyler, Tex.)

each watt of electrical energy put in (COP of 1). When heat source temperatures are relatively warm [about 50°F (10°C)], a COP as high as 3 is possible for heat pumps. In other words, for each watt of electricity used, 3 W of heat energy is available for heating—1 W from the unit, 2 W from the source. The energy cost for heating would therefore be one-third that of normal electric resistance heating.

As might be expected, a warm heat source will provide higher efficiencies from a heat pump. As the temperature of the heat source declines, the efficiency drops until a COP of 1 is reached at about 10°F (−12.2°C), providing little advantage for the user over traditional electric resistance heating. For cooling, the unit's efficiency again drops when the source temperature exceeds 90°F (32°C). A heat pump will therefore be advantageous if the medium with which it is coupled remains above 10°F (−12.2°C) and below 90°F (32°C). The fact that a heat pump can provide both heat in winter and cooling in summer makes it an ideal system in moderate climates, where a single system can provide for both needs. Water source units consequently provide the best COP for most systems on a year-round basis since groundwater is always warmer than air in winter and cooler in summer.

Heat pumps can also be advantageously coupled with active solar systems. Both liquid and air solar collector loops can be used. Each system would require storage, from which the heat would be removed as required. As might be expected, geographical areas with considerable winter cloudiness would not benefit from this coupling nearly as much as would those with more sunshine.

The cost of heat pump units is comparable to that of conventional central heating systems. Air source units are the least expensive and the easiest to install, while high expenses must often be incurred to acquire ground coupling or a water source. This additional expense is often 10 to 50% higher than the cost of the heat pump unit. High-quality equipment is currently being marketed and local service is available from the dealer or most qualified commercial refrigeration companies.

A heat pump can provide economical temperature and humidity control year-round. It will also amortize itself in a few years compared to conventional centralized heating and air-conditioning systems, through reduced energy bills. COP values of up to 5.4 have reportedly been attained, although most in-field units function between 2 and 4. Savings of 20 to 54% have been reported by people who have installed retrofit units. Solar-assisted units have achieved COP values as high as 10.5.

Some leaders in the field suggest that the best measure of a heat pump system's performance is the seasonal performance factor (SPF). The SPF is the ratio of the total seasonal demand to the energy required:

$$\frac{\text{total kWh demand}}{\text{total kWh required}}$$

and includes supplementary heat necessary when temperatures are low. An average house in the Boston area with a 2½-ton air source heat pump required 13,900 kWh to supply 26,500 kWh worth of heat for the total heating season, resulting in a SPF of 1.91 [Gilman, pp. 110-111]. McGuigan describes how these ratings vary from a high of 2.61 in Miami and Los Angeles to a low of 1.22 in Minneapolis.

Table 14-2 provides a comparison of calculated annual heating costs for a typical home using various conventional heat sources and heat pumps. The prices of energy and the heat loss of buildings varies nationwide according to climate.

Although recent sales have increased and the market appears brighter for the future, heat pumps have not yet become popular,

TABLE 14-2 Comparison of Yearly Cost of Usable Heat for Various Types of Available Energy

Energy Source	Energy Unit		Cost/Unit		Cost/Year
Electric resistance: $\dfrac{93{,}118{,}694 \text{ Btu/hr}^a}{3413 \text{ Btu/kW} \times 100\%}$	=	27,284 kW	×	\$0.06 per kWh	=	\$1637.04
Liquefied petroleum gas: $\dfrac{93{,}118{,}694 \text{ Btu/hr}}{91{,}800 \text{ Btu/gal} \times 0.75}$	=	1352 gal	×	\$1.00 per gal	=	\$1352.48
No. 2 fuel oil: $\dfrac{93{,}118{,}694 \text{ Btu/hr}}{140{,}000 \text{ Btu/gal} \times 0.80}$	=	831 gal	×	\$1.20 per gal	=	\$ 997.70
Natural gas: $\dfrac{93{,}118{,}694 \text{ Btu/hr}}{100{,}000 \text{ Btu/therm} \times 0.75}$	=	1242 therms	×	\$0.606 per therm	=	\$ 752.65
Heat pump (air source): $\dfrac{93{,}118{,}694 \text{ Btu/hr}}{3413 \times \text{COP of } 2.4}$	=	11,368 kW	×	\$0.06 per kWh	=	\$ 682.09
Natural gas (high efficiency): $\dfrac{93{,}118{,}694 \text{ Btu/hr}}{100{,}000 \text{ Btu/therm} \times 0.90}$	=	1035 therms	×	\$0.606 per therm	=	\$ 627.21
Heat pump (ground-coupled): $\dfrac{93{,}118{,}694 \text{ Btu/hr}}{3413 \times \text{COP of } 3.4}$	=	8025 kW	×	\$0.06 per kWh	=	\$ 481.47

[a] Based on a typical central New York home with a calculated heat loss of 49,084 Btu/hr at $-15°$F. Average winter air temperature $35°$F.

Source: Adapted from a Table by D. Christy, Inland Supply, Inc., Syracuse, N.Y.

especially in the residential sector. This can be attributed to many factors, including the mystery of how they work, the lack of advertising, the economic recession, and the fact that fewer new homes are being built. There are high initial costs, high electricity costs, and the "low" cost of traditional fossil fuels. Few manufacturers and contractors are familiar with heat pumps. Other inhibitors include the fact that the availability of water for the more efficient water source units is limited, the water is not usable in some geographic areas, and much of the northern United States has air source temperatures which do not provide a high COP value. In addition, unless the climate is such that both heating and cooling are necessary, requiring year-round use, the system may not be economical.

LEGISLATION

The initial 1974 efforts toward national energy self-sufficiency and legislation since 1976 have strengthened the expansion of cogeneration systems. The National Energy Act of 1978 addressed cogeneration and encouraged the commercial, industrial, residential, and community sectors to become involved. It suggested that communities could develop municipal solid waste projects which would produce electricity and industrial process steam and at the same time reduce their refuse disposal problems. Industries and independents were also encouraged to become generators of electrical power by ensuring that electrical energy not required for their own use would be purchased by the local utility company.

Very little new encouragement to conserve energy has been advanced under the Reagan administration. There were even discussions to terminate the energy tax credits which provide financial incentives to corporations and individuals who initiate energy conservation efforts. Early 1984 news reports carried tidings of a potential energy surtax which could be placed on all fossil fuels. This tax, if initiated, could create a significant increase in the use of cogeneration and heat reclamation systems, as conservation would suddenly become economical in all sectors.

ENVIRONMENTAL IMPACT

Extracting heat from liquids and gases before they are discharged into the environment means less thermal pollution. This in turn will permit a return to normal marine plant growth rates. If heat-recovery units were used on very large units that discharge huge amounts of water vapor together with the heat, a return to nonmodified weather could occur. Cooler emissions could also cause a greater concentration of toxins and pollutants, since less natural convective dispersal action would be present.

Water source heat pump systems that do not reinject the used water back into the aquifer could have a negative effect on the area's water table. Some systems simply dump the circulated water into the sewage system, while others use ponds or supply water for irrigation or other uses.

More efficient use of our resources would cause a slower rate of increase in exploration, drilling, mining, refining, transporting, and their associated impacts on the environment. Although this may create a slower growth of employment opportunities in these areas, a dramatic growth in employment can be expected in fields related to the production and installation of cogeneration and heat reclamation systems. A net economic gain for society can be expected together with an improved life style and quality of life. See Chapter 16 for further discussion of the employment impact of renewable energy technologies.

COGENERATION AND HEAT RECLAMATION POTENTIAL

One of the reasons why cogeneration has not grown rapidly has been the lack of available prepackaged systems. Galvin reports that until 1982, each installation required research and development efforts to design and build an appropriate system. Now, prepackaged cogeneration systems in the range of 45 to 150 kW are available for small commercial and industrial applications at installed costs that are up to two-thirds less than those of custom-engineered systems. This cost reduction is attributed to the mass production of system components. Installed costs range from $600 to $1000 per kilowatt and the systems are projected to pay for themselves in 1 to 3 years, depending on local electrical rates and the amount of waste heat and electricity that can be used on site. This compares with custom-designed large commercial systems for which it can take up to 20 years to recover the initial investment.

A 1976 Dow Chemical Company study projected that by 1985, industry with sincere involvement could meet one-half of its electrical needs through cogeneration systems. At that time only 14% were being met, with the balance purchased from the local utility. John N. Eustis, branch chief of DOE's Office of Industrial Programs, has projected that cogeneration could save approximately 490,000 barrels of oil per day and cut the needs for oil imports by as much as 15%. In New Hampshire, where the industrial sector uses one-third of all the state's electricity, wood-fired cogeneration units are being advanced especially for those companies using wood as a raw material, such as the paper industry.

Research is continuing on heat exchangers and cogeneration systems design problems. The focus has been on reducing flow-induced vibration, fouling, and corrosion, and on producing more durable materials that can affect cost and performance for both low- and high-grade heat exchangers. Mass-produced components and prepackaged systems are expected to have a considerable impact and result in a positive response in industrial and commercial acceptance. Bottoming-cycle units will also become popular as mass production and escalating fuel costs impose their respective economic influences.

Heat pipes are being seriously considered as recovery devices because they are more efficient than many current heat exchangers. Ceramic heat pipes using a liquid-metal working fluid are being tested in the hope of minimizing problems of thermal stress. As early as 1979, ceramic heat pipes using sodium as the working fluid were successfully tested at 1742°F (950°C). A new furnace, using a bank of heat pipes as a heat exchanger, was introduced late in 1983. Other manufacturers can be expected to follow.

Heat-transfer studies of fluidized combustion units have demonstrated that up to twice as much heat can be recovered from burning waste products and volatiles than from conventional incinerator units. Small industrial and commercial producers of waste products are finding that use of these units is economically beneficial compared to current disposal methods.

The use of heat pumps is expected to increase dramatically in all sectors as the cost of fossil fuels increase. This will occur especially in the southern half of the United States where both heating and cooling are required. In Sweden, the use of heat pumps increased from approximately 1000 in 1975 to 30,000 in 1982. Although U.S. adoption may not be as rapid, there will be substantial growth, especially of ground-coupled and water source units. As of 1977, nearly 1.5 million homes had heat pumps; in 1981, a total of more than 3 million units were in service. Solar-assisted heat pumps are becoming popular in the sunnier areas of the country for residential, commercial, and industrial applications. Housing contractors are becoming familiar with heat pumps and are recommending them for both space and water heating. Nationwide, heat pumps are being installed in nearly half of the new homes using electric heat. The Air Conditioning and Refrigeration Institute projected that heat pumps would be installed in 27 to 29% of all new housing built in 1983.

Low-grade heat can be stripped from industrial wastewater and air-exchange systems to provide space heat and hot water for use in-house or in adjacent buildings. Applications in agriculture and aquaculture are also being expanded. Cooling can be provided from nearby waterways.

Many utility companies have initiated advertising campaigns to inform their customers about heat pumps and the potential for retrofitting into existing systems. These "add-on" units provide heat except in times of coldest temperatures, when the existing (conventional) heating system is actually more efficient. This is especially timely as energy rates escalate and consumers seek ways to reduce energy costs.

New chemical heat pumps are emerging through studies funded by DOE at Brookhaven National Laboratory. These systems will save millions of barrels of oil annually when used with solar systems or waste industrial process heat. They are based on a reversible heat absorbing/heat releasing (endothermic/exothermic) chemical reaction initiated by a low-temperature heat source. One system dissociates calcium chloride and methanol using solar energy. Once separated, the chemicals represent stored potential energy that can be recovered as the chemicals are recombined. Since solar energy is intermittent, the use of waste industrial process heat for this operation appears more attractive and closer to commercialization.

As new industrial plants are built or old facilities remodeled and old equipment replaced, energy-efficient units with heat reclamation and cogeneration capability will be installed. This will be especially true with large buildings and energy-intensive industries. Institutional barriers will be reduced so that energy formerly wasted will be available for use in adjacent facilities—whether public or private. Economics will be the driving force.

Future development is expected to include even greater energy conversion efficiencies and the use of fuel cells and thermionic converters. These systems convert source energy directly to electricity without the intermediate mechanical energy (turbine and generator rotation) step (see Chapter 15).

SUMMARY

All of our machines, in converting one form of energy to another to do work, lose some energy in the form of heat. The more a machine loses, the less efficient it is. Since more efficient use of our energy resources is imperative, cogeneration and heat recovery units will help us get "better mileage" from our energy conversion systems. Some of the newer industrial cogeneration systems can convert 75 to 85% of the energy in fuel into useful forms.

Cogeneration, which involves the simultaneous generation of electrical energy and the production of thermal energy from a single source, is growing in acceptance as mass-produced, more economical units become available and energy costs escalate. Combined-cycle, topping-cycle, and bottoming-cycle systems are available commercially but bottoming-cycle units are still too expensive to provide a realistic return on investment.

Thermal energy, which in the past has been released into the environment, is increasingly being captured through the use of heat reclamation systems and used to reduce total energy requirements in many sectors. This "waste" heat is being used to preheat incoming air, water, or materials for processing and can provide the only energy source needed to run a complete electricity-generating system.

Heat pumps are devices that work like refrigerators but in reverse. They effectively take low-grade thermal energy and increase it to usable temperatures for both space and water heating and/or cooling. The popularity of these units is increasing, especially in the southern half of the United States where both winter heating and summer cooling is required. Heat pipes and exchangers are also assisting us to acquire the greatest amount of thermal energy from our fuel conversion systems.

Efforts to remove the technical, economic, and institutional barriers to efficient energy use are continuing. New, more effective systems and equipment are emerging which are providing both faster and greater financial return on investment. This is encouraging both industrial and commercial involvement. The benefits of heat recovery are the reduction in energy costs and reduction in equipment size and cost. The bottom line, however, should be to eliminate avoidable energy loses—to reduce energy consumption.

ACTIVITIES

1. Determine several sources of waste heat in your home. Measure the temperatures of these sources. Tell how this energy might be reclaimed and used. Indicate whether the sources are practical or impractical for realistic use.
2. Review literature on the environmental effects of industrial thermal pollution in our waterways. Describe how this energy might be used instead of wasted. Find out why companies are wasting rather than using this energy. Describe incentives necessary for encouraging conservation.
3. Determine if there are any cogeneration or heat pump users in your immediate area. Tour these facilities after additional study of systems types and functions.
4. Conduct a home energy audit to determine the total Btu requirements during a typical heating season (see Activity 2 in Chapter 13). Using Table 14-2 as a reference, obtain local energy costs and calculate the costs to heat your home using each of the energy sources.

BIBLIOGRAPHY

Angrist, Stanley W., *Direct Energy Conversion*. Boston: Allyn and Bacon, 1982.

Christy, Dan, "Comparison of Yearly Cost of Usable Heat for Various Types of Available Energy." Syracuse, N.Y.: Inland Supply, Inc., 1983.

Friends of the Earth Foundation, advertisement for *Soft Energy Notes*, 1981.

Gadomski, Chris, "New Chemical Heat Pumps Emerging from Three Separate Research Efforts," *Solar Times*, May 1981.

Galvin, Cindy, "Pre-packaged Cogeneration Units Now Available for Small Users," *Energy User News*, December 12, 1983.

Gannon, Robert, "Ground-water Heat Pumps," *Popular Science*, February 1978.

Gilman, Stanley F., ed., *Solar Energy Heat Pump Systems for Heating and Cooling Buildings*. University Park, Pa.: Pennsylvania State University, 1975.

Hill, Richard F., *Energy Technology VI—Achievements in Perspective*. Rockville, Md.: Government Institutes, Inc., 1979.

Hunt, V. Daniel, *Handbook of Energy Technology*. New York: Van Nostrand Reinhold, 1982.

Kim, Yong, H., "Dairy Farm Model of Energy Efficiency," *Renewable Energy News*, June 1983.

Krenz, Jerrold H., *Energy: Conversion and Utilization*. Boston: Allyn and Bacon, 1976.

Lindsley, E. F., "Heat Pumps Go Underground," *Popular Science*, October 1983.

McGuigan, Dermot, *Heat Pumps—An Efficient Heating and Cooling Alternate*. Charlotte, Vt.: Garden Way Publishing, 1981.

McMullan, J. T., and R. Morgan, *Heat Pumps*. Bristol, England: Adam Hilger, 1981.

Reiter, Sydney, *Industrial and Commercial Heat Recovery Systems*. New York: Van Nostrand Reinhold, 1983.

Roth, Robert, *All about Heat Pipes*. Barrington, N.J.: Edmund Scientific Co., 1973.

Swenson, S. Don, *Heating Technology—Principles, Equipment, and Application*. North Scituate, Mass.: Breton, 1983.

Turner, Wayne C., ed., *Energy Handbook*. New York: Wiley, 1982.

Chapter 15
Energy for the Future
From Fossil Resources toward a Mix of Renewables

CONCEPTS

1. We burn the past to heat, cool, illuminate, and provide mobility for the present.
2. Future projection is difficult due to the multitude of unknowns.
3. Energy sources for the future will have to be renewable or capable of being converted from plentiful supplies.
4. All energy sources except nuclear and geothermal come from the sun.
5. We are not running out of energy; rather, some traditional sources are being depleted.

GLOSSARY

PLASMA—a gaseous state in which the particles are electrically charged (ionized) rather than neutral.

THERMIONIC CONVERTER—a device that changes high-temperature heat energy directly to electrical energy.

INTRODUCTION

Projecting into the future has been attempted by some people through the use of a crystal ball. This device may be the only way that we can foresee the future because of the many dynamic interacting forces that bear upon events to come.

Future energy sources and supplies are dependent on several factors which are entirely outside our realm of influence or control and range from new discoveries through individual responsibility. The events that occur in an unsettled world—political climate; national, state, and local government interactions; the business community; prices; and individual use patterns—will all influence the energy future.

You have read in previous chapters how excessive use of limited resources has affected their costs and how these increasing costs have stimulated conservation efforts. Through conservation and the resulting reduction of demand growth rates, you have also seen how a blossoming nuclear power industry has become seriously financially burdened. Just as these events were not accurately appraised, the continuation of cost escalation, conservation, and the substitution of alternative and renewable energy forms are sure to provide an unpredictable and exciting future. If scientific forecasts related to the future shortage of pure water are accurate, it will add yet another strong influence to the growing uncertainty.

OVERVIEW

The cost of energy has a direct bearing on our standard of living and the overall economic health of our nation. This has been most evident during the recent recession. The 1973-1974 oil embargo, followed by the 1977 national gas shortage and the still higher prices and limited supplies of 1979, were followed by an oil "glut" with a slight price decline. Energy prices and supplies are expected to remain somewhat stable through the balance of the 1980s, but who can be sure—with the world's volatile political situation and our continued dependency on imported oil?

Governmental agencies rely on private industry to provide information on which public policy is determined and future projections made. The contribution of renewable energy sources has frequently been omitted from the data and projections have been made on energy-use trends of the past.

We have been lead to believe that we cannot survive as a nation unless we continue to develop and use nuclear fission energy or continue to consume large quantities of fossil fuels. Perhaps this is true if we maintain the position that the vast majority of our energy must be in the form of electricity—that we must be an all-electric society. Lovins and Lovins tell us that an analysis of U.S. energy needs shows that 35% is necessary for low-temperature applications, 23% for high-temperature applications, 34% for liquid fuels for vehicles, and 8% for electricity. We do not really need a nuclear reaction to heat our homes to 68°F (20°C)! Is it not more intelligent to use the energy form and source that is most appropriate, economical, and efficient to perform a desired task?

Energy forecasters expect the U.S. population to increase to approximately 300 million by the year 2000. This expanding population will require energy to meet its basic needs and enjoy an acceptable standard of living. Virtually all projections indicate an increase in energy demand between now and 2000 but at a reduced rate compared with the rapid growth of the past. The National Academy of Sciences, Resources for the Future, and the Ford Foundation project that even with serious and successful conservation, energy demand will increase approximately 30% by 2000 if the economic growth rate is held at mid-1970 levels. This would mean an increase from 78 quads to about 100 quads during the next 15 years. Others project an increase only half as large.

President Carter's message on June 20, 1979, indicated that the national goal for 2000 was to have 20% of our total energy needs met by solar and other renewable sources. Since he left office and the emphasis on nonfossils changed, some studies project that no more than 5% of the total energy needs of the United States will be provided by renewable sources. These latter studies are already inaccurate due to the recent increases in fossil fuel prices, the resulting reduction in fuel utilization, and the continued movement toward renewable forms of energy.

The transition from a highly industrialized society to one emphasizing high technology and service could result in a further decline in the projected energy growth rate. The reason is that heavy industry requires large quantities of energy, while high-tech requires less. Some studies project a decline in electricity use after 2000 due to conservation and life-style changes, while others indicate that twice as much electricity will be needed by 2020 as in 1983.

Betts reports on a major House subcommittee hearing in February 1984, where a debate was conducted on two conflicting studies. A DOE study called for a $1 trillion electrical plant building program necessary to avert brown-outs in the 1990s, because electricity demand was expected to grow along with the recovering economy. A Congressional Research Service study indicated that supply could meet the demand through increased efficiency, conservation, cogeneration, load management, alternative production methods, and only about one-tenth the new output projected by the other study. Which study is right? In which direction should our leaders point us? Because of the long time required to build electrical power plants, decisions need to be made soon. If the wrong decision is made, the consequences will undoubtedly be devastating. Which would you choose?

ENERGY SOURCES OF THE FUTURE

What will the energy picture be like in 2000 or 2010? How will you heat your home and your domestic water? What fuel will power your transportation?

There is little doubt that there will be an energy mix, with many sources contributing to the total needs of society. Petroleum and natural gas, the resources we will have to rely on through the turn of the century, will be needed as feedstocks for products and will therefore be too scarce and valuable for continued use as boiler fuels. Are coal and nuclear energy our only other choices? At this time, if you think only of large, centralized electrical technologies to supply the largest share of our energy needs, you are probably correct.

In 1977, Amory Lovins coined the phrase "soft energy paths" and introduced the concept of using technologies that are appropriate for each energy task. This concept suggests that through wise energy selection and efficient use, there would be less waste, reduced environmental impact, no decline in the standard of living, and an increase in the national wealth. By selecting the most appropriate energy source for the job, we would no longer be transporting Middle East Oil halfway around the world so that we could burn it and use 1.75% of its energy value to make an incandescent light bulb glow. (The efficiency of a fossil-fueled electricity generating plant = 35%; the efficiency of an incandescent bulb = 5%; therefore $0.35 \times 0.05 = 1.75\%$ = the efficiency of burning oil to light a bulb.)

Appropriate energy utilization means the use of electricity for the processes that really require electricity and the use of other energy forms where they best apply. The operation of motors, illuminating devices, and your stereo set are best done with electricity. Burning wood for heat rather than for illuminating is more appropriate. The effort that we are already observing is energy engineering or the matching of the most effective energy source and form with the desired task. It is already too expensive and wasteful to do it any other way.

Because fossil fuels and uranium have very limited availability, renewable energy sources will have to be used, and they are available in great variety. This concept is new to many people, but remember that very few people really thought about it until we were awakened by the Arab oil embargo just a few years ago. Continuing to use the world's remaining fossil fuels without sincere development and application of the renewable sources will only hasten economic demise and political instability.

The term "alternative energy sources" is often used when discussing nonconventional energy, but this concept is somewhat misleading. If you have a true alternative, you have a choice in selecting options. Continued use of finite petroleum and natural gas supplies does not leave us any true alternative. We must absolutely use other energy sources if we are to survive with any quality of life style intact. Living without adequate energy to meet our needs is not high-quality living.

Our society is demonstrating a shift to renewable energy sources as the price of traditional fuels escalates. The declining supply of fossil fuels and the increasing demands for energy clearly project that our future economy will have to be based on renewables if it is to continue. Reliable studies that report energy-use data are now including the contributions made by renewable sources, but the input of many small installations is nearly impossible to calculate accurately. Some renewable energy use is being classified as "conservation." Replacing one energy source with another is not true conservation even though it does actually save a resource for future use.

Diversity and combinations of energy sources will be the pattern for our energy future. Since many of the renewable energy sources are site-specific, one geographic area may be taking advantage of geothermal resources while another will be using solar, biomass, wind, or tidal energy. Combinations of sources are necessary because of the intermittent nature of some renewables. We are actually quite familiar with energy combinations and substitutions already, so it will not be a new approach. When a nuclear-fueled electricity-generating plant fails or is shut down, oil, coal, or hydropower must be substituted until the unit is returned to production. The same is true for renewable energies. Stormy, cloudy weather which is bad for direct solar acquisition is generally good for wind energy applications.

Some progressive utility companies are already involved in an energy mix. Long reports that Southern California Edison has announced that 30% of its new electricity-generating capacity over the next 10 years will come from renewables. Electricity is expected to be produced from nine primary sources, including oil, natural gas, coal, nuclear, wind, geothermal, biomass, solar, and hydropower. Fuel cells and cogeneration systems are also expected to make significant contributions. The New England Electric System is not yet so diversified, but its emphasis is away from costly central power station construction. It is aggressively pursuing an alternative program which includes hydro, wind, coal, and biomass conversion purchase agreements together with conservation and system load management. Earth Resources Research Ltd. of Great Britain forecasts that up to 60% of their nation's needs can be met by 2025 using renewable energy sources.

Energy decisions generally must be implemented gradually, and major shifts have taken decades. Many low-technology renewable energy applications can be implemented quite rapidly, especially when traditional fuels become unaffordable. The more sophisticated systems require research, experimentation, and long-term testing before commercial application. Conscientious efforts should be undertaken now by all sectors before we end up helplessly hitchhiking down the road with an empty fuel can in our hands. There are no easy choices, and each will be accompanied with its respective problems and environmental impacts.

Two major factors that inhibit rapid adoption of renewable energy sources are: (1) their expense, compared with some cheaper fossil fuels; and (2) the general American attitude. The cost of using renewables is high initially because in most cases they cannot meet the total energy demands of the user and therefore must be supported by a backup system. This forces the user to incur the expense of the "extra" system. Even with federal and state tax credits and a payback from savings, renewables have not enjoyed a dramatic rush in demand. Even though the prices for fossil fuels are high, most people simply complain more but persist in their consumption.

The affluence of the American citizenry and the complacent attitude that it demonstrates is not found in many nations of the world. Due to the high standard of living and abundance of material wealth, many Americans simply do not become seriously involved in or concerned with the world and national energy situation. As long as they can afford to pay the prices demanded in the marketplace and there is no apparent shortage, they simply do not seem to care. This attitude is very evident in the reluctance of many people to become involved with appropriate conservation measures in their homes, the dramatic resurgence to purchase larger, fuel-inefficient automobiles and recreational vehicles, and reluctance to use available mass transit systems.

The source of the complacent attitude of the citizenry can also be identified as emanating from government and industry. A true national energy policy and program to make this nation self-sufficient simply does not exist. The government apathy projects little concern for the future and appears content to be reactive rather than proactive as energy crises appear one after another. Industry has a great controlling influence on government, and if you observe that many of our larger, more affluent corporations are in the energy field, this apathy can be understood. Regular fuel deliveries are not necessary to people who have solar systems. These same corporations, however, are investing their fossil fuel profits heavily in the renewable energy fields. Why?

Government involvement in the energy picture will undoubtedly increase. Although deregulation is permitting the market forces of supply and demand to influence fuel prices, large profits by domestic suppliers and price/supply manipulation by our foreign suppliers will continue to be problems. Since Americans seem to quickly forget the lines, "No Gas" signs, and fights at service stations when someone tries to get ahead in line, some groups are proposing a heavy federal tax on gasoline (as in Europe) to encourage Americans to conserve. Revenue from this tax could be used to advance energy technologies and conservation. Federal government funding for energy development has been reduced and probably will remain at a lower level until our next energy crisis. Until that time, state and private investors will have to continue with minimum assistance. The DOE will hopefully continue to support, provide incentives, and assist in removing institutional barriers to continuing development. Distribution of funding that is available through federal sources is still directed primarily and inequitably toward the development of high-energy technologies.

Oil and natural gas will remain the principal sources of energy in the United States well beyond 1995 because of our dependency on them, but their contribution will continue to decline as the price escalates and the supply becomes depleted. Since people normally chose the lowest possible price, the conversion to renewable energy forms is inevitable. The dollar runs the marketplace. If oil and natural gas are the most economical fuels, people will use them until they are gone. Natural, not manipulated, markets and prices will dictate the demand.

No matter which energy sources we select, there will be risks and negative environmental impact. National self-sufficiency would help alleviate the impending international tensions driven by energy shortages as each nation attempts to protect its interests. Will the people of the world permit the relaxation of environmental standards so that coal can be used to replace oil, natural gas, and nuclear fission power? Will they "freeze to death in the dark"? As the risks and benefits for each energy source are weighed, the interplay of the economic, political, social, and environmental aspects will determine their rise or fall in popularity.

FUTURE PROJECTIONS FOR EACH OF THE ENERGY RESOURCES

(These projections assume no wars, economic collapse, or political disruptions.)

Coal: Bridge to the Energy Future

Coal is seen by some people as the solution to our energy problem until those magical unlimited, nonpolluting, economical energy sources of the future are discovered and developed. With petroleum and natural gas supplies dwindling and prices increasing, a return to the use of coal is occurring and along with it the social and environmental side effects from which we tried to escape in the past. These effects are expected to have a significant impact on the success of the return to coal and may make impossible a commitment to a long-term coal economy.

The American Petroleum Institute projects a dramatic increase in coal use through the 1980s. DOE projects an output of about 9.5 million barrels of oil per day equivalent (mbde) while the National Coal Association projects 11.6 mbde through 1990. An accelerated case would require 13.3 mbde, nearly double our 1979 output. Others project that we must double, triple, or quadruple our 1984 coal output by 2000, use other sources, and diligently conserve to meet the energy needs for moderate economic growth. A conservative consumption growth rate of 1.75% per year is expected after 1985 [Ruedisili and Firebaugh], while others project an increase of 5 to 7% per year. Coal's major use will be for producing electrical power.

Coal exports have been rising dramatically since 1978. During that year, the United States exported 40,000,000 tons of coal, and this figure grew to 90 million tons by 1980. Continued export growth is projected to 120 million tons in 1990 and 255 million tons in 2000 (Office of Technology Assessment). The United States, due to its extensive coal resources, could become the Saudi Arabia of coal as world petroleum and natural gas supplies decline.

After 1990, synthetic oils and gases from coal and peat are expected to increase dramatically for both fuel and feedstock applications. In-situ techniques will reduce the environmental impact and permit reclamation of deposits that would otherwise be inaccessible. The first commercial in-situ process is expected between 1985 and 1987.

Petroleum and Natural Gas: The Fuels that "Made" the American Economy

Inflation is directly tied to the prices of oil and natural gas because our economy is so greatly dependent on these declining sources of energy. No longer are these resources easily found and extracted to provide economical energy for our industrial society. When speaking on the topic of the war between Iraq and Iran in February 1984, President Reagan said that our oil supply lines will be kept open "no matter what it takes," proving that oil is a weapon in international politics and that oil is worth almost any cost. Approximately half of the world's supply of oil is under direct control of governments. Proven existing oil reserves are found predominately in the OPEC countries.

The decline of the oil and natural gas contribution to the total U.S. energy picture from nearly 75% in the mid-1970s to 69% (oil, 40%; natural gas, 29% [Dorf]) in the early 1980s is demonstrating that we can no longer depend on finite fossil fuels to power our economy. At 1984 consumption rates, oil especially will be in such short supply and so expensive by 2020 that it will no longer be used for combustion purposes except in specialized cases. How will we heat our homes and power our cars?

Domestic oil production is expected to remain relatively stable through the 1980s, with the quantity of imports increasing due to its lower price. Approximately 26% of our 1983 oil was imported [American Petroleum Institute] and with increases expected to begin in late 1984 or early 1985, we will be dragged back toward greater dependency on unstable suppliers. Since demand will continue to grow slowly, the efforts made since 1973 toward oil self-sufficiency will virtually be lost. Alaskan oil production will start to decline after 1985 unless new oil is discovered. A dramatic shift to coal cannot be expected as long as imported oil is "economical." We can therefore expect another oil crisis in the 1990s, followed by still another effort to become energy self-sufficient.

The price of natural gas has been controlled since the National Gas Act of 1938, but prices have been increasing and total control is expected to be lifted in 1985. Once the price stabilizes after decontrol, prices will continue to increase but since greater gas reserves are available, they are not expected to increase as rapidly as oil prices. The 4700-mile (7562-km) natural gas pipeline from Alaska's North Slope to the Midwest and California began in 1981 and was expected to be finished in 1986. With the present delays, completion is currently speculative. When the natural gas resources are exhausted, methanol, from the conversion of Alaska's vast peat resources, can be transported using this line.

Much of the nation's future oil and natural gas supply lies offshore, on the outer continental shelf and in marshes or wetlands. Some inland and coastal recovery is presently being conducted and drilling barges and other specialized equipment are being developed to minimize environmental impact. Although very large deposits are unlikely, other promising discoveries are anticipated in some of the western states and Alaska.

Enhanced oil recovery (EOR) methods are being developed to facilitate a greater volume of recovery from oil and gas fields. Recovery by pumping allows only 30 to 33% of the crude oil in a deposit to be removed. Thermal, hydro, biological, chemical, CO_2, and other tertiary extraction methods are expected to allow the recovery of up to 50% of the oil and 75% of the natural gas available from a deposit, depending on the permeability of the structure. Experimental data taken from only a few examples show that perhaps 90% recovery is possible using CO_2 techniques [Ruedisili and Firebaugh].

In 1983, Union Oil finished building the first stage of its oil shale processing plant. Under the Reagan slowdown, the progress of the synthetic fuels program has slowed dramatically and with the stabilized prices and excess production of oil in the early 1980s, continued com-

Figure 15-1 Drilling for oil in the hostile environment of Alaska's North Slope. (Courtesy of Texaco, Inc.)

mercialization is expected to be delayed. Canada, with extensive tar sand deposits, is also progressing toward commercial levels of processing but at a more brisk pace.

Oil company mergers and diversity of investments in other fields are causing concern. Control of the market by a few very large companies that have bought out the competition is the impending problem. Investments in these mergers and those outside the oil industry reduce financial backing for continued petroleum and natural gas development.

Nuclear Fission: The Interim Fuel for Producing Electricity

Confidence in nuclear fission as a feasible energy source for the future is continually being shaken. One day an operating utility is fined $40,000 for operating a plant with its cooling system shut down [Associated Press, "Con Ed . . ."], six days later another is fined $180,000 for violations of NRC regulations [Gallagher, "Niagara Mohawk . . ."], and ten days later the cost of Nine Mile 2 jumps $900,000,000, to $5.1 billion with further increases expected [Gallagher, "N M 2 . . ."] and the Seabrook plant is canceled [Associated Press, "Owners . . ."]. Ratepayers are being required to pay for discontinued plants which will never produce power, while investor profits are being assured by the public utilities commissions. The rapidly increasing electricity use rates are exceeding all other escalating prices and tending to reduce the standard of living for everyone.

The Office of Technology Assessment reports that the United States may need more nuclear plants in the future but will not build them without major changes in the technology, management, and level of public acceptance. The cost overruns, rate hikes, construction problems, operating mishaps, and long, costly power outages have eroded the confidence of investors, consumers, regulators, and the utilities. Basically, the fission nuclear industry is dying from these related problems, as accurately forecast by Lovins in 1977. Many of the plants now under construction will be completed and placed on-line, but some will be canceled or converted to coal-combustion units. Industrial and government studies project a demand increase of 2.5 to 3.2% per year from 1984 to 2000, but their forecasts have always proven high.

MIT's Department of Nuclear Engineering suggests that improved light-water reactors can satisfy future utility needs, but other leaders in the field propose the modular high-temperature gas-cooled graphite reactor (HTGR). Graphite reactors are smaller, safer, more efficient, and have been used in Great Britain for nearly 30 years. Two of these units have been built in the United States, but the Fort St. Vrain facility in Colorado is currently the only one operating. International cooperation on the continued development of breeder reactors is expected, with commercial units to be ready in about 40 years.

We are in an energy dilemma for at least the near future if we persist in believing that the best way to supply electricity is through larger and larger centrally located facilities. With oil and natural gas in declining supply, the only fuels remaining that will fit the mega philosophy are coal and nuclear fission, and neither is economically or environmentally acceptable. The use of nuclear energy will continue, despite costs, unless a major accident occurs. There is no reasonable alternative if we continue to waste and inappropriately use electricity for every energy need.

Solar: The Energy Source We Left Behind

According to the *Domestic Policy Review*, solar energy could supply as much as 18.1 quads, 13% of total U.S. energy, by 2000. Other groups project from a low of 7% to a high of 25%, with an average of 17%. Both space and water heating using solar energy are becoming popular, especially in the Sun Belt. With prices of conventional energy increasing, business, industry, and individual homeowners are taking advantage of this available "free" energy source. Federal and state tax credits, as well as a good return on the financial investment, are also encouraging greater interest and involvement in both new and retrofit solar applications. Solar tax credits are expected to be available for only a few more years, especially as solar begins to make major energy contributions after 1990.

The use of solar energy is downplayed by some large commercial energy suppliers, with their advertising projecting "tomorrow" as the time for solar. This is true for large commercial-scale conversion of solar energy to electricity, but any adoption of solar will cut into their sales and consequently reduce profits. Is this why Arco, Amoco, and Exxon rank first, second, and third, respectively, in ownership in the photovoltaic industry?

Large central plants to produce electricity from solar energy are presently in the experimental and testing stages. Solar One (Figure 15-2), located near Barstow, California, uses 1818 computer-controlled heliostats (mirrors) to focus the sun's heat on a boiler at the top of a tower. The steam thus produced is used to drive a turbine.

Another commercial venture involves the use of solar-powered "utility" satellites. These satellites will provide electrical energy for space stations and manufacturing plants planned for the 1990s. The need to provide and transport power systems for each station would therefore be eliminated. Other utility satellites, located in geosynchronous orbit, will transmit energy via microwaves to antennas on earth.

Figure 15-2 Solar One. (Courtesy of Southern California Edison.)

According to Peter Glasser, energy expert for Arthur D. Little, solar utility satellites would orbit earth at 22,300 miles (35,881 km) and supply up to 5000 MW per unit. At this altitude, the panels would always be in sunlight except during an eclipse. Figure 15-3 is an artist's concept of a solar cell array that could measure 3 miles by 6 miles by 29 miles (5 km × 10 km × 47 km) and send electricity to earth.

The nation's largest industrial solar generating facility is in Shenandoah, Georgia (Figure 15-4). In a joint effort with DOE, the Georgia Power Co.'s 5-acre plant uses 114 parabolic dish solar collectors which track the sun to obtain 750°F (400°C) fluid temperatures which drive a steam turbine. The electricity generated provides about 50% of the energy needs of the Bleyle of America knitware factory. Many more of these solar energy applications to meet industrial needs are expected as electricity costs escalate.

Photovoltaic research and development efforts continue with thin-film amorphous and polycrystalline materials, resulting in decreasing costs and increasing efficiencies. Amorphous silicon cells now have an efficiency of about 10%, with a 12% theoretical limit; other materials may approach 20% efficiency. Mass-production techniques are being explored which will further reduce the price if high-quality production can be maintained. As the price comes down, more photovoltaic panel displays will be seen, as on the Citicorp Center in New York City and at the Epcot Center at Disney World. Currently at $5 per watt, the DOE,

Future Projections for Each of the Energy Resources 385

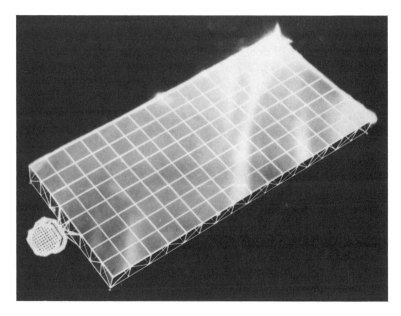

Figure 15-3 Solar-powered utility satellite (From the National Aeronautics and Space Administration.)

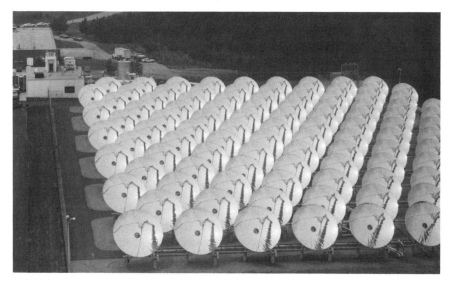

Figure 15-4 Solar total energy project. (Courtesy of Georgia Power Company.)

Dorf, and others expect photovoltaic systems to be competitive by 1986. When this happens, the demand for conventional, commercially generated electricity will decrease.

The first section (6 MW) of a 16.5-MW photovoltaic power plant was initiated in northern California in the spring of 1984. This plant, when completed, will replace the 1-MW unit in southern California as the world's largest on-line photovoltaic facility.

Two other major research efforts are under way with photovoltaic systems. One, at the Dallas–Fort Worth Airport, uses a combination lineal Fresnel lens and photovoltaic cell array, resulting in a combined photovoltaic/photothermal efficiency of 47%.

The second research effort involves replication of the photosynthesis process. Joseph Katz and others at the Argonne National Laboratory have shown that an artificial chlorophyll membrane, under the influence of light, will separate positive and negative charges and produce an electrical current. This "synthetic leaf" converts solar energy directly into electricity and may become economically competitive within 4 to 5 years because the organic materials of the membrane are very economical.

Other industrial applications of solar energy include desalinization of seawater, the electrolysis of water to produce hydrogen fuel, and use in a hybrid fuel cell.

Residential solar applications include space heating and cooling and domestic water heating. Passive space heating is economical in all areas of the country, but active systems are not making high returns in the cloudy areas. Active water heating systems are economical in all areas. More than 12% of all homes in Australia have solar domestic hot water systems. As American homeowners face higher and higher energy bills, a greater move toward these systems is expected, but what is necessary is proof that solar works and is dependable. Prices are expected to come down as mass-production techniques are used to manufacture the units. Photovoltaic systems will be economical for residential applications around 2000, when the prices will make the solar-powered home affordable.

New storage systems can be expected that will permit both storage and transportation of solar heat energy. Dehydrated metal salts are being developed that work similar to phase-change materials, that is, they release heat when water is added. By using solar energy to remove the moisture from these anhydrous salts, the heat energy can be stored until needed. Additional developments are expected in self-activated glazing, heat-transfer mediums, solar-powered refrigeration units, flat-plate phase-change units which drive turbines connected to generators, collectors that use photosynthesis to produce oils and electricity, and many others.

Figure 15-5 Residential passive solar space heating: new or retrofit. (Courtesy of Energy Research Inc., Novi, Mich.)

Intelligent use of solar energy for low and moderate heating needs conserves valuable fossil fuels for necessary uses such as vehicular fuels, pharmaceuticals, and feedstocks. The technology tends to be safe, dependable, and easily applied at the individual and community levels.

Wind: The Rapidly Growing Energy Source for Electricity Production

The use of wind to drive electricity generators has surpassed OTEC, bioconversion, and photovoltaic systems. During 1983, massive numbers of investors stimulated rapid growth of the industry, which is making California the world's leader of wind-powered electricity production. Production more than doubled in 1982, exceeded 150 MW in 1983, and may double or triple again in 1984. Projections for the future are especially encouraging in California as long as the tax credits remain in place. Cost to the consumer is expected to be less than 10 cents per kilowatt-hour.

The concept of large, heavy units which have been successfully serving Europe and Sweden during the last decade has been replaced by smaller, multiple, high-performance units grouped in "wind farms." This concept, now proven, is spreading throughout the United States and is being considered in all the coastal areas, the Great Plains, the Midwest, the Southwest, the Great Lakes region, and others. For example, a 100-MW wind "park" is planned for Medicine Bow, Wyoming. Small, single, residential-size units, although increasing in popularity, are not expected to contribute significantly due to the high initial investment required by the homeowner.

New units and concepts are constantly being developed. Since wind energy is more cost-effective when applied to pumping water or compressing air, hydraulic coupling with electric generators is expected to gain in popularity. Single-blade rotors, cylindrical arms that use the Magnus effect, and shrouded turbines are new. There is even a system that uses the wind to drive a fine mist of water through an electrically charged grid, thus producing an electrical potential on collector electrodes.

Solar photovoltaic and wind/electric systems make a good combination. Bright sunny days generally have low wind velocities, and windy days are generally cloudy. Together they will make an impressive contribution to our energy needs.

Geothermal Energy: Energy from the Heat within the Earth

As discussed in Chapter 9, California currently leads all other states in geothermal electric development and is expected to retain this position. Exploratory wells are being drilled in several states, but few high-temperature sources are expected to be found. Some developers propose to use low-temperature sources by circulating fluids with low boiling points and expanding them through low-pressure turbines.

Favorable interest and investment continues with municipalities becoming involved in producing their own electricity. The Navy Air Weapons Training Complex in Fallon, Nevada, expects to complete its 75-MW plant about 1990. Sandia National Laboratory is conducting experiments and feasibility studies using magma rather than just high-temperature fluids from geothermal sources.

Significant cooperative international programs have been established with Italy, Mexico, and Germany. In addition to the exchange of information and cooperative research and development activities, reservoir stimulation, computer modeling, environmental control studies, and seismic studies are continuing.

The Geothermal Energy Program projects continued efforts in collecting data, defining resources, and developing discovery, production, and application techniques for both electrical and thermal applications. Economical technologies for exploiting normal-gradient geothermal heat are expected after 2000. Demonstration projects are needed to provide evidence of the profitability of nonelectric applications.

TABLE 15-1 Potential Geothermal Electrical Development, by State

	MW of Electricity	
	By 1990	By 2020
Alaska	?	?
Arizona	200	230
California	5,570	10,270
Colorado	0	0
Hawaii	70	920
Idaho	400	4,950
Louisiana	300	1,200
Montana	50	50
Nevada	850	3,700
New Mexico	350	1,500
Oregon	250	1,150
Texas	725	4,675
Utah	500	3,000
Washington	50	50
Wyoming	0	0
	9,315	31,795

Source: Interagency Geothermal Coordinating Council, Washington, D.C.

Waste and Biomass: From Garbage Power to Grow-Your-Own

Several leading energy forecasters project that organic wastes and biomass will provide the greatest contribution of all renewable energy sources to our future energy needs. The average American family discards about $100 worth of groceries per year in the form of food wastes. This computes to 20% of the food going into the home, for a cost of $16 billion per year for the entire nation [Ripley]. Total urban wastes amount to 125 million tons per year, and some municipalities and independent power producers are beginning to take advantage of the energy that we discard so easily. Europe has been taking advantage of this resource for years.

Since 2.5 lb (1.1 kg) of prepared garbage has the energy equivalent of 1 lb (454 g) of coal, burning the waste to provide process heat or generate electricity is attractive. Garbage-to-electricity or process heat plants can be found in New York, California, Arkansas, and other states, with Massachusetts leading the pack. Interest is high and plants are proliferating, with 35 operating, 20 under construction, and more than 75 in the planning stages in 1983. New combustion techniques are being developed to reduce emissions and postcombustion ash. Some units are already 85 to 95% efficient.

Methane gas from landfills and sewage treatment plants is also being used increasingly in many parts of the country. Dumping sites that no one wanted are now going through litigation to determine who has the right to extract and benefit financially from the "newly found" energy. The average methane-fueled plant can pay for itself in about 5 years, with the supply from most city landfills expected to last 20 to 30 years. Methane from sewage treatment facilities can be used on site for process heat and generating electricity. Since the removal of the methane reduces the fire and explosion hazard that often has to be dealt with, most large landfills and treatment plants are expected to be exploited. The use of gasified fuels is expected to remain strong in the areas of space heating and electric power generation.

Although the synthetic fuel program of the federal government has slowed under the Reagan administration, progress is being made in producing fuels from sawdust and other wood wastes, nutshells, seaweed, agricultural wastes such as cornstalks, municipal sewage sludge, and nearly any easily available cellulosic waste. Methane, methanol, and a form of crude oil are all being extracted, but commercial production is not expected before 1990 and significant contribution not before 2000. Proposals are being made for small reactor units that could be transported to sites where biomass is easily available, thereby reducing expensive feedstock transportation.

In 1983, Ford produced 587 methanol-powered Escorts to be used as fleet and test cars. Recent studies have shown that leaded gasoline is a major cause of human poisoning, and rapid strides are being taken to use ethanol and methanol as octane enhancers, especially in the larger cities.

Debate has been and will continue to be strong if agricultural food crops such as corn are converted to ethyl alcohol or fuel oils so that Americans can drive their cars while others in the world go hungry. Although fuel alcohol crops are not expected to affect the production of food crops, the driving versus hunger controversy will frequently make the headlines. Some farmers are producing assorted crops which are being used in experimental and developmental processes. The U.S. Department of Agriculture has screened about 600 native crops that can be profitably grown for energy use. Of these crops, 80 are promising and 12 have been selected for further intensive study.

Hydropower: Returning to the Power of Colonial Days

The rush to hydropower is on! Using the power of flowing water is the most economical and least environmentally disturbing of all other conventional methods of generating electricity. A limited number of large hydropower sites remain undeveloped, so major efforts are toward developing or retrofitting moderate and small sites. Large utilities, municipalities, small companies, and independent persons are all involved.

Canada has extensive undeveloped hydroelectric resources and is pursuing sales agreements with the states adjacent to its southern border. The financial and environmental costs of nuclear-, oil-, and coal-fueled plants make the purchase of economical imported hydroelectric power attractive, but such purchases continue to disrupt the balance of international payments just as the purchase of imported oil does. The balance of payments deficit set a new record in 1984, $101.6 billion.

Litigation and demonstrations by environmental groups can be expected as developers file environmental impact statements for their intended facilities. Since the facilities will be smaller there will be more of them and consequently a great number of people will be affected.

Holland is initiating an interesting hydropower/windpower project. Large wind-powered generators produce electricity for the national grid, and any excess is used to drive Kaplan turbines, which pump water to a large reservoir. During calm periods the water stored behind the dam drives the turbines to generate electricity. A pilot plant is now under construction, and the decision whether to build a 3000-MW plant to be located near central Holland will be made in 1985. The final project completion, if approved, is projected for 2000.

Due to its economy, hydroelectric power utilization can be expected to increase slowly but steadily through 2000.

Ocean Energy Systems:
A Mixture of Energy Extraction Techniques

The energy available in ocean thermal, wave, current, tide, and salinity differentials is attracting attention. A problem with each of these energy sources is that they are site-specific and therefore limited by the practicality of transporting the converted energy form to demand centers.

Great Britain and Japan are working on wave-powered electricity generating units and the first prototype of the British air-driven turbine is expected to be complete in 1986. Located off the coast of the Island of Lewis in the Outer Hebrides, Scotland, the new design incorporates precast concrete structures with air chambers. Waves, acting like pistons in the air chambers, compress the air, which drives the turbine connected to a generator. This design eliminates many of the problems encountered previously with "nodding ducks," flexible rafts, floating turbines, and other mechanical devices.

The first modern tidal energy electricity generating power plant came on-line in 1984 at Annapolis Royal, Nova Scotia, and the second on Passamaquoddy Bay in the Bay of Fundy is in the initial stages of construction. Where tidal elevation differentials are adequate and demand centers nearby, development of this resource can be expected.

Experiments have been successful, and demonstration ocean thermal energy conversion (OTEC) units for use in tropical waters are expected in the early 1990s. These units are calculated to be competitive with oil-fired electricity-generating plants, but government subsidy has been relaxed and independent investors are a bit leary of speculating. States along the Gulf coast can expect energy from the ocean in the late 1990s.

Electricity generated from the flow of ocean currents is not expected until the mid-1990s, although experiments and a demonstration unit will be initiated in the late 1980s. Again, there are few population centers near enough to adequate currents to make this technology a priority.

Salinity gradient technology is still experimental. Membrane materials and manufacturing techniques need to be developed before a demonstration can even be projected. Unless dramatic breakthroughs occur in the very near future, the first unit can not be expected before 2000.

Cogeneration: Gleaning "Extra" Energy from a Source

Cogeneration systems attempt to make maximum use of every possible unit of energy by using waste heat from electrical power generation to serve other energy requirements. With the price of electricity continually increasing, industries that require large quantities of heat or electricity are installing their own units in the attempt to remain competitive.

Some users of cogeneration systems use the energy entirely themselves, while others use what they need and sell the balance. Another method of operation is to sell all the energy produced to the local utility company and purchase what is needed as a normal customer.

Cogeneration systems are expected to affect the demand rates of local utilities. This reduction in demand is projected by several experts in the field to eliminate any real need for large electricity-generating plants.

Conservation: Key to the Future

The patriotic and moral concepts of energy conservation for future generations appear to have little meaning to many Americans. Whether the inhabitants of the United States in 35 years have adequate fuels to meet their basic needs seems of little importance. The attitudes that

"Our energy problem is a gigantic scheme by the oil companies to get rich," "By then we will have found more oil or a new, endless energy source," or "Let them worry about it" demonstrate a serious misunderstanding of the world energy situation. Is it any wonder that approximately 50% of Europe's fuel prices are government-imposed taxes levied to force conservation? Does the United States need to establish an energy user's tax, as some conservationists suggest, so that we can protect our own future?

American society runs on the dollar interacting with supply and demand. If the price is low, there is little concern, but if the price is high, we become concerned enough to conserve. Efficient use of energy is not synonymous with human suffering or deprivation. It is simply the wise use of the resources that you do use (i.e., wasting as little as possible). Many energy users have already completed conservation measures that are the least costly and easiest to implement. Further conservation efforts will be more costly per unit of energy saved. Industry has done more than any other sector to conserve because reducing expenses demands less waste, resulting in a better competitive position, but estimates for the entire sector indicate only a 15% reduction. Homeowners have done the least to conserve. Where possible, some have installed devices such as wood-burning stoves to displace higher-cost fuels, but they have not implemented energy-saving measures. Replacing one energy source with another more economical source is not conservation.

Continued high levels of conventional energy use will bring higher prices and government mandates, regulations, and controls, while reductions will provide a better energy future for the world. Recycling of paper, steel, and aluminum provides a 70%, 75%, and 93% savings in energy, respectively, over the use of raw materials, but only a small effort has been made toward accomplishing these savings. Energy demand is projected to increase at the rate of 0.5 to 1.2% per year through 1990, even with additional effort to conserve. Economics will force us to conserve. Continued inefficient use and inappropriate application of energy sources will be expensive not only to ourselves but to the environment and future generations.

IMPLEMENTING THE ENERGY MIX

There appears to be no argument that several energy sources are required to replace declining fossil fuel supplies. For implementation of the more economical alternatives, there is little necessity to provide incentives. An example is the great proliferation of wood-combustion units since

1973. Adoption of the more expensive alternatives, however, is a problem, especially for people of limited financial resources.

The federal and many state government energy tax incentives have assisted in encouraging conservation and renewable energy systems adoption, but federal continuance beyond 1985 is questionable. Without tax incentives, low-interest loans, or other support, adoption of the more expensive renewable energy systems will be delayed and the impact on the general economy and standard of living could be devastating. Projections are for continued but perhaps reduced levels of federal incentives until a negative impact becomes evident.

When we look at the development and use of smaller renewable energy production systems, the scene is encouraging. The 1978 PURPA and energy tax credits are providing the initiative for the "little guys" because the large utility companies are required to purchase electricity offered to them. Utility reluctance has been high in many areas of the country, but with the price of large new fossil-fueled or nuclear generating facilities financially and/or environmentally prohibitive, becoming an energy "middleman" has great potential. Utilities are seen as having an increasing role as energy distributors. The use of district heating may increase as utility companies find that it is financially more attractive to sell waste heat to area customers than to expel it into the environment.

Innovative third-party financing, insurance coverage of energy production, no-risk installations, easily acquired loans, and other initiatives are expected to continue the move toward renewable adoptions and national energy self-sufficiency. Palo Alto, California, is considering a solar equipment leasing system, and Santa Clara and Phoenix already have companies that rent solar domestic water heating units. The county of Kauai in Hawaii requires all new multifamily structures of 10 units or more to install domestic solar water heating systems. San Diego has mandated that solar space and water heating be installed in all new buildings since 1979. Several state and local building codes are requiring energy-efficient construction techniques on all building, and solar access is being protected. Montana has a tax on coal which is used to provide grants to installers of renewable energy systems. Watch your local news media for more applications and innovations for implementing renewable energy systems.

Each geographic area of the United States will be taking advantage of those renewable energy assets which are most abundant in the region. The Northeast and Northwest will be using more hydro power, while the Sun Belt will be using solar. Ocean energy systems will be used on the Pacific and Gulf coasts, and wind will contribute in the Midwest and many other areas. The energy mix that will occur will be interesting, exciting, and challenging to our formerly fossil-based society. America's security and economic health depend greatly on adequate energy at reasonable prices.

Table 15-2 shows the 1983 estimated cost range for electricity produced by various energy sources and projects these costs for 1990. It also indicates the average percentage of change during that period. Costs are sure to affect our energy source selection.

TABLE 15-2 Estimated Cost of Electricity (Cents per Kilowatt-hour) from New Plants, 1983, with Projections for 1990 and Percent Change[a]

Energy Source	1983	1990	Percent Change
Nuclear power	10-12	14-16	+36.4
Coal	5-7	7-9	+33.3
Small hydropower	8-10	10-12	+22.2
Cogeneration	4-6	4-6	0
Biomass	8-15	7-10	−26.1
Wind power	12-20	6-10	−50.0
Photovoltaics	50-100	10-20	−80.0
Energy efficiency	1-2	3-5	

[a]Costs in 1982 dollars.
Source: Christopher Flavin, Worldwatch Institute, Washington, D.C.

FUTURE ENERGY TECHNOLOGIES: A CRYSTAL BALL VIEW

There are several new and experimental energy technologies which inventors, scientists, and engineers are trying to perfect for future commercialization. Some of the technologies are advancements of current practices, while others are totally new concepts that may appear inconceivable.

Fuel Cells

A fuel cell is a device that converts chemical energy directly to electricity (see Figure 15-6). Although this is true for dry cells and batteries, fuel cells differ in that they operate continuously as long as fuel and an oxidizer are supplied from an external source. Inside a fuel cell (Figure 15-7), hydrocarbons such as coal, oil, and natural gas are converted to a gaseous mixture of hydrogen and CO_2 possessing a positive charge at one electrode. Air is fed to the other electrode, which creates a negative charge. The resulting imbalance is separated by an electrolytic solution such as phosphoric acid. Connecting the electrodes to a load causes current to flow because of the electrochemical reaction of hydrogen combining with oxygen to form water. Direct-current

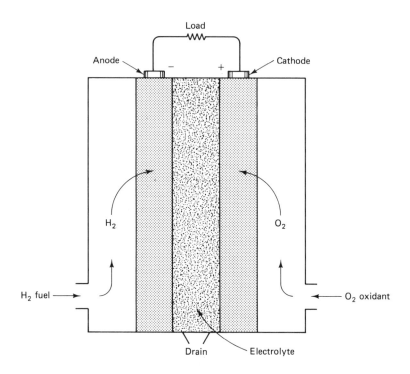

Figure 15-6 Fuel cell schematic.

electricity and heat are resultant energy forms. The dc can be inverted, transformed, and transmitted as required.

An advantage of fuel cells is that they convert fossil fuels to electricity with minimum waste. Efficiencies range from 60 to 70%, with hydrogen/oxygen cells at room temperatures operating at about 94% efficiency [Angrist]. Another advantage is that fuel cell power plants can be built near demand centers, thereby reducing energy transportation losses and environmental effects of extensive distribution systems. Principal environmental emissions are warm air, water, and carbon dioxide, thereby making fuel cells ideal for urban locations.

United Technologies, Inc., has successfully tested a 1-MW fuel cell in Windsor, Connecticut, and has developed two additional larger-scale experimental fuel cells, the first sponsored by Consolidated Edison (see Figure 15-8) and the other by Tokyo Electric. The unit in Japan has been on-line since 1983, but the New York City 4.8-MW unit has had some problems with delays and internal leaks. The San Diego Gas and Electric Company installed two units in its service area in 1984.

Several smaller (40-kW) units have been designed and successfully tested for use in remote locations or in commercial and industrial

Future Energy Technologies: A Crystal Ball View 397

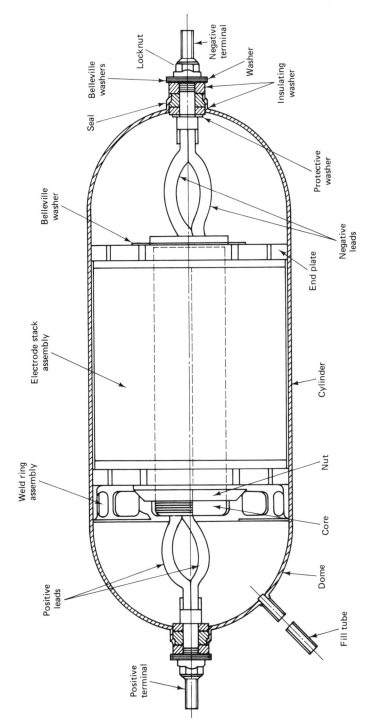

Figure 15-7 Miniature fuel cell: cross section. (From the National Aeronautics and Space Administration.)

Figure 15-8 Consolidated Edison's fuel cell power plant. (Courtesy of Consolidated Edison.)

facilities (see Figure 15-9). Other applications are to provide power for ships, trains, electric vehicles, and spacecraft.

Experiments are continuing with various materials and electrolytes, including molten carbonate, naphtha, calcium/oxygen, nickel/hydrogen, and hydrogen/halogen. Feedstocks from which hydrogen can be acquired, in addition to those previously mentioned, include methane, LP gas, alcohols, industrial waste gases, petroleum distillates, and coal-derived gases and liquids. Canada currently leads the research and development effort with a $13 million governmental grant to the University of Toronto.

Texas Instrument has developed a combination solar cell and fuel cell. Small silicon cells are placed in a solution of hydrobromic acid. When light strikes the cells, they electrolyze the acid, forming hydrogen gas and bromine ions. Compartmentalized in a fuel cell, the bromine stays in solution and the hydrogen is stored as a metal hydride. The two are recombined to produce electricity.

Commercial marketing for fuel cells is expected to begin in 1987, with the units providing the electrical base load and 300°F (149°C) heat for the owners. These cogeneration units can be manufactured in modules and shipped to the site, thus reducing overall costs. Cost per kilowatt-hour is projected to be 6 to 8 cents, with the thermal energy costing less than one-half what it would using conventional fuels. This thermal energy, when not required by the owner, can be sold to neighbors for space or process heating.

Figure 15-9 Small, remote fuel cell. (Courtesy of Gas Research Institute.)

Hydrogen: The Fuel of the Future

We constantly encounter the word "hydrogen" when considering various forms of energy. Most often it is associated with hydrocarbons in the form of fossil fuels such as coal, petroleum, or natural gas.

"A world that runs on hydrogen-electricity is inevitable," says David Scott, director of the University of Toronto Institute for Hydrogen Systems. Hydrogen is the most common element in the universe. Water, which covers two-thirds of our planet, consists of two atoms of hydrogen bonded to an atom of oxygen (H_2O). When separated, hydrogen is very combustible and can be used to operate turbine generators, heat homes, or power automobiles. When burned, hydrogen oxidizes to water, which can be electrolyzed back to hydrogen. There is negligible environmental impact, and the base resource is easily acquired and plentiful.

Why are we "fooling around" with finite, polluting petroleum fuels, coal, and nuclear energy when such an ideal fuel exists? The bonds that hold water molecules together are very strong, and large amounts of energy are required to break them to release the hydrogen. There are several ways to do this, but the most promising method is electrolysis.

Electrolysis is the process of passing electricity through a medium, in this case water. This process separates the hydrogen from the oxygen atoms. When hydrogen has been needed, fossil-fueled electricity gen-

Figure 15-10 Small electrolysis hydrogen generator. (Courtesy of Billings Energy Corporation.)

erating plants have provided the energy, but this makes the process uneconomic. The problem has been that with economical, plentiful fossil fuels available, hydrogen has cost more to produce than the fuel it would replace.

Harnessing the sun's energy to produce hydrogen appears promising. Whether converted to heat or directly to electricity by photovoltaic cells, solar energy is likely to be the major hydrogen-producing energy form of the future. Photovoltaic electrolysis has reached 8% efficiency in experiments at Texas A&M University. Photosynthesis or photochemical splitting using a photoelectrochemical cell has been done at the University of California at Berkley, but only 1% efficiency has been obtained thus far.

Whether from solar-powered electricity generators, photovoltaic electrolysis, or photochemical or thermochemical reaction, hydrogen from water will be an important fuel of the future. When produced, it can be transported through steel pipes, stored in underground chambers, compressed or changed to metal hydrides, and used in either gaseous or liquid form.

Nuclear Fusion: The "Pot of Gold at the End of the Rainbow"?

Many people are looking toward nuclear fusion energy as the salvation to our energy problem. These people may be right, but there were many people in the 1950s that proclaimed that nuclear fission was the ultimate in answers until economic, safety, and environmental factors reduced its competitive position.

Fusion nuclear energy is the energy of the sun, stars, and the hydrogen bomb. Fission involves the splitting apart of uranium nuclei to release energy, but fusion is the joining together of hydrogen nuclei in order to release energy. Since hydrogen is available from seawater, the "fuel" for fusion is inexhaustible.

Experiments with nuclear fusion began in 1932 but the science was not taken seriously until 1952, after the hydrogen bomb was developed. In 1958, the Atoms for Peace program initiated a cooperative effort among the United States, Great Britain, and the USSR toward finding an economical, unlimited energy source for the world. Since then other European countries and Japan have joined in the international effort with the supposition that if all nations had unlimited energy to meet their needs, world peace would be sustained.

There are several possible fuels that can be used in the fusion process, but they must be among the lightest elements. Hydrogen, lithium, helium, boron, and beryllium are all contenders, but the isotopes deuterium and tritium have been selected for the first generation of fusion reactors. Different forms of a substance are called "isotopes," and deuterium and tritium are both positively charged isotopes of hydrogen.

The fusion process requires extremely high temperatures and pressures that must be sustained for periods of time. If any of these three conditions are not maintained, the reaction terminates. We therefore have a controlled process.

A thermonuclear reaction involves an atomic structure change. During the process, the positive-charged deuterium and tritium are heated to 90,000,000°F (50,000,000°C), which changes them to gases. At this temperature, the electrons surrounding each atom become free and the gas turns into plasma—a collection of free nuclei and electrons. Pressure within the confinement structure forces pairs of positively charged deuterium and tritium nuclei into single nuclei of the gas helium and a neutron is ejected. Since the atom of helium is lighter than the sum of the two original isotopes, the lost mass becomes energy according to Einstein's equation, $E = mc^2$. Continuing the reaction therefore releases vast amounts of energy.

Fusion nuclear energy is still experimental. Scientists are attempting to attain adequate plasma density, temperatures, pressures, and time to sustain a reaction. As might be expected, attaining and retaining 90,000,000°F is a bit tricky. Our technology has not yet created a material that will hold such a temperature without melting. The reason that deuterium and tritium are being used as first-generation fuels is that all others require even greater temperatures.

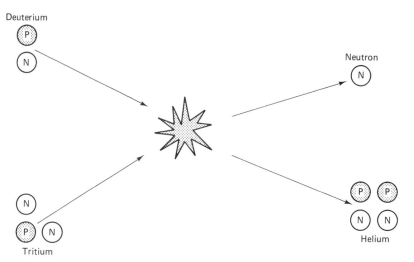

Figure 15-11   Fusion reaction.

Since there is no "material" that can contain the temperatures necessary for fusion, considerable experimentation has been done using magnetic confinement. A leading design for magnetic confinement is a large doughnut-shaped unit known as a tokamak (a Russian acronym for a toroidal chamber) (see Figure 15-12). Confinement was successfully attained with a tokamak in 1975. Further advances occurred in December 1982, and in November 1983 full density and confinement time were achieved for the first time in the United States.

Figure 15-12   Tokamak fusion test reactor. (Courtesy of Princeton Plasma Physics Laboratory.)

Magnetic containment simulates the conditions of the sun. Nuclear fusion tends to blow the sun apart, but the sun's own gravitational force restrains it, creating a stable mass of extremely hot plasma. Because the plasma is a good electrical conductor, the magnetic field acts like gravity, holding the force of the fusion reaction within the structure of the unit.

Inside the center of the toroidal magnetic confinement reactor is an evacuated chamber where the deuterium and tritium fuels are injected and subjected to the heat and pressure that turn them into plasma (see Figure 15-13). Surrounding this chamber is a 3-ft (0.914-m) "blanket" which absorbs the neutrons that carry the energy away from the fusion reaction. As the blanket becomes very hot, coolant is circulated through it, conducting the heat away to produce steam to drive a turbine.

A major new confinement experiment is the Scyllac device, which confines the plasma to a "theta-pitch" configuration within the toroidal unit. Figure 15-14 shows scientists working inside the magnetic field structure as they assemble the unit for confinement experiments.

Another new unit, called the Advanced Toroidal Facility, is being built at Oak Ridge National Laboratory. This unit also differs in the geometry of the magnetic field used to confine the plasma. The potential for steady-state operation rather than the pulsing mode of other toka-

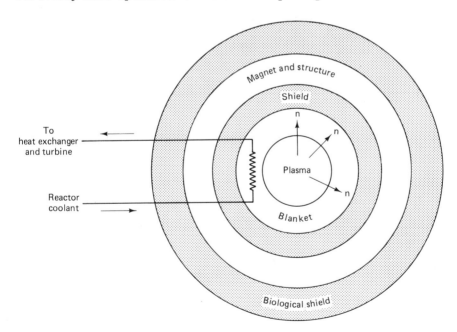

Figure 15-13 Schematic of a magnetic fusion confinement unit.

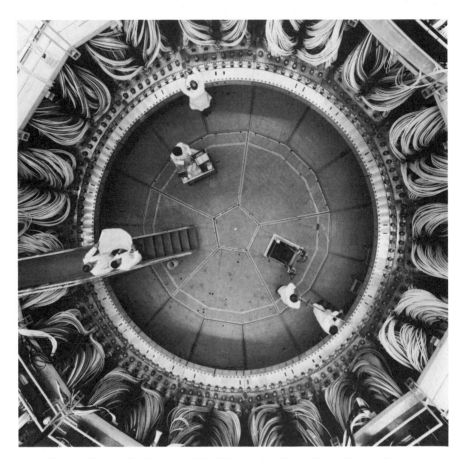

Figure 15-14 Scyllac toroidal "theta-pitch" configuration unit: top view. (Courtesy of Los Alamos National Laboratory.)

maks is encouraging. The USSR is experimenting with a fusion–fission hybrid which they hope will produce enormous amounts of plutonium. The potential dangers of plutonium were discussed previously in relation to fission breeder reactors.

Other possibilities for containment include "mirror machines," in which the plasma is guided along straight lines in a cylindrical container with strong magnetic fields at each end. The magnetic fields "mirror" the fusion plasma toward the center, thus oscillating it back and forth between the fields.

Confining the fusion reaction is also being tested using inertia. With this system, deuterium and tritium pellets, so tiny that many would fit on the head of a pin, are heated and compressed long enough for a reaction to occur. The heat and pressure are caused by intense short-wavelength laser light or energetic particle beams (charged atoms) directed at the pellets from several different directions. The beams are focused with sufficient intensity to crush the pellets, causing them to implode with tremendous pressure and heat. The compression and temperature ignite the fuel, causing fusion. A huge surge of energy is released before the pellet expands. The concept uses the pellet's resistance to change in motion (inertia) to hold the pellet in place before it blows apart from internal pressure.

The use of lithium ions rather than protons is a new beam concept which is proving easier to focus and more economical. However, lasers provide much more power to initiate the reaction. New designs for banks of capacitors are permitting the release of very short, high-power pulses of about 60 to 100 nanoseconds.

Significant strides have been made in fusion-system engineering. The first commercial-scale demonstration unit may be ready around 2000, which means that commercial contribution would be around 2020 at the earliest. The acceptability of fusion, as with any other energy source, depends on technical feasibility at affordable prices. The technology is complex, and commercial units are expected to be very expensive. The technology will not be accepted by industry unless the costs are less than or at least competitive with fission, solar photovoltaics, and others. A big advantage is that the fuel cost will be negligible and no long-lived, heavy-element isotopes are produced. Accident potential in comparison with fission is greatly reduced because a meltdown is not possible. If conditions for fusion cannot be maintained, the reaction simply does not occur. The radioactive materials that are produced are less toxic. All fuel processing can be done on-site, and there is no need to produce, handle, or transport radioactive fuel elements or plutonium as with fission. Estimates of 90% efficiency are projected for the fusion reaction, but 40 to 60% is expected from conversion to electricity in the final commercial units.

Figure 15-15 OMEGA 24-beam laser facility. (Courtesy of the Laboratory for Laser Energetics, College of Engineering and Applied Sciences, University of Rochester.)

Figure 15-16 Laser implosion at KMS Fusion. (Courtesy of Atomic Industrial Forum, Inc.)

Magnetohydrodynamics (MHD): Electricity Without a Generator

Since basic science class you have known that electricity is generated when a conductor moves perpendicular to a magnetic field. You are also aware that in a conventional electricity-generating plant, a turbine is rotated by an external energy force which in turn rotates a coil of wire through a magnetic field to generate electricity. A thermionic converter uses MHD and converts heat directly into electricity by passing a conducting fluid through a magnetic field. An electric current is established in the fluid.

Although experimental engineering began as recently as 1976, significant progress has been made in proving and demonstrating the feasibility of the MHD process. Devices have been developed which can produce in excess of 50% more electricity from a given quantity of heat than conventional turbine-generator systems. In the process, the ionized hydrogen isotopes of deuterium and tritium in the form of gasses are moved through the magnetic field, and a current is produced within the gas; electrodes projecting into the unit collect this current, allowing it to be conducted away for use. Since very hot gasses (2000 to 3000 K) are required for the process, MHD is ideal for use as a topping cycle in a cogeneration system, especially coal-fired units.

Research is continuing on MHD units and components. Coal-based

Future Energy Technologies: A Crystal Ball View   407

Figure 15-17   Experimental MHD unit. (Courtesy of Nooter Corporation.)

gas, electrode material, pollution abatement, design configuration, and other experimentation is being conducted. Expectations are high that MHD, with its elevated efficiencies, will contribute to base load, peak load, cogeneration, and retrofit applications of commercial electricity generation. The principle of MHD is also proposed for space-vehicle propulsion.

Preparations are now under way for the design and testing of a new 10-kW unit which is said to make solar-generated electricity cost-effective. This system, without moving mechanical parts, involves liquid-metal magnetohydrodynamics. Raloff explains that

> mercury circulates through a solar collector, acting as its heat-transfer fluid. Upon entering a mixer, a volatile liquid—in this case, freon—is injected as droplets. Heat transferred directly to the freon causes the droplets to boil, creating a two-phase liquid-metal and vapor-bubble flow. The continuing exchange of heat between the metal and the vapor expands the bubbles. This swelling action mimics miniature pistons and imparts a mechanical energy to the circulating mixture. It actually drives the liquid through a converging nozzle and into the MHD channel. Here the high-velocity flow of metal through a magnetic field generates an electric current; electrodes tap the current. As the dual-phase mixture exits the channel, the metal and vapor are separated. The vapor is routed through a heat exchanger—where it condenses,

releasing heat to warm air or water—and then on to the mixer. Mercury circulates from the separator back to the solar collector. [Raloff, p. 109]

Other Energy Sources for the Future

Investigations are being conducted in the use of lasers as a means for propelling aircraft and spacecraft. The vehicles would carry water and liquid nitrogen as fuel and receive laser pulses from space satellites.

Aneutronic energy may soon be a new word in your vocabulary. Again, this investigation involves isotopes of hydrogen. In this process a beam of atomic nuclei is bent by magnets into a figure 8 path at the center of which the nuclei collide. The enormous amount of heat that is created can be used to produce steam to drive a turbine, but efforts toward converting this heat directly to electricity, as in MHD, are under way. Research scientists project that electricity may be generated for one-sixth of the present cost of conventional methods. Small units may power trains, ships, and even aircraft, and small communities could produce their own economical, safe nuclear power. If continued funding can be found, initial commercialization might begin about 1993.

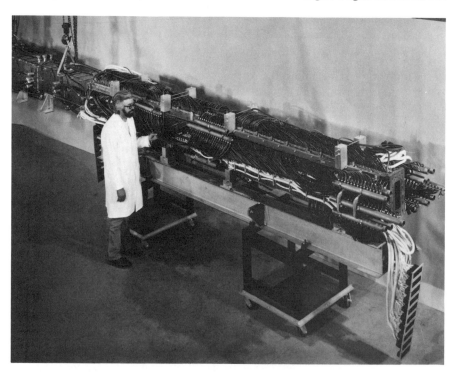

Figure 15-18 Electromagnetic coil, a major component of a MHD unit. (Courtesy of Avco Everett Research Laboratory.)

Thermoelectric generators that produce heat from the spontaneous disintegration of plutonium 238 are imminent. As in MHD and aneutronics, this heat is converted directly into electricity. These "nuclear batteries" are projected for cardiac pacemakers, satellites, and other large and small applications, A disadvantage is the necessity of handling plutonium rather than hydrogen isotopes.

Batteries have been improving rapidly over the years. In 1900, 38% of U.S. automobiles were powered by electricity. Zinc-nickel, zinc-chlorine, and lithium-sulfur batteries are all being projected for use in cars, commercially, within the next 3 to 5 years. As might be expected, the utility companies are eager to see this new source of demand materialize. The use of fuel cells and thermoelectric units, when developed further, may eliminate the weight and limited driving range associated with batteries.

Another energy source that could be available by 1993 but probably will not, due to public pressure, is the use of hydrogen-bomb heat energy. In this proposal, two small bombs each day would be detonated underground to deliver superheated steam which would drive turbines.

SUMMARY

Our energy future has so many variables that it is very difficult to predict. When will our fossil fuels become depleted to the point that they are no longer affordable? How soon will inexhaustible, economical energy sources be developed? The critical years will be between 1985 and 1995 because technology will not "be there" with the answer when it is needed if funding and conscientious effort is not continued.

The stability of our economy and quality of life depend greatly on energy and its wise, efficient use, but the demonstrated attitude of both our country's leadership and citizenry has been unresponsive. Total energy demand will continue to grow slowly due to an improving economy and increasing population.

Nuclear, coal, and solar energies hold considerable promise for the future, but some aspects require scientific breakthroughs to make them feasible or commercially available. Time to research and develop new energy technologies can be gained through conservation. There will be a trend toward smaller, decentralized energy production facilities with an "energy mix" blending high and low technology to meet our energy needs. Geothermal, solar, wind, and biomass energies will make considerable contributions.

Keeping up with newspaper articles, current scientific and journalistic reports, and electronic communication media presentations will help keep you aware of near-future changes, developments, research, and discoveries. Your decisions will play an important role in deciding the energy future, the severity of the impending socioeconomic impact, your quality of life, and the life you leave your children. The only forecast of which we can be reasonably sure is a continuing increase in energy costs and a decline in the availability of fossil fuels. America's energy future will be significantly different from its past.

ACTIVITIES

1. After reading various scientific reports, energy projections, and periodicals related to energy and the future, report or discuss your projections for future energy sources. Address specifically the energy forms and sources projected for application in the residential space heating and domestic water heating, transportation, and other sectors. Compare with your colleagues.
2. Conduct research as in Activity 1, addressing the environmental impacts of each of the energy sources.
3. Discuss how the interaction of political and economic factors affect the development of various energy sources.
4. Discuss how the American life style and standard of living has affected attitudes toward energy use and conservation. What effects do you see concerning the changing energy picture and future life styles?

BIBLIOGRAPHY

American Petroleum Institute, *Two Energy Futures*. Dallas, Tex.: American Petroleum Institute, 1983.

Anderson, Jack, "Oil Crisis Waiting to Happen," *Syracuse Post-Standard*, November 30, 1983.

Angrist, Stanley W., *Direct Energy Conversion*. Boston: Allyn and Bacon, 1982.

Associated Press, "Con Edison Fined $40,000 for Nuclear Safety Lapse," *Syracuse Post-Standard*, March 15, 1984.

Associated Press, "Owners of New Hampshire Nuke Plant Agree to Cancel Troubled Twin Reactor," *Syracuse Post-Standard*, March 31, 1984.

Betts, Mitch, "Elec. Debate: More Plants or Conservation?" *Energy User News*, February 13, 1984.

Dorf, Richard C., *The Energy Factbook*. New York: McGraw-Hill, 1981.

Flavin, Christopher, "Market Forces Mothball Nukes," *Renewable Energy News*, March, 1984.

Gallagher, John, "Nine Mile 2 Estimate Reaches $4.6B Ceiling," *Syracuse Post-Standard*, March 16, 1984.

——, "Niagara Mohawk Fined 180G," *Syracuse Post-Standard*, March 22, 1984.

——, "N M 2 is $5.1B; Study to Weigh Plant's Future," *Syracuse Post-Standard*, March 31, 1984.

Gipe, Paul, "Wind's Tough Dreamers," *Solar Age*, February 1984.

Kaku, Michio, and Jennifer Trainer, *Nuclear Power: Both Sides*. New York: W. W. Norton, 1982.

Landsberg, Hans H., "Relaxed Energy Outlook Masks Continuing Uncertainties," *Science*, Vol. 218, December 3, 1982.

Lidsky, Lawrence M., "The Trouble with Fusion," *Technology Review*, October 1983.

Long, Pat, "What's Next for California?" *Solar Age*, November 1982.

Lovins, Amory B., *Soft Energy Paths—Toward a Durable Peace*. Cambridge, Mass.: Ballinger, 1977.

Lovins, Amory B., and L. Hunter Lovins, *Brittle Power—Energy Strategy for National Security*. Andover, Mass.: Brick House, 1982.

Meador, Roy, *Future Energy Alternatives*. Ann Arbor, Mich.: Ann Arbor Science, 1978.

Punwani, Bodle, Rader, and Tarman, "SNG Production from Peat." Institute of Gas Technology Paper, Miami International Conference on Alternative Energy Sources, December 5-7, 1977.

Raloff, Janet, "Liquid-Metal Solar Power," *Science News*, Vol. 120, August 15, 1981.

"Ripley's Believe It or Not." Television program, WIXT, February 5, 1984.

Ruedisili, Jon C., and Morris W. Firebaugh, *Perspectives on Energy: Issues, Ideas, and Environmental Dilemmas* (2nd ed.). New York: Oxford University Press, 1978.

Schefter, Jim, "Energy Experts Pick 5 Wild Windmills," *Popular Science*, June 1983.

Scott, David, "Wave Power Spins a Ram-Air Generator," *Popular Science*, March 1984.

Thomsen, Dietrick E., "Economy of Fusion," *Science News*, Vol. 125, January 7, 1984.

Chapter 16
Energy-Related Employment and Careers
Where the Jobs Are

CONCEPTS

1. Renewable energy systems are more labor intensive than non-renewable energy systems.
2. Renewable energy employment can generate more personal income for each unit of energy produced than can nonrenewable energy systems.
3. Renewable energy systems can make use of skills already available in the labor force.
4. People who are unemployed or underemployed can be employed more easily in the renewable energy field than is possible in centralized, nonrenewable energy systems.

GLOSSARY

DIRECT EMPLOYMENT—used to identify jobs specifically associated with an industry, work site, or occupational category.
HVAC—heating, ventilation, and air conditioning

INDIRECT EMPLOYMENT—jobs created by supporting industries which provide goods and services to those in "direct" employment, as well as other positions created by money spent or invested by those so employed.

INTRODUCTION

Job opportunities in the renewable energy field often bear a close relationship to traditional occupations. Many of these "new" jobs are in reality the result of training and experience from occupations that have existed for several decades. An example of how previous work can be directed to new technologies is that of the aerospace engineer who now works on the development of large-scale wind turbines.

A major emphasis needs to be placed on the need for flexibility in the preparation of persons who choose to work in the rapidly growing field of renewable energy. Not only is flexibility needed for advancement in the field, but it is also essential as the industry struggles through initial growth pains. As technologies develop, certain materials and procedures are found to be more suitable to construction and manufacturing. Some systems succeed and others are found to be unworkable, despite the early optimism regarding their suitability.

The renewable energy field is much more labor intensive than is the nonrenewable energy sector. Most noticeable needs for personnel are in the areas of manufacturing, sales, and service, and in the development and installation of equipment for solar and wind energy systems. People who have been trained and have experience in carpentry, construction, welding, plumbing, and electrical work can make the transition to the renewables field with relative ease, since the skills needed are in most instances very similar to those needed for traditional careers.

Many of the same construction and installation skills developed by people in the conventional building trades can be applied quite easily to the installation of insulation, whether in new construction or in retrofitting operations. The installation of energy-efficient windows, for example, requires much the same skill and technical knowledge as that for non-energy-efficient construction.

Energy technology programs for the preparation of technicians, engineers, and researchers are currently in place in more than 100 colleges and universities in the United States. Completion of a program may lead to certification as an energy auditor, or may lead to the A.A.S., B.S., M.S., or other graduate degrees in one of many energy-related fields. Many of the degree programs are interdisciplinary—a recognition of the need to prepare people who are versatile and adaptable.

OCCUPATIONS IN RENEWABLE ENERGY FIELDS

Engineers and Scientists

Recent studies of employment distribution in the renewable energy field indicate a high percentage of engineers employed in research and development work. This is to be expected in a new and expanding group of technologies. The job or employment opening most frequently mentioned in the engineering group is that for mechanical engineers. They plan and design products and systems, develop testing procedures, and evaluate pilot projects for their application in providing functional and profitable equipment such as heat-transfer systems or electromechanical controls. Other engineering specialists needed for energy research and development include applications engineers, who develop new applications for products and systems and provide technical support in sales and marketing, as in the development of more efficient solar collector panels.

Commercial production and installation work with renewable energy products employs many other types of engineers. Civil engineers plan, design, and direct construction of structures such as towers for WECS. Electronic engineers are responsible for the design, manufacture, and testing of electronic components used in almost every renewable application, and chemical engineers may be involved in experimentation with materials to make photovoltaic cell materials more efficient.

Environmental engineers are expected to be in greater demand because of the increased concern of most citizens regarding the danger to the environment with increased consumption of fossil fuels. In particular, the projected increase in the use of coal raises serious questions about higher and higher levels of sulfur in the atmosphere and subsequent degradation due to acid rain. Environmental impact statements are a part of the planning for essentially every new facility being constructed. Opportunities for environmentally related work will increase the demand for biologists, chemists, and biochemists.

The recognition of the greatly increased potential for the use of water power as a component of overall electricity-generating capacity increases the need for hydraulic engineers. They design and supervise projects which control and use the energy: canals, conduits, reservoirs, and control and conversion systems. Ocean energy systems, which include wave and tidal energy, will make use of hydrologists, oceanographers, marine biologists and hydrophysicists, and mechanical engineers.

Industrial engineers coordinate the work done in production facilities, working with organization charts, work distribution, process flow, and plant layout. Manufacturing engineers direct manufacturing processes in plants, estimate production times, and determine staffing for production schedules. Solar, biomass, and other renewable energy industries make use of their skills in developing systems and plant layouts for new energy products. Many renewable energy conversion systems can be manufactured by conventional production methods and transported to the site where they will be installed. Large, centrally located fossil fuel conversion systems most often require custom-designed features and must be constructed on-site as one-of-a-kind devices, increasing the cost to the ultimate consumer—you and me.

Engineers and scientists are assigned a wide variety of positions, and many change or expand their duties several times during their working lifetimes.

Scientific positions include those of aquatic biologist, air quality analyst, atmospheric scientist, or mathematical modeler. Operations/ systems researchers are in much demand as plants are remodeled for new products and manufacturing systems. Geologists, geophysicists, and chemists are sought after in the geothermal field. Wind energy study requires meteorologists and physical scientists.

There is a great demand for persons with training and experience in conventional energy fields (Figures 16-1 through 16-4). Chemical, mechanical, petroleum, and nuclear engineers (especially those who can provide solutions to nuclear waste disposal) are needed as well as physical scientists, economists, and operations research specialists. The demand for accountants, computer programmers, and analysts continues strong. Geologists and geophysicists are expected to remain in strong demand for exploration and development of fossil fuels. They would be able to make the transition to the geothermal field with ease.

College/university teachers with the background and preparation to teach others are difficult to find and to hire. Energy study and research tends to cut across traditional academic lines and requires someone who has both a broadly based academic background and experience in teaching. Those who have had practical experience as well as one or more college degrees are in demand both for teaching and for the manufacturing and production sectors of industry.

Teachers are needed in the school systems where young people are making choices about their further educational and/or employment plans. Those who have training and experience in the physical sciences and technology and have an interest and knowledge in energy systems and conservation are in demand to prepare those who will be entering the job market as the next century begins.

The range of occupational choices is too broad to be covered here

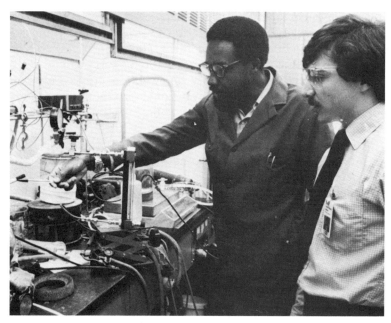

Figure 16-1 Research chemist. (From Career Awareness Kit; courtesy of Chevron USA, Inc.)

Figure 16-2 Chemical engineer. (From Career Awareness Kit; courtesy of Chevon USA, Inc.)

Figure 16-3 Mechanical engineer. (From Career Awareness Kit; courtesy of Chevron USA, Inc.)

but should be explored with the help of counselors and publications specifically devoted to job skills and requirements.

Managerial, Administrative, and Service Jobs

Both large and small companies and corporations employ administrative personnel (Figures 16-5 and 16-6). In smaller companies, one person may need to wear many hats (in the interest of economy), whereas a larger company may have one or more persons for a single position, although the percentage of administrative positions may not be as large in proportion to overall employment.

Some of the occupations in this category include utility rate analyst, solar energy market researcher, real estate lawyer, customer relations representative, technical publications writer, public relations representative, advertising aide, salesperson, and industrial relations representative.

Technical Personnel

The ratio of technicians to engineers and scientists in renewable energy fields is not as high as in better-established areas. Much of this

Figure 16-4 Geologist. (From Career Awareness Kit; courtesy of Chevron USA, Inc.)

Figure 16-5 Tax attorney. (From Career Awareness Kit; courtesy of Chevron USA, Inc.)

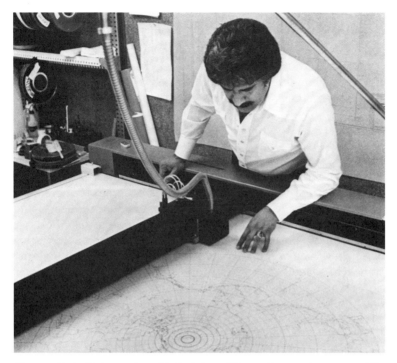

Figure 16-6 Information Technician. (From Career Awareness Kit; courtesy of Chevron USA, Inc.)

is undoubtedly due to the higher levels of research and development work presently being undertaken. As these new technologies become better established, larger numbers of technicians will be needed to supervise the highly sophisticated systems. Those employed in solar energy as it is now developing include chemical, electrical/electronic, and mechanical technicians. Most work being undertaken in solar development is in the commercialization of solar space and water heating needs. Some employment is available in solar cooling research, using well-known principles similar to those in the HVAC industry.

Skilled Crafts

Every skill used by carpenters and other construction workers is adaptable in both passive and active solar construction. Modifications in plans are required, but standard construction techniques have proven to be effective in building energy-efficient structures. Earth-sheltered designs require several different skills, but one owner-builder used state-approved bridge building techniques to plan and construct an attractive, yet energy-efficient home.

Those persons with training and experience in plumbing and HVAC are often able to transfer those skills to the installation, repair, and servicing of either air- or liquid-type solar systems; both use liquid flow principles in their operation. Other skilled crafts in demand include machinists, mechanics, sheet-metal workers, and welders/flamecutters as well as boilermakers, millwrights, pipefitters, and electricians.

RENEWABLE ENERGY, EMPLOYMENT, AND THE ECONOMY

Almost 25% of all energy used in the United States is consumed in heating, ventilating, and air conditioning. A program to increase the use of renewable energy forms for these purposes could reduce the nation's energy consumption by as much as 2 or 3 billion barrels of oil per year. The reduction in balance of payments for foreign oil we would not purchase would be a positive factor in the gross national product. Experience in several states has shown that dollars saved through conservation efforts are in turn spent on domestic goods and services. This provides an impetus to the local and regional economy and an increased potential for purchase of goods and employment.

Levy and Field found in a 1980 study of 518 businesses engaged in solar energy that the largest group of employees were involved in production and installation of equipment for space heating and cooling and water heating. Nearly as many persons were employed in research and development, indicating this field is still dealing with the initial growth of basic research and its applications.

In the same study, all other renewable energy fields, photovoltaic conversion, industrial process heat, thermal power, wind conversion, and biomass conversion employed several times more persons in research and development than in commercial and installation work. Research and development generally requires higher levels of education than does production and installation work. The shortage of those with the requisite levels of experience was indicated in the report by a "difficulty-to-hire" ratio. In "most difficult to hire" categories were atmospheric scientists, operations systems researchers, and college/university teachers. By comparison, jobs in the skilled craft and operative positions, and in clerical and unskilled jobs, were relatively easy to fill.

Those individuals with training and experience in the latter fields can make the transition to similar jobs in the renewable energy field with little additional training. Those with little or no training in the traditional crafts may be able to find employment with small companies and receive on-the-job training as the company grows.

The ratio of jobs to capital investment can make a very strong case for further expansion of the renewable energy sector. The construction of a typical nuclear plant of 1000 MW, for example, costing $5.5 billion, would create about 13,000 jobs over a 10- to 14-year period. This would provide a ratio of about one construction job for each $400,000 invested, or one job for each 77,000 kW of installed capacity. Construction at central sites tends to require large numbers of a few very specific craft skills at one stage of construction, followed by layoffs, then the hiring of another large group of one or more crafts. Local communities usually cannot supply large numbers of any one craft, so skilled labor must be "imported." Each bulge of employment followed by a subsequent layoff causes disruptions in the employment picture as well as in the need for housing, food, recreation facilities, schools, and other public services. There may be other serious consequences for the local economy as well, particularly when there are large shifts in population in what have traditionally been relatively small, stable communities.

Construction jobs at large centralized energy production units are almost entirely unionized. Although efforts have been made to expand the range of personal qualifications for acceptance into apprenticeship positions, unions have traditionally hired only those with a minimum of a high school education and the ability to pass tests based largely on mathematics skills. This strong emphasis on formal educational skills has tended to prevent many school dropouts (predominently from minority groups) from entering apprenticeship positions.

By contrast, many solar and other renewable energy jobs can and are being done by small firms, with more flexible employee entrance requirements. The work done by these companies is done at thousands of sites rather than at centralized sites. Employment is maintained and is more stable over a period of years. Some manufacturing, such as that for efficient wood-burning heating equipment has enjoyed rapid growth and increased employment as individuals and corporations attempt to reduce energy costs. The rapid growth in the WECS industry has increased employment in many electrical and mechanical production facilities, and has created a demand for new plants.

Another comparison can be made of the ratio of operational jobs made available by nonrenewable energy plants versus those created by renewables. A nuclear plant of 1000 MW is estimated to need about 500 employees to operate the plant, a ratio of 2 MW of installed capacity for each employee, or $11 million of invested capital for each position. Over one-third of these employees require at least a college degree plus highly specialized training in nuclear plant operations. Most of the remaining employees require at least a high school diploma. The highly automated petroleum refining industry provides one job for each

$100,000 in capital investment, again with the majority of the positions being those demanding high levels of education, skill, and long periods of training.

Renewable energy investment is more labor intensive than capital intensive. Solar development can provide 2.5 times the number of jobs for the same amount of energy capacity produced than can nuclear energy development. Another comparison may be made in the ratio of trade employees for each engineer or technician employed. For nuclear the ratio is two trade persons for each engineer; for solar, the ratio is nine to one.

A report by the Jet Propulsion Laboratory compared jobs in the liquefied natural gas field specifically with that of solar employment. Their conclusions were that 50 to 80 times more jobs would be created by solar technologies at an overall cost of 1.5 times less than that of the use of LNG. At the time of the report, solar was the only renewable energy resource with much visibility. The addition of other renewable energy sources since then would tend to support the general conclusions of that study.

Recent developments in the construction of wind farms, particularly in California, have created thousands of jobs and resulted in a 400-MW electrical generating capacity from that technology alone. Monroe reported that the operation and maintenance of large WECS in New York State would employ two to four times the labor force of an equivalent nuclear- or coal-fired system. Wind systems would also provide employment in existing generator and electrical component manufacture, steel fabrication plants, and in associated industries, all of which have experienced lowered demand for their products and subsequent losses of jobs in recent years. Many of the renewable energy conversion systems can be manufactured by plants already in existence. They can be readily transported to the site, while the larger, centrally located fossil-fueled conversion systems require customized, on-site construction at higher costs.

WECS have been shown to provide electrical energy for systems ranging from a few hundred watts to several megawatts. Wind farms have been shown to be highly desirable energy producers at relatively isolated sites on the west coast, providing jobs in the manufacture of electrical generators, inverter and control systems, and in the manufacture of turbine blades, gearboxes, and support towers. The acquisition of land leased for the construction of wind farms provided realtors, lawyers, and insurance people with a "wind-fall" business.

Employment increases estimated to be developed by renewable technologies are almost entirely in occupations that are already well established or in fields with moderate-to-high unemployment. The direct effects of employment in these fields to develop renewable energy systems is twofold: to produce components and services for which skilled personnel have already been trained, and to place many of the unemployed into productive, meaningful jobs. The results would be a decrease in unemployment benefits paid out, an increase in consumer spending, and therefore an improvement of the economy in general as well as a decrease in reliance on imported petroleum products.

The "advantages" of large, centralized energy utilization should be seen in comparison with decentralized uses. There are economies of scale to be considered when producing energy for large population centers and for large energy consumers in the commercial/industrial sector, particularly when efficiencies of larger energy-producing equipment are considered. However, centralization (putting all your eggs in one basket) brings with it concentrations of population, with traffic congestion, pollution, and other problems of urban living which seem to have few solutions. Centralization also creates the threat of major power blackouts due to equipment failure and an increasing potential for sabotage. Decentralization can provide many opportunities for employment: manufacturing and assembly, construction, small and medium-size plants, and self-employment, as well as working for someone else.

Renewable energy technologies can generally be characterized as less centralized than conventional energy production, yet they offer the potential of supplementing that system. Solar collectors and WECS can be installed as individual units or in sets, with no connection to an external power grid; or they can be tied through an interconnect system with the utility and provided with automatic cutout protection in the event of the failure of either. Homeowners as well as businesses can make use of these systems to provide lower-cost energy than is currently available via the utility grid. Concentrations of population, traffic congestion, and pollution can be lessened with the infusion of more decentralized energy-producing equipment.

Hydroelectric generation has been shown to be much less costly for installation and operation than coal-fired or nuclear power plants. Hundreds of sites with the potential for development have been identified in many parts of the United States and in many other countries. Our discussion of hydropower and ocean energy systems indicates that when more fully developed, these systems will be of small and medium scale, yet can be treated as large WECS are—as separate units or interconnected with other utility systems. Employment opportunities are just beginning to develop and should increase in the manufacture of components, construction of facilities, assembly, and service of this equipment.

Waste and biomass systems lend themselves well to the solution of major industrial and commercial as well as urban/suburban environmental disposal problems. Conversion of waste materials to energy is an excellent example of the wise use of resources. Plants have been built in many municipal areas to reclaim metal, glass, and other recyclable materials and to convert other waste materials to steam for electrical production. The construction and operation of resource recovery units employs local people and makes a positive contribution to the local economy and the environment.

The heat pump industry is one of the fastest growing industries today, particularly in the southern half of the United States. Engineers are at work on "super" heat pumps which will operate effectively at temperatures well below the freezing point of water (currently about the lower limit for effective heat pump operation). This development will make heat pumps feasible for all but the most frigid regions in most of the populated areas of the world. The expanding need in this field is for installers and service persons to meet local demands for the installation and maintenance of new units. Sales of these units have been discouraged by the lack of service/maintenance personnel in local areas.

Conservation of energy is the least costly measure which can be taken to reduce energy demands. Conservation consists not only of reducing demands through lower thermostat settings in winter and higher in summer, but many other measures as well. Insulating buildings and heat-transfer systems, installing heat recovery systems, and providing information services to residential and commercial consumers all develop employment. A greater demand for insulation materials has increased the number of employees needed in the manufacture and installation of several types of insulation materials. A study done for the AFL-CIO by Brian Turner and Iris Lav predicted that implementing energy efficiency standards for homes and commercial structures between now and the year 2000 would create 200,000 jobs per year.

Companies that produce window units have developed windows which are not only more energy efficient but more easily cleaned and maintained. The manufacture, sales, and installation of new units all provide employment, and when the installation is complete, energy is saved. The production and installation of solar water and space heating systems would provide 400,000 jobs per year.

Utility companies have hired new employees and retrained existing employees as energy auditors to comply with legislation developed to encourage conservation. Retraining provides another avenue for upward mobility and increased responsibility and its (usually) corresponding salary increase.

EDUCATION AND TRAINING PROGRAMS

The National Solar Energy Directory published by SERI describes energy education programs to be found in colleges and universities in the United States. Every state has at least one program, as do Guam, Puerto Rico, and the District of Columbia. There are programs listed for five levels of preparation. Those with shorter training requirements include certificate programs for preparing personnel for positions such as air conditioning, heating, and refrigeration; plumbing and pipefitting—solar specialization; solar energy technician; and wind energy technology. Associates degrees follow much the same pattern; in fact, some of the titles are the same, indicating that there is not complete agreement on the amount and type of preparation needed for various positions. More than 50 programs are offered in each of the certificate and associate degree programs.

Bachelor's and master's degrees in energy education are available in more than 50 colleges and universities. The preparation is often found to be interdisciplinary. Departments of engineering and physics predominate in offering energy-related courses and programs. Some programs are heavily involved with business or management courses, and frequently include social and political issues as a primary or secondary focus.

Doctoral programs emphasize such approaches as energy engineering, energy management systems, mechanical engineering (emphasizing direct energy conversion), energy conversion and resources, space solar power research, and solar energy in environmental studies.

Graduate programs in energy-related fields more often tend to be based on traditional degree programs such as those found in mechanical, electrical, and chemical engineering. Those programs leading to certificates and associate degrees seem to be more flexible in type and range of their offerings. The latter are also more often available from community colleges and technical institutes, both public and private.

Schiff takes universities to task for not responding more quickly to the energy problems that were perceived by the public-at-large several years before energy programs were instituted at these larger institutions. Community colleges were developed to respond to the needs of a more local constituency, and rely on those groups for continued support. They have been able to respond more quickly and easily to economic and social trends than have four-year colleges and universities. Large size tends to foster inertia, and programs once in place are not easily modified, particularly when such modification means that faculty and students already enrolled in existing programs might suffer disruption. There is, however, the continued need to develop additional courses and programs in many areas related to energy.

As the recognition of this fact grows, persons with strong preparation in the physical, biological, and environmental principles which underlie the use of energy resources will be in demand as instructional staff. Those with training in both the traditional disciplines and in applications of newly developed technologies will be in great demand.

An area that has grown rapidly, responding quickly to people's needs, is that of informal, noncredit instruction. These workshops and seminars, often consumer oriented, are frequently offered as short courses—sometimes as short as a single meeting. They may be as mundane as quick pointers on no-cost energy-saving tips (lowered settings on thermostats, turning off unused lights, driving slower or with more planning to reduce the number of stops), or low-cost energy-saving techniques of insulating, caulking, or putting up storm windows.

A large number of one-day seminars on constructing, installing, and servicing solar panels for domestic hot water and space heating have been conducted. The reasons for attending a seminar may be to gain enough knowledge to make a more intelligent purchase, to learn enough to construct a system, or to determine if an investment in one type of system is economical for a particular location or purpose.

Institutes are frequently offered for periods of one or two weeks and some for as long as a month or more. Energy-saving construction techniques, log cabin construction, techniques in developing self-sufficiency skills (and the problems of "going it on your own") are among the topics covered. Many are hands-on workshops incorporating both the technical information content and supervised skill development.

Employment opportunities are listed in Job Service Offices (job banks) in most metropolitan areas by the U.S. Employment Service and may also be found in the U.S. Labor Department's monthly publication "Occupations in Demand." Technical journals published by various professional fields are good sources of up-to-date job information. Examples of this are *Oil and Gas Journal*, *Offshore Magazine*, *Energy User News*, *Solar Age*, and *Solar Engineering and Contracting*.

SUMMARY

Traditional occupations parallel those found in the growing renewable energy field. Jobs in the skilled and semiskilled trade areas should be enhanced by renewable energy technologies. Some new positions have been created, and others will undoubtedly appear as new technologies are developed. The most likely similarity between existing and new occupations will be the need to remain flexible with regard to training and retraining.

Engineering and scientific positions will be those most needed for the short term, with the technician and trade jobs following as expansion occurs in renewable energy. Research and development work will require the skills of more than one speciality, implying that those who seek employment as researchers and developers should have training which is both broad and in depth. This could mean longer preparation time or a person who can maintain a regular work load and be involved in constant upgrading of skills and knowledge.

With the increase in awareness of the technical aspects of society, it will become increasingly difficult to find employment without skills and knowledge; there will be few places for the undereducated. Some preparation is required. It need not be a formal college degree but may be the type of training acquired at workshops, institutes, and certificate programs.

There will be a low ratio of technical personnel to engineering and scientific employees as each technology matures, but the opportunity for a wide range of positions exists in every renewable energy field. Those who seek employment in crafts or trades which are traditionally unionized should expect to have at least a high school diploma and work/school experience in a technical field. Renewable energy technology does not mean a return to the simple, nontechnical life.

Increased employment is very real with the development of additional energy technologies. Not only is the job ratio increased for the level of energy produced, but the range of positions is widened as well. Centralized energy-producing equipment requires a high proportion of very highly skilled trades persons, technicians, engineers, and administrators. Greater reliance on smaller, more decentralized energy-producing systems will increase the employment opportunities for those with less skill and less training. Since 1975, smaller companies have been shown to have provided more new jobs than have large corporations. One industry affects employment opportunities in another. For example, the increase in demand for flat-plate collectors has increased the need for employment in the glass industry, specifically for specialty (low-iron) glass.

Education and training for all levels of employment in both the conventional and renewable energy fields can be found in every state in the United States and is also widely available in most of the industrialized nations of the world. The exchange of information and expertise that characterizes the "information society" will increase the rate at which knowledge of developing technologies will be dispersed.

ACTIVITIES

1. Identify one occupational category that would be of interest to you. Review literature on the skill and preparation needed to meet requirements for entry-level employment. Assess wages/salary levels, opportunity for advancement, working conditions, and other factors related to employment in this job.
2. Review the Help Wanted columns of several local and regional newspapers as well as listings in energy-related journals. What types of positions are offered? What training or educational requirements are there? Are the jobs available in your community or region, or would the position require you to relocate? How do the salaries compare with those in other jobs?
3. What job-searching services are available in your school or community? What can you do to improve the types of information available?
4. What energy-related jobs are available in your community? (electricity-generating plants, gasoline truck drivers, propane suppliers, transportation of oil and coal, construction of new energy facilities, and others).
5. Review the classified sections of local or regional telephone directories for energy-related services.

BIBLIOGRAPHY

Bechtel National Inc., "Projected Annual Resource Requirements at the National and Regional Level, for the Department of Commerce Energy Forecast 1985 and 2000." San Francisco, 1977.

The College Blue Book, 18th ed., "Degrees Offered by College and Subject," New York: Macmillan, 1982. (12 states; 17 programs in Energy Management.)

Fermoselle, Rafael, *Energy Occupations in Demand*. Arlington, Va.: R. F. Associates, 1980.

ILO Information, International Labor Office, Washington, D.C.

Jet Propulsion Laboratory, *Solar Energy in Buildings*. Pasadena, Calif.: JPL, 1977.

Levy, Girard W., and Jennifer Field, *Solar Energy Employment and Requirements 1978-1985*. Columbus, Ohio: Battelle Columbus Laboratories, April 1980 (for U.S. Department of Energy).

Monroe, Jackson, et al., *Energy and Employment in New York State*. A report to the New York State Legislative Committee on Energy Systems, Albany, N.Y., May 3, 1975 (draft).

Peterson's Annual Guide to Undergraduate Study, 13th ed. Princeton, N.J.: Peterson's Guides Inc., 1983. (47 programs.)

Quammen, David, *Appropriate Jobs, Common Goals of Labor and Appropriate Technology*. Brief 3. Butte, Mont.: National Center for Appropriate Technology.

Renewable Energy News (various issues), CREN Publishing Ltd., Ottawa, Canada.

Schiff, Gary S., *The Energy Education Catalog, Programs in American Colleges and Universities*. New York: Academy for Educational Development, American Council on Education, 1981.

———, "The State of Energy Education in American Colleges and Universities," paper presented to the International Energy Information Forum and Workshop for Educators, 1982 World's Fair, Knoxville, Tenn., June 1982.

Settlemire, Mary Ann, *Energy Education Programs: Perspectives for Community, Junior and Technical Colleges*. Washington, D.C.: American Association of Community and Junior Colleges, 1981.

Solar Energy Research Institute, *National Solar Energy Education Directory*, 2nd ed. Golden, Colo.: Solar Energy Research Institute, January 1980.

Uleck, Ronald B., editor-in-chief, *Energy Jobs Handbook*. Gaithersburg, Md.: Prospect Press, 1981.

Energy User News (various issues).

Solar Age (various issues).

Technical Education News (various issues).

Appendix A
Energy Units and Conversion Factors

Multiply	By	To Obtain
British thermal units (Btu)	0.2530	kilogram-calories
British thermal units (Btu)	777.5	foot-pounds
British thermal units (Btu)	1054	joules
British thermal units (Btu)	2.928×10^{-4}	kilowatt-hours
Btu per min	0.02356	horsepower
Btu per min	0.01757	kilowatts
calories per gram	1.8	Btu per lb
centimeters	0.3937	inches
centimeters	0.01	meters
cubic centimeters	10^{-6}	cubic meters
cubic centimeters	10^{-3}	liters
cubic feet	1728	cubic inches
cubic feet	0.02832	cubic meters
cubic feet	7.481	gallons
cubic feet	28.32	liters
cubic feet per minute	0.1247	gallons per sec
cubic feet per minute	0.4720	liters per sec
cubic inches	5.787×10^{-4}	cubic feet
cubic meters	35.31	cubic feet
cubic meters	264.2	gallons
cubic meters	10^3	liters
cubic meters per sec	2118.9	cu ft per min
decimeters	0.1	meters
degrees (angle)	60	minutes

Multiply	By	To Obtain
degrees (angle)	0.01745	radians
dekameters	10	meters
feet	0.3048	meters
feet per min	0.508	cm per second
feet per sec	30.48	cm per second
feet per sec	18.29	meters per minute
foot-pounds	1.28×10^{-3}	British thermal units
foot-pounds	5.050×10^{-7}	horsepower-hours
foot-pounds	3.766×10^{-7}	kilowatt-hours
foot-pounds per minute	1.286×10^{-3}	Btu per minute
foot-pounds per minute	0.01667	foot-pounds per sec
foot-pounds per minute	3.030×10^{-5}	horsepower
foot-pounds per minute	2.260×10^{-5}	kilowatts
foot-pounds per second	7.712×10^{-2}	Btu per minute
foot-pounds per second	1.818×10^{-3}	horsepower
foot-pounds per second	1.356×10^{-3}	kilowatts
gallons (U.S. liquid)	3785	cubic centimeters
gallons (U.S. liquid)	0.1337	cubic feet
gallons (U.S. liquid)	231	cubic inches
gallons (U.S. liquid)	3.785×10^{-3}	cubic meters
gallons (U.S. liquid)	4.951×10^{-3}	cubic yards
gallons (U.S. liquid)	3.785	liters
gallons (U.S. liquid)	8	pints (liq)
gallons (U.S. liquid)	4	quarts (liq)
gallons per minute	2.228×10^{-3}	cubic feet per second
gallons per minute	0.06308	liters per second
grams	10^{-3}	kilograms
grams	10^{3}	milligrams
grams	0.03527	ounces
grams	0.03215	ounces (troy)
grams	2.205×10^{-3}	pounds
horsepower	42.44	Btu per min
horsepower	33,000	foot-pounds per min
horsepower	550	foot-pounds per sec
horsepower	1.014	horsepower (metric)
horsepower	0.7457	kilowatts
horsepower	745.7	watts
horsepower-hours	2547	British thermal units
horsepower-hours	1.98×10^{6}	foot-pounds
horsepower-hours	0.7457	kilowatt-hours
kilograms	10^{3}	grams
kilograms	2.2046	pounds
kilograms	1.102×10^{-3}	tons (short)
kilogram cal per kg	1.8	Btu per lb
kilogram cal per kWh	3.9685	Btu per kWh
kilogram per sq cm	14.223	lbs per sq in.
kilometers	10^{5}	centimeters
kilometers	3281	feet
kilometers	10^{3}	meters
kilometers	0.6214	miles

Multiply	By	To Obtain
kilometers	1093.6	yards
kilometers per hour	54.68	feet per minute
kilometers per hour	0.9113	feet per second
kilometers per hour	16.67	meters per minute
kilometers per hour	0.6214	miles per hour
kilowatts	56.92	Btu per min
kilowatts	3413	Btu per hr
kilowatts	4.425×10^4	foot-pounds per min
kilowatts	737.6	foot-pounds per sec
kilowatts	1.341	horsepower
kilowatts	10^3	watts
kilowatt-hours	3412	British thermal units
kilowatt-hours	2.655×10^6	foot-pounds
kilowatt-hours	1.341	horsepower-hours
kilowatt-hours	3.6×10^6	joules
liters	0.2642	gal (U.S. liq)
liters	1.057	quarts (U.S. liq)
lb per sq ft	4.8824	kg per sq m
lb per sq in.	0.0703	kg per sq cm
meters	100	centimeters
meters	3.2808	feet
meters	39.37	inches
meters	10^{-3}	kilometers
meters	10^3	millimeters
meters	1.0936	yards
meters per minute	3.281	feet per minute
meters per minute	0.05468	feet per second
meters per minute	0.06	kilometers per hour
meters per minute	0.03728	miles per hour
meters per second	196.8	feet per minute
meters per second	3.281	feet per second
meters per second	3.6	kilometers per hour
meters per second	0.06	kilometers per min
meters per second	2.237	miles per hour
meters per second	0.03728	miles per minute
miles	5280	feet
miles	1.6093	kilometers
miles	1760	yards
miles per hour	0.44704	miles per second
ounces	29.573730	cubic centimeters
ounces	28.35	grams
pounds	7000	grains
pounds	453.6	grams
pounds of water	0.01602	cubic feet
pounds of water	27.68	cubic inches
pounds of water	0.1198	gallons
square feet	144	square inches
square feet	0.09290	square meters
square feet	3.587×10^{-3}	square miles
square feet	0.111	square yards

Multiply	By	To Obtain
square inch	6.4516	square centimeters
square meters	2.471×10^{-4}	acres
square meters	10.764	square feet
square meters	3.861×10^{-7}	square miles
square meters	1.196	square yards
square miles	640	acres
square miles	2.788×10^7	square feet
square miles	2.590	square kilometers
square miles	3.098×11^6	square yards
square yards	2.066×10^{-4}	acres
square yards	9	square feet
temp (°C) + 273	1	abs. temp. (°K)
temp (°C) + 17.8	1.8	temp. (°F)
temp (°F) −32	0.555	temp. (°C)
tons (long)	1016	kilograms
tons (long)	2240	pounds
tons (metric)	10^3	kilograms
tons (metric)	2205	pounds
tons (short)	907.2	kilograms
tons (short)	2000	pounds

Appendix B
Average Heat Value of Fuels

Fuel	Energy Content (Btu)
1 kilowatt-hour electricity	3,413
1 therm gas (natural, LPG, propane)	100,000
1 cubic foot natural gas	1,025
1 cubic foot LPG or propane	2,500
1 cubic foot low-Btu synthetic gas	200
1 cubic foot med-Btu synthetic gas	270
1 cubic foot hydrogen gas	333
1 cubic foot high-Btu synthetic gas	970
1 gallon LPG	91,000
1 pound LPG	21,500
1 gallon fuel oil	138,700
1 gallon kerosene	135,000
1 gallon methanol (methyl alcohol)	60,000
1 gallon gasoline	115,340
1 ton peat	3,500,000
1 ton lignite coal	12,300,000
1 ton subbituminous coal	19,800,000
1 ton bituminous coal	24,050,000
1 ton charcoal	24,000,000
1 ton anthracite coal	30,000,000
1 ton dung	15,000,000
1 ton garbage and trash	10,000,000
1 ton agricultural wastes	12,000,000
1 cord dry softwood	15,000,000
1 cord dry mixed wood	17,000,000
1 cord dry hardwood	19,000,000
1 pound uranium	215,000,000

Appendix C
Heat Value of Various Wood Species

Species	Btu/lb[a]	Weight/cord[b]	Btu/cord[c]
Alder	6720	2540	17,068,800
Apple	7890	4400	34,716,000
Ash	8300	3440	28,552,000
Aspen	5980	2160	12,916,800
Birch, white	7470	3040	22,708,800
Cedar, western	6720	2060	13,843,200
Cherry	7470	3200	23,904,000
Cottonwood	5980	2160	12,916,800
Dogwood	8300	4230	35,109,000
Elm	7470	2260	16,882,200
Fir, Douglas	7470	2970	22,185,900
Fir, grand	5980	2160	12,916,800
Hemlock	6720	2700	18,144,000
Hickory	9500	4240	40,280,000
Juniper	7470	3150	23,530,500
Larch, western	7890	3330	26,273,700
Madrone	8300	4320	35,856,000
Maple, red	7885	3200	25,232,000
Maple, sugar	7890	3680	29,035,200
Oak, red	8300	3680	30,544,000
Oak, white	9500	4200	38,900,000
Pine, lodgepole	5980	2610	15,607,800
Pine, ponderosa	6720	2240	15,052,800
Pine, white	6720	2250	15,120,000
Pine, yellow	7890	2610	20,592,900
Poplar	5980	2080	12,438,400
Redwood	7470	2400	17,928,000
Spruce	5980	2070	15,462,900
Spruce, Norway	5980	2240	13,395,200

[a] Approximate heat value, Btu/lb, for air-seasoned (20% moisture) wood.
[b] Approximate weight, lb/cord, for average density of air-seasoned wood.
[c] Approximate Btu/cord for air-seasoned (20% moisture) wood and for average density (1 cord = 128 ft^3).

Appendix D
Power Consumption by Appliances

	Average Wattage	Kilowatt Hours Consumed Annually	Annual Cost at $0.07 per Kilowatt Hour
Food			
Range	8200	1175	82.25
Oven, self-cleaning	4800	1146	80.22
Oven, microwave	1500	300	21.00
Deep fryer	1448	83	5.81
Broiler	1436	100	7.00
Dishwasher	1201	363	25.41
Frying pan	1196	186	13.02
Toaster	1146	39	2.73
Waffle iron	1116	22	1.54
Waste disposer	445	30	2.10
Freezer (frostless) 15 ft^3	440	1761	123.27
Blender	386	15	1.05
Refrigerator (frost free) 12 ft^3	321	1217	85.19
Refrigerator (not frost free) 12 ft^3	153	580	40.60
Mixer	127	13	0.91
Carving knife	92	8	0.56
Laundry			
Clothes dryer	4856	993	69.51
Water heater (standard)	2475	4219	295.33
Iron (hand)	1008	144	10.08
Washing machine	512	103	7.21

	Average Wattage	Kilowatt Hours Consumed Annually	Annual Cost at $0.07 per Kilowatt Hour
Comfort			
Air conditioner	1566	1389	97.23
Heat (portable)	1300	176	12.32
Dehumidifier	257	377	26.39
Electric blanket	177	147	10.29
Humidifier	177	163	11.41
Fan (rollaway)	171	138	9.66
Heating pad	65	10	0.70
Health			
Hair dryer	381	14	0.98
Sun lamp	279	16	1.12
Shaver	14	1.8	0.13
Toothbrush	7	0.5	0.04
Entertainment			
TV (color)	332	502	35.14
TV (black and white)	237	362	25.34
Radio-record player	109	109	7.63
Radio	71	86	6.02
Housewares			
Vacuum cleaner	630	46	3.22
Floor polisher	305	15	1.05
Sewing machine	75	11	0.77
Clock	2	17	1.19

Notes: Electric knife consumes about 56 cents in electrical power per year. Electric water heater consumes about $295.33 in electrical power per year. A microwave oven will *save* 75% in energy costs.

Source: Adapted from Patrick J. Galvin, *Book of Successful Kitchens* (Farmington, Mich.: Structures Publishing Co., 1974), p. 64.

Appendix E
Home Energy Audit

INTRODUCTION

This home energy audit will guide you in calculating the heat loss of your residence. Included in this audit are heat gains from various features as well as item F (on the summary page), which will allow you to compare your home with the energy use of others. You will also be able to compare costs of heating your home using various fuels by using Section II and item G. Analysis of your findings will indicate which home features are losing the greatest amount of heat and which features should receive your efforts toward energy conservation.

DIRECTIONS

Page 1 A. Fill in *each* of the blanks in Section I.
          B. Determine the local cost per unit for *each* of the fuel sources in Section II.
          C. Calculate the cost per MBtu delivered to the living space.
Page 2 A. A sketch of your home including the area of each exterior surface is recommended.
          B. There are two copies of page 2, allowing calculation of heat losses for up to 12 rooms.

C. For *each* room of your home with exterior, heat-dissipating surfaces, fill in the data requested:
   1. Provide name of room.
   2. Provide the surface areas of all heat-dissipating surfaces.
   3. After referring to the Factors sheet (page 3) note the surface type at ① and the corresponding factor at ② for each surface.
   4. Multiply the surface area by the factor ② to determine the Btu/hr loss.
   5. Add the Btu/hr losses for each room to determine the total room loss.

Page 4
A. Add the Btu/hr losses of *all* windows and record under Sum of Rows.
B. Add the Btu/hr losses of each other feature and record under Sum of Rows.
C. Calculate infiltration losses:
   1. Select K factor from the factor sheet, item 6.
   2. Transfer volume of heated space from page 1.
   3. Multiply 0.018 × 70 × K × volume of heated space.
D. Add all losses (Sum of Rows) to determine the total peak hourly loss (Btu/hr).
E. Calculate the percent of total heat loss for each surface type (total loss of each type ÷ total peak hourly loss × 100).
F. Calculate losses due to heat transfer (B).
G. Calculate heat gains (C).
H. Calculate peak net heat loss (D).
I. Calculate seasonal heating needs.
J. Determine total heating cost by multiplying item E by the cost/MBtu for the fuel used to heat your home (see Section II on page 1).
K. Compare your calculated heating season cost with your actual heating season cost.
L. Calculate Btu/DD/ft^2 and compare with heating requirements of other homes.

Name: _____

BUILDING HEAT-LOSS CALCULATIONS ΔT = 70°F

I. Location of building: _____ (city); _____ (county)

    Latitude: _____ (to nearest whole degree)   Degree-days: _____

    Type of structure: _____ house   Number of stories: _____

    (check)   _____ apartment/duplex

                _____ mobile home

Type of construction — explain. (Example: concrete block foundation. Wood frame with 2″ × 4″ studs, $3\frac{1}{2}″$ fiberglass insulation, cedar shiplap siding)

_____

_____

_____

Total amount of heated space: _____ square feet (floor area)

                              _____ cubic feet (volume)

Type of heat source(s): _____
(oil, natural gas, electricity, LP, wood, etc.)

Type of heating system: _____
(Example: forced hot air, baseboard hot water, electric baseboard, radiant, etc.)

Rating of heating system: _____ Btu/hr (output)
(See specification plate on unit)

General information: 1. All dimensions should be measured and not estimated.
                            2. The surface heat loss is equal to the factor you choose for that surface multiplied by the surface area.
                            3. Selection of a living unit with a self-contained heating system is preferred.
                            4. A sketch will assist in recording layout, locations, and sizes of various features.

II. Fuel/energy costs in your area × units/MBtu ÷ conversion efficiency = $/MBtu delivered to living space

Electricity: $ _____ /kWh × 293 kWh/MBtu ÷ 1.00   =

Oil (No. 2): $ _____ /gal × 7.14 gal/MBtu ÷ 0.70   =

Natural gas: $ _____ /therm × 10 therm/MBtu ÷ 0.80   =

LP gas: $ _____ /gal × 11 gal/MBtu ÷ 0.80   =

Coal (bit.): $ _____ /ton × 0.039 ton/MBtu ÷ 0.65   =

Wood:* $ _____ /cord × 0.035 cord/MBtu ÷ 0.60   =

*Full cord = 128 ft^3

Page 1

Building Heat-Loss Calculations ΔT = 70°

Name: _____

Room: _____

Room: _____

Conduction losses from heat dissipating (exterior) surfaces (square-foot area)	Area	Surface Type Factor	Btu/hr Loss
Windows Conduction loss		① ②	
Doors Conduction loss		① ②	
Walls Conduction loss (minus window and door areas)			
Roof or ceiling			
First floor/slab/ crawl space/ basement floor			
		Total room loss	

Room: _____

Conduction losses from heat dissipating (exterior) surfaces (square-foot area)	Area	Surface Type Factor	Btu/hr Loss
Windows Conduction loss		① ②	
Doors Conduction loss			
Walls Conduction loss (minus window and door areas)			
Roof or ceiling			
First floor/slab/ crawl space/ basement floor			
		Total room loss	

Room: _____

Room: _____

Conduction losses from heat dissipating (exterior) surfaces (square-foot area)	Area	Surface Type Factor	Btu/hr Loss
Windows Conduction loss			
Doors Conduction loss			
Walls Conduction loss (minus window and door areas)			
Roof or ceiling			
First floor/slab/ crawl space/ basement floor			
		Total room loss	

Room: _____

Conduction losses from heat dissipating (exterior) surfaces (square-foot area)	Area	Surface Type Factor	Btu/hr Loss
Windows Conduction loss			
Doors Conduction loss			
Walls Conduction loss (minus window and door area)			
Roof or ceiling			
First floor/slab/ crawl space/ basement floor			
		Total room loss	

Building Heat-Loss Calculations ΔT = 70°

Name: _____

Room _____

Conduction losses from heat dissipating (exterior) surfaces (square-foot area)	Area	Surface Type Factor	Btu/hr Loss
Windows Conduction loss		① ____ ② ____	
Doors Conduction loss		① ____ ② ____	
Walls Conduction loss (minus window and door areas)			
Roof or ceiling			
First floor/slab/ crawl space/ basement floor			
		Total room loss	

Room _____

Conduction losses from heat dissipating (exterior) surfaces (square-foot area)	Area	Surface Type Factor	Btu/hr Loss
Windows Conduction loss			
Doors Conduction loss			
Walls Conduction loss (minus window and door area)			
Roof or ceiling			
First floor/slab/ crawl space/ basement floor			
		Total room loss	

Room _____

	Area	Surface Type Factor	Btu/hr Loss
		Total room loss	

Room _____

	Area	Surface Type Factor	Btu/hr Loss
		Total room loss	

Page 2

Factors for heating loss calculations (includes composite construction, $\Delta T = 70°F$)

Surface Type	Factor (Btu/ft^2/hr)
1. Windows and Sliding Glass Doors	
a. Conduction losses	
Single pane	74
Double pane ($\frac{1}{2}$" air space, use for storm)	45
Double pane ($\frac{1}{2}$" air space)	39
Double pane ($\frac{13}{16}$" air space)	34
Triple pane ($\frac{1}{4}$" air space)	32
Triple pane ($\frac{1}{2}$" air space)	25
2. Doors (excluding sliding glass doors)	
a. Conduction losses	
Wood – $1\frac{1}{2}$" no storm	34
$1\frac{1}{2}$" w/storm	20
Steel – $1\frac{3}{4}$" Pella	9
$1\frac{3}{4}$" Therma-Tru	5
3. Exposed walls	
Concrete block (8")	24
Brick (4")	24
Uninsulated frame (wood)	16
Insulated frame (wood)	
$3\frac{1}{2}$" fiberglass	5
6" fiberglass	3
8" cellulose	2
For heated underground basement walls, use factors above and multiply by 0.5	
4. Roof/Ceiling	
No insulation	22
$3\frac{1}{2}$" fiberglass	5
6" fiberglass	3
13" cellulose	1.5
5. Floor	
Over unheated basement or crawl space vented to outside*	
Uninsulated	21
6" fiberglass insulation	3.5
Over sealed, unheated, total underground basement 6" fiberglass	1.5
On concrete slab	
No insulation	6
Perimeter insulation	1.4

*Multiply percent of exposed basement wall above frostline (3 ft below ground level by 2.5 (insulated floor) or 20 (uninsulated) and add 1.0. (Omit windows.)

6. Air exchanges per hour = K

Old, uninsulated, not maintained	K = 6.0
Old, uninsulated house, maintained	K = 3.0
Average, insulated house, well maintained	K = 1.5
New, well-insulated house	K = 0.5
New, superinsulated (12" walls)	K = 0.2

Note: Interpolation of factors may be required for surfaces not included in this table.

Summary page   Name: _____

III. A. For all rooms, total the heat loss per surface type (sum of rows):

Conduction Losses	Sum of Rows	% of Total Heat Loss
Windows		
Doors		
Outside walls		
Roof/ceiling		
Basement floor/slab		
Infiltration losses (see at right)		
Total — peak hourly loss (Btu/hr)		100%

Air exchanges per hour (see item 6 on factor sheet)

Factor chosen: K = _____ Btu/hr/ft^3

Volume of heated space = _____ ft^3

Infiltration loss = heat capacity of air $\times \Delta T \times$ air exchanges/hr $\times$ volume of heated space

Infiltration loss = 0.018 $\times$ 70 $\times$ K $\times$ volume of heated space

B. Losses due to heat transfer
   Duct losses (if forced hot air system)
   Pipe losses (if hot water baseboard)
   Multiply total hourly heating loss by 1.10 if ducts/pipes are insulated, 1.20 if not.
   (No correction for electric baseboard heat)   _____ Btu/hr

C. Gains (yearly averages)
   Appliances and lighting: subtract 2500 (2000 if water heater in
   unheated space)   _____ Btu/hr
   People: subtract 400 for each occupant (averaged)   _____ Btu/hr
   Solar gain: subtract 35 $\times$ area of south-facing windows _____ Btu/hr
                                              Total gains   _____ Btu/hr

D. Peak net heat loss: subtract total gains (C) from total under item B = _____ Btu/hr

E. Seasonal heating needs
   Multiply item D by number of degree days in your area $\times \frac{24}{70} \times \frac{1}{1,000,000}$ = _____ MBtu

F. Total heating season *costs* based on calculations above using fuel/energy costs in *your* area
   (calculations based on $ _____ /MBtu for _____ [kind of fuel or energy used]) = $ _____

G. Actual heating season costs taken from your bills (subtract domestic water heating costs if
   you use the same type of energy source for both space and water heating — 20% of total
   bill) = $ _____

H. For comparison purposes, divide item D by the *area* of heated space and multiply by
   $\frac{24}{70}$ =   _____ Btu/DD/ft^2

Page 4

Source: Hinrichs/Salvagin/Schneider, State University of New York, College at Oswego.

Appendix F
Estimate Your Annual Exposure to Radiation

	Common Sources of Radiation	Your Annual Dose* (mrems)	Total
Where you live	Cosmic radiation at sea level:	26	
	For your elevation: add 1 mrem for every 100 ft above sea level		
	For example: Atlanta is 1050 ft, so add 10 Chicago is 600 ft, so add 6 Denver is 5300 ft, so add 53 Pittsburgh is 1200 ft, so add 12		
	Ground radiation (U.S. average):	26	

	House construction:
	If your house is wood, add 30 mrems
	                  brick, add 50 mrems
	                  concrete, add 70 mrems		

	Total		
What you eat, drink, and breathe	Water, food, and air radiation (U.S. average):	25	
	Nuclear weapons testing fallout:	4	
	Total		29
How you live	Medical diagnosis: If you had a chest x-ray this year (newest x-ray equipment), add 10 mrems per film If you had an intestinal x-ray, add 500 mrems Or use U.S. average of 92 mrems		
	Jet plane travel: For each 1500 miles, add 1 mrem		
	Color TV viewing: For each hour per day, on average, add 0.15 mrem		
	If you sleep with another person, add 1 mrem:		
	Total		
How close you live to a nuclear plant	If you live 1 mile from a nuclear power plant, add 0.02 mrem for each hour you are typically home per day:		
	If you live 5 miles away from a nuclear power plant, add 0.002 mrem for each hour you are typically home per day:		
	If you live over 5 miles away from a nuclear power plant, add 0 mrems:		
	Total		
	Total annual dose:		

Note: The average annual dose in the United States is 180 mrems.

*These numbers will vary slightly in different publications

Adapted from "Personal Radiation Dose Chart" by the American Nuclear Society and revision by Alcoa in *Alcoa News Reprint*, January 1982.

Appendix G
World Use of Renewable Energy
1980, 2000, Change Factor, and Potential

Source	1980	2000	Change Factor	Long-term Potential (exajoules)
Solar energy; passive design	<0.1	3.5–7	+52.5	20–30
Solar energy; residential collectors	<0.1	1.7	+17	5–8
Solar energy; industrial collectors	<0.1	2.9	+29	10–20
Solar energy; solar ponds	<0.1	2–4	+30	10–30++
Solar photovoltaics	<0.1	0.1–0.4	+2.5	20++
Wood	35	48	+1.4	100++
Crop residues	6.5	7	+1.1	—
Animal dung	2	2	—	—
Biogas: small digesters	0.1	2–3	+25	4–8
Biogas: feedlots	<0.1	0.2	+2	5++
Urban sewage and solid waste	0.3	1.5	+5	15++
Methanol from wood	<0.1	1.5–3.0	+22.5	20–30++
Energy crops	0.1	0.6–1.5	+10.5	15–20++
Hydropower	19.2	38–48	+22	90++
Wind power	<0.1	1–2	+15	10++
Geothermal energy	0.3	1–3	+6.7	10–20++
Total	63.5	113–135		334–406++

Notes: + indicates a positive change factor. ++ indicates that technical advances could allow the long-term potential to be much higher; similarly, a range is given where technical uncertainties make a single estimate impossible.
< means less than.

Source: Adapted from Worldwatch Institute, Washington, D.C.

Appendix H
U.S. Energy Production and Consumption by Source

Source	1973	1979[a]	1983
		Energy Production (%)	
Coal	22.45	27.43	28.29
Crude oil	31.44	28.40	29.98
NGPL	4.14	3.58	3.60
Natural gas	35.79	31.49	26.87
Nuclear	1.47	4.35	5.29
Hydro	4.62	4.60	5.75
Other[b]	0.07	0.14	0.22
		Energy Consumption (%)	
Coal	17.39	19.02	22.50
Natural gas	30.34	26.23	24.80
Petroleum	46.95	47.10	42.25
Nuclear	1.22	3.55	4.58
Hydro	4.06	3.99	5.47
Other[b]	0.06	0.11	0.19

Note: Totals for each year may not be 100.0% due to rounding.
[a] Year of peak U.S. consumption.
[b] Includes geothermal, electricity from wood, waste, and wind.
Source: Energy Information Administration, *Monthly Energy Review*, February 1984, pp. 5, 7.

Index

Acid rain, 25, 68-70, 90
Agricultural wastes (see Wastes)
Airfoil (see Wind Energy Conversion Systems)
Alaska, North Slope, 2, 82, 380
Alcohol (see also Ethanol; Methanol)
　from biomass, 242, 251-56
　fermentation process, 251-56
　from grain, 252
　net energy production, 254
Algae (see Biomass)
Altamont Pass, 16
American Petroleum Institute (API), 83, 379-80
Anacazi Indians, 15
Anaerobic digestion (see Biomass)
Anapolis River, 287-88
Aneutronic energy, 408
Anthracite coal (see Coal)
Appliances, power consumption, 436-37
Argonne National Laboratory (ANL), 111, 386
Ash from coal-fired power plants, 60, 70
Atomic energy (see Nuclear energy)
Atomic Energy Commission (AEC) (see Nuclear Regulatory Commission)
Atomic Industrial Forum (AIF), 105, 107, 108-9, 111-12
Azimuth, 135, 139

Bagasse, 233
Bag house, 69
Baltimore Canyon, 16, 96
Batteries
　applications, 174, 190
　types, 409
Bay of Fundy, 286-87, 392
Bergius process, 64
Betz coefficient, limit of, 181, 200
Biomass (see also Energy farms; Wastes)
　from agricultural crops, 228, 256-58
　from agricultural waste, 18, 248-49
　from algae, 257
　anaerobic digestion of, 227, 244-49
　composting, 249-51
　future of, 389-90
　gasification, 238-41
　from ocean, 311-12
　substitute natural gas, 242-49
Bituminous coal (see Coal)
Boiling water nuclear reactors (BWR) (see Nuclear reactors, designs)
Bottoming cycle (see Cogeneration)
Breeder reactors (see Nuclear reactors, designs)
Bristol cylinder, 306-7

Carbon dioxide emissions, 6, 69
Catalytic combustion units (see Wood, combustion units)
Chain reaction (see Nuclear reactors, designs)

Clean Air Act, 70
Coal, 43-73
　anthracite, 46, 49
　bituminous, 46, 49
　culm, 62
　direct combustion, 59
　distribution/transportation systems, 57, 71
　fluidized-bed combustion (see Fluidized-bed combustion)
　formation of, 44
　as a fuel, 12, 14, 17, 22, 23, 25, 44, 59, 71, 379
　gasification, 59, 64-66
　heat content, 12
　lignite, 46, 49
　liquefaction, 59, 63-64
　mining (see Mining)
　peat (see Peat)
　production, 25-27
　properties of, 48
　ranks of, 46, 49
　recovery (see Mining, coal)
　reserves, 48-50
　subbituminous, 46
　sulfur content, 49
Coal Mine Health and Safety Act, 58
Coal mines (see Mining)
Coefficient of performance (COP), 37, 343, 362-66
Cogeneration, 97, 343-51, 368, 392, 398, 406-7
　bottoming-cycle systems, 350
　topping-cycle systems, 348-49
Combined-cycle power generation, 62
Compost, 249-51
Conservation, 7-8, 319-24, 326-32, 392-93
　effect on fossil fuel utilization, 24, 80, 374
　of energy, 7, 29-30, 320-22, 324-26
　funding for, 378
Contour rafts, 304
Core samples, 81
Coriolis
　effect, 292
　turbine, 292
Crops for energy, 254-56 (see also Energy farms)
Crude oil, 76-77, 88 (see also Petroleum)
Crude Oil Windfall Profits Tax Act (COWPTA), 275, 338
Current energy (see Ocean current energy)

Darrieus rotor (see Wind Energy Conversion Systems)
d'Arsonval, Jacques, 294, 296
da Vinci, Leonardo, 136
de Coriolis, Gaspard Gustave, 292
Degree day, 314, 330-31
Dehydrated metal salts (see Thermal storage)

448

Delta T (ΔT), 135, 151-54, 159, 284, 294
Demonstrated reserve base (DRB), 48, 49, 51
Department of Energy (DOE), 11, 17, 22, 275, 288, 297, 348, 350, 367, 369, 374, 378
Deregulation, 378
  natural gas, 338
Devonian shales, 98
Doubling time, 32, 39-40

Earth Day, 6
Economic Recovery Tax Act (ERTA), 275, 338-39
Economizer, 353
Education/training programs, 425-26
Efficiency, 37-39, 154, 318-19, 323
Efficiency of Energy Conversion Systems, 352
Einstein, Albert, 170, 401
Electric power generation
  cost estimates, 359
  efficiencies, 38
  growth rate, 27-28
  production, 27-29, 44
Electrostatic precipitator, 69
Employment (*see also* Education/training programs)
  direct, 412, 420-24
  indirect, 413
  job titles
    engineers/scientists, 414-16, 420
    managerial/administrative/service, 416
    technical, 416-19
    skilled crafts, 419-20
Energy
  classifications, 35
  defined, 33
  forms, 33-34
  kinetic, 34, 262-63, 268
  potential, 34
  sources, 34-35
Energy, Department of (*see* Department of Energy (DOE))
Energy audit, 438-44
Energy content of fuels, 447
Energy conversion efficiency (*see* Efficiency)
Energy conversion factors, 430-33
Energy efficiency (*see* Efficiency)
Energy farms
  aquatic, 257
  tree, 235-37
  wind, 193-94
Energy Information Administration (EIA) 117-18
Energy options for the future (table), 5
Energy Policy and Conservation Act, 332
Energy production by source, 447
Energy resources, United States
  coal, 20, 25-27
  geothermal, 18, 20, 212
  natural gas, 17, 20-26
  nuclear, 17, 19-20
  petroleum, 17, 20-26, 78-82
  uranium, 113
  water/hydro, 28, 271
  wind, 17
  wood, 17-18
Energy resources, world
  coal, 48-49
  geothermal, 214-15
  natural gas, 20
  petroleum, 26, 78-80
  wood, 17
Energy savings (*see* Conservation)

Energy storage
  batteries (*see* Batteries)
  compressed air, 190
  by electrolysis, 190
  with flywheels, 190
  photovoltaic systems, 174
  pumped hydro, (*see* Pumped-storage hydroelectric plants)
  for solar systems, 151, 157-63
  with wind energy, 189-91 (*see also* Wind energy)
Energy Tax Act of 1978, 338
Energy units, 35, 430-33
Energy use, 20-25
  commercial, 315, 317
  industrial, 315-18, 322-24
  residential, 315, 319, 326-32
  transportation, 315, 318
Energy value of fuels, 434
Engineers (*see* Employment)
Environmental concerns (*see* Pollution)
Environmental degradation, 6, 67-68 (*see also* Pollution)
Environmental Protection Agency (EPA), 7, 49, 69-70, 126
Ethanol (ethyl alcohol), 228, 252-56, 390
Eutectic salts (*see* Thermal storage, phase-change materials)
Exponential growth, 39

Federal Energy Regulatory Commission (FERC), 274
Fermentation (*see* Biomass)
Fischer-Tropsch process, 14, 64, 72, 256
Fission (*see* Nuclear energy)
Flat-plate solar collectors (*see* Solar collectors)
Fluidized-bed combustion, 60-62, 64
Fly-ash, 60, 69-70
Force, 35
Forestry management for wood production, 235-37
Fort Union Formation, 49
Fractionation tower, 88-89
Fresnel lens, 156
Fuel cells, 72, 369, 395-99
Fuels
  heat value of, 434
  imported, 4, 19, 25
Fusion (*see* Nuclear energy)

Gas (*see* Natural gas)
Gasification of coal (*see* Coal)
Gasoline, 34, 239, 251, 254-56
Gengas unit, 239
Geological formations
  for geothermal energy, 209-15
  for oil and natural gas, 77
Geothermal energy, 14, 18-19, 208-26
  388-89 (*see* Geysers, The)
  conversion technology, 216-20
  future for, 388-89
  geological features, 18, 215-16
    geopressurized, 215-17, 222
    hot dry rock, 215, 217, 222-23
    hydrothermal, 215, 217, 222
    sites for, 213-15
Geysers, The, (California), geothermal plants, 14, 18-19, 211, 213, 220
Glaubers salts (*see* Thermal storage, phase-change materials)
Grande Coulee Dam, 271, 280
Greenhouse (*see* Solar heating/cooling)
Greenhouse effect, 27, 43, 69

Head, 262
Heat exchangers, 146, 353-55
Heating costs for homes, 364-65
Heat pipe, 355-56, 368
Heat pump, 343-44, 356-66, 368, 370
Heat recovery systems, 350-56
Heat transfer methods, 40-41, 325
Heat value of fuels, 434
Heliostats, 156, 383
Home energy audit, 331, 438-44
Home heating costs, comparisons, 364-65
Hoover Dam, 272
Hubbert, M. King, 24, 95, 97
Hydraulic head, 262
Hydroelectric power generation, 14, 16, 19, 261-83, 291, 390-91 (*see also* Electric power generation; Turbine designs, water; Water wheel designs)
Hydrofracturing, 215-16, 223
Hydrogen, 221, 398-401, 406, 408
Hydrogen-bomb heat energy, 409
Hydrothermal resources (*see* Geothermal energy)

Imperial Valley, (California), 212
Industrial wastes (*see* Wastes, industrial)
Insolation (*see* Solar radiation)
Institute of Nuclear Power Operations, 128
International Energy Agency, 306
Inverter, 189-90, 279

Jet Propulsion Laboratory (JPL), 422

Kaimei, 306
Kelvin, Lord, 357
Kinetic energy, 34, 262-63, 268
Known geothermal resource area (KGRA), 212

Lancaster Flexible Bag, 307
Lanchester Clam, 307-8
Landfills (*see* Methane)
Laser confinement (*see* Nuclear energy, fusion)
Life-cycle cost, 321
Lignite (*see* Coal)
Liquefaction (*see* Coal; Wood)
Los Alamos Scientific Laboratory, 355
Lurgi gasifier, 14

Magnetic confinement (*see* Nuclear energy, fusion)
Magnetohydrodynamics (MHD), 72, 406-8
Methane
 from biomass, 244-49
 from coal, 58, 64, 99
 from farms, 245-49, 355
 in geothermal fluids, 221
 from landfills, 244-49, 390
 from municipal sewage, 390
 in natural gas, 94, 244
 from ocean, 311
Methanol (methyl alcohol)
 from biomass, 228, 239, 252, 254-56
 from coal, 64, 72
 from peat, 380
Millirem, 126
Mining
 coal, 49, 51-56, 67, 71
 uranium, 121

Mouchot, Augustin, 137, 154
Municipal solid waste (*see* Wastes)

National Academy of Sciences, 374
National Aeronautics and Space Administration (NASA), 17
National Energy Act, 366
National Energy Conservation Policy Act, 337
National Energy Plan, 275
National energy policy, need for, 5, 6, 377
National Gas Policy Act, 94, 380
Natural gas
 composition, 94
 distribution, 94
 drilling for, 83
 as a fuel, 12, 17, 19, 22-23, 58, 76, 95, 379-82
 exploration for, 77
 production, 95
 properties, 12, 76, 94
 recovery processes, 85-88
 reserves, 95-99, 375
 storage, 94-95
 substitute natural gas (SNG) (*see* Synthetic natural gas)
 transportation, 76, 95
Natural Gas Policy Act, 338
Newton, Sir Isaac, 35
Nodding duck, 305
Noise pollution (*see* Pollution)
North American Reliability Council, 28
North Slope (*see* Alaska, North Slope)
Nuclear electric power plants
 costs, 20, 114-21
 decommissioning, 21, 121, 124
 major plants
  Clinch River Breeder Reactor, 112
  Nine Mile II, 118
  Seabrook, 118
  Shoreham, 118
  Three-Mile Island, 125, 127-28
  Washington Public Power Supply System, 118
 safety, 125-29
Nuclear energy (*see also* Nuclear electric power plants; Nuclear waste)
 chain reaction, 106-7
 as an energy source, 14, 17, 19, 22-23, 58
 fission, 102-33, 374, 382-83
 fuel costs, 115
 fuel cycle, 113-16
 fuel fabrication, processing, 113, 119, 121
 fuel reprocessing, 104-5, 115, 121-22
 fusion, 14, 103, 400-405
  confinement methods, 401-5
Nuclear Power Operations Institute of, 128
Nuclear radiation dose (*see* Radiation)
Nuclear reactors, designs, 107-13
 Breeder reactor, 108-9, 111-12
 Boiling-Water Reactor (BWR), 107-8, 114, 130
 Graphite-Controlled Breeder Reactor (GCBR), 130
 High-Temperature Gas-Cooled Reactor (HTGR), 107
 Light-Water Breeder Reactor (LWBR), 107, 130
 Liquid-Metal Fast Breeder Reactor (LMFBR), 107, 109, 112, 130
 Pressurized-Water Reactor (PWR), 107-8, 130
Nuclear Regulatory Commission (NRC), 103, 117-20, 127-28
Nuclear Safety Analysis Center, 128

Nuclear waste
 disposal, 21, 122-24
 environmental implications, 125-29
 (see also Pollution)
 geological formations for storage, 121-23

Oak Ridge National Laboratory (ORNL), 403
Ocean current energy, 291-94, 392
Ocean salinity gradient, 309-11, 392
Ocean thermal energy conversion (OTEC),
 295-302, 311, 392
 Mini-OTEC, 296, 298
 Project OTEC-1, 297
Ocean wave energy, 302-9, 391
Offshore oil production (see Petroleum,
 production)
Oil (see also Petroleum)
 from biomass, 237-38
 embargo, 373, 376
 shale, 93
On-shore oil production (see Petroleum,
 production)
Organization of Petroleum Exporting
 Countries (OPEC), 4, 79
OTEC (see Ocean thermal energy conversion)

Pacific Gas and Electric Company, 211,
 218-20
Particulate matter (see Pollution, air)
Passive solar heating (see Solar heating/cooling)
Peat, 17, 46-47, 380
Pelton turbine, 267-68
Petroleum
 catalytic cracking, 76, 88-90
 composition, 77
 distillation units, 88-90
 distribution, 90-91
 drilling for, 82-83
 exploration for, 77, 81-83
 as a fuel, 12, 14, 17, 19, 22-23, 25, 58,
 76, 78, 379-82
 origin, 77-78
 polymerization, 90
 production, 12, 26, 80-81, 83, 95
 properties, 76, 85
 recovery
  primary, 85
  secondary, 85, 87-88
  tertiary, 85-88, 381
 refining, 88-90
 reserves, 79-80, 82, 375
 storage, 90-92
 transportation, 76
Phase-change materials (see Thermal storage)
Photoelectric systems (see Photovoltaic
 systems)
Photovoltaic effect, discovery of, 170
Photovoltaic systems
 applications, 14, 141, 400
  industrial, 176
  remote, 175
  residential, 175
 cells (see Solar cells)
 collector orientation, 139-40
 costs, 175
 electrical output, 176
 focusing collectors (see Solar collectors)
 storage systems (see Energy storage)
Plasma, 401
Plutonium, 107, 109, 128-29, 404, 409
Pollution, 91
 air, 6, 49, 62, 90, 93, 229, 333-36
 from coal-fired utilities, 59
 from geothermal power plants, 220-23
 noise, 204, 220-21
 from oil, 91-94
 particulates, 59, 69
 sulfur dioxide, 59, 70
 thermal, 366
 visual/aesthetic, 161, 203-4
 water, 67-68, 93
 from wood combustion units, 236-37
Polymerization in oil refining (see
 Petroleum)
Power, defined, 36, 49
Prices, average national, selected energy
 sources, 2
Public Utilities Regulatory Policies Act
 (PURPA), 174, 182, 184, 191, 275-76,
 338
Pumped-storage hydroelectric plants, 190
 272-73, 281, 303 (see also
 Hydroelectric power generation)
Pyrolysis
 of biomass, 228, 237-38
 of coal, 64

Rad (see Radiation, nuclear)
Radiation
 dose levels, 126, 445
 geothermal, 224
 nuclear, 125-29
 solar, 137
Radioactive wastes (see Nuclear waste)
Radon gas, 221, 334-35
Rance River, 286
Rankine cycle, 296, 297, 350
Reactor systems (see Nuclear reactors, designs)
Refuse-derived fuel (RDF), 241-45
Rem, 126
Renewable energy use, 446
Reprocessing of nuclear fuel (see Nuclear
 energy)

Salinity gradient, 309-11, 392
Salter duck, 305
San Gorgonio Pass (California), 17
Scientists (see Employment)
Seasonal performance factor, 364
Semiconductor, 170-72
Shale, 93
Smith-Putnam wind turbine, 190-91
Solar cells, 170-77 (see also Photovoltaic
 systems)
 amorphorous, 175
 band-gap, 171
 efficiencies, 173, 175
 silicon, 173
Solar chimney, 15, 165
Solar collectors, 135, 142-50, 174
 batch heaters, 142-43
 flat-plate, 143, 164, 166
 focusing
  parabolic trough, 137, 154, 156
  point focus, 154
  tracking, 157
 mirror-type, 136-37
 stagnation, 144
 thermosiphoning, 146
Solar constant, 135, 138
Solar energy
 applications, 140-41
 electrical power stations, 383-84
 nature of, 134-80, 383-87
 photovoltaic systems (see Photovoltaic
  systems)

Solar energy (cont.)
  space heating systems, 15, 157-63
  water heating systems, 141-57
Solar heating/cooling
  active systems, 140-42, 146-50
    air systems, 164-66
    cooling, 164-66
    liquid systems, 146-50, 153-57
    system sizing, 150-54
  greenhouses (sunspaces), 141, 163-64, 353, 355
  passive systems, 140-42, 157
    cooling, 164-65
    direct-gain systems, 146-50, 158-63
    liquid systems, 146
    Trombé wall, 135, 159-61
    water wall, 161
  sunspaces, attached, 135, 163-64
  thermal storage (*see* Thermal storage)
Solar insolation (*see* Solar radiation)
Solar One, 383-84
Solar photovoltaic energy (*see* Photovoltaic systems)
Solar radiation, 135, 138-39, 151-52, 154, 170 (*see also* Solar constant)
Solar roof pond, 161-62
Solar thermal systems (*see* Solar heating/cooling)
Solar utility satellite, 383-85
Solvent-refined coal (SRC), 64
Southern California Edison, 212, 376
Specific heat, 158-59
Subbituminous coal (*see* Coal)
Substitute natural gas (*see* Synthetic natural gas)
Sulfur dioxide (*see also* Acid rain; Pollution)
  emissions, 6, 60, 62, 68-70, 90, 93
Sulfuric acid (*see* Acid rain)
Sunspaces (*see* Solar heating/cooling)
Synchronous inverter, 189-90, 279
Synroc process, 122
Synthetic fuel, 22, 44, 62-66, 72, 237-39, 381, 390 (*see also* Synthetic natural gas)
Synthetic Liquid Fuels Act, 63
Synthetic natural gas (SNG), 72

Tar sands, 93, 382
Tax credits/incentives, 6, 203, 275, 301, 337-39, 353, 366, 377, 383, 394
Tehachapi Pass, 16, 194
Thermal pollution (*see* Pollution)
Thermal storage, 167-70 (*see also* Energy storage, for solar systems)
  masonry/rock, 158-61, 167-70
  phase-change materials, 167
  roofs, 161-63
  Trombé wall, 135, 159-61
  walls, 158-61
  water, 166-68
Thermionic converter, 369, 372, 406
Thermodynamics
  definition, 39
  first law of, 39, 40
  second law of, 41
Thermoelectric generator, 299
Tidal energy, 14-15, 285-91, 391-92
Tokomak Fusion Test Reactor, 402-4
Topping cycle (*see* Cogeneration)
Tracking collector (*see* Solar collectors)
Tree farms, for biomass fuel, 235-37
Trombé wall, 135, 159-61

Tround drill, 82
Turbine designs, water, 267-70, (*see also* Waterwheel designs)
  Coriolis, 292, 294-95
  cross-flow, 267, 269
  efficiency of, 270
  Francis, 267-70
  impulse, 267
  Kaplan, 267-68
  Pelton, 267-68
  reaction, 267
  Turgo, 267
Turbine designs, wind
  control mechanisms, 187-88, 194-95
  drag type, 197
  experimental types, 198-203, 292
  horizontal axis, 191-95
  vertical axis
    cyclogyro, 198
    Darrieus rotor, 195, 197
    Savonius rotor, 195-97
United States energy needs, 374
United States Geological Survey (USGS), 212
Units of energy, 11-12, 430-33
Uranium (*see* Nuclear energy, fuel)
Urban waste (*see* Wastes, municipal solid)

Waste-heat recovery, 350-55
Wastes
  agricultural/biomass, 228, 258
  energy from, 19, 29, 229, 237
  industrial, 228, 237
  municipal solid, 228, 237-45, 389-90
Waste, radioactive (*see* Nuclear waste)
Water power, history of, 15-16
Waterwheel designs (*see also* Turbine designs, water)
  breast, 264-65
  overshot, 264-65
  Poncelet, 263, 265
  undershot, 263-64
Wave energy (*see* Ocean wave energy)
West Valley, 122-23
Wind energy
  applications of, 281
  available to machine, 199
  conversion systems (*see* Wind Energy Conversion Systems)
  density, U.S., 186
  future of, 387-88, 391
  history of, 14-16, 19, 182
Wind Energy Conversion Systems (WECS), 183-205
  airfoil, 181, 183, 185
  blades/designs, 185-87, 195-204
  controls for, 187-88
  environmental effects, 204-5 (*see also* Pollution)
  generators for, 188-89
  power vs. tip-speed ratio, 182
  television reception, effects on, 187
  tower design, 184-85
Wind farms, 193-95
Wind turbine designs (*see* Turbine designs, wind)
Wood
  Btu by species, 233, 435
  combustion units, 230-35
  as fuel, 12, 14, 17-19, 229-35, 257
  gasification, 238-41
Work, defined, 33, 36